T0146063

Nature's Path

Nature's Path

*A History of Naturopathic Healing
in America*

SUSAN E. CAYLEFF

Johns Hopkins University Press
Baltimore

Johns Hopkins University Press
2715 North Charles Street
Baltimore, Maryland 21218-4363
www.press.jhu.edu

Library of Congress Cataloging-in-Publication Data
Cayleff, Susan E., 1954– , author.
Nature's path : a history of naturopathic healing in America /
Susan E. Cayleff.
p. ; cm.
Includes bibliographical references and index.
ISBN 978-1-4214-1903-9 (hardcover : alk. paper) — ISBN 1-4214-1903-3
(hardcover : alk. paper) — ISBN 978-1-4214-1904-6 (electronic)
— ISBN 1-4214-1904-1 (electronic)   I. Title.
[DNLM: 1. Naturopathy—history—United States. 2. History, 20th
Century—United States. 3. Holistic Health—history—United
States.   WB 935]
RZ440
615.5'350973—dc23   2015018664

A catalog record for this book is available from the British Library.

Special discounts are available for bulk purchases of this book.
For more information, please contact Special Sales at 410-516-6936
or specialsales@press.jhu.edu.

Johns Hopkins University Press uses environmentally friendly book
materials, including recycled text paper that is composed of at least
30 percent post-consumer waste, whenever possible.

To Dr. Sue Gonda, storyteller, writer, guide, partner
You make every day a gift

Contents

Acknowledgments

Often in acknowledgments, one of the last lines thanks someone whose support, work, and belief in the author kept the project going. I am putting first that one key person: Dr. Sue Gonda, of Grossmont College in El Cajon, California. Her insights as a historian, editor, writer, compatriot, and stalwart coworker helped bring this mammoth project to light. The usual thank-you is not enough.

Jackie Wehmueller, at Johns Hopkins University Press, has been an ally and a valued colleague. Joanne Allen improved this manuscript with her smart, meticulous, and insightful editing. And I thank my colleagues in the Department of Women's Studies at San Diego State University for their ongoing support and encouragement.

I am indebted to the following people who have shared their expertise and knowledge with me over the years: Friedhelm Kirchfeld, archivist at the National College of Naturopathic Medicine Library in Portland, Oregon, and coauthor of the seminal *Nature Doctors: Pioneers in Naturopathic Medicine*, who gave generously of his expertise and access to uncataloged sources when I consulted them; he later corresponded with me as information emerged. I also am indebted to Sussanna Czeranko, ND, curator of rare books at the NCNM. Lauren Paschal Proctor, daughter of Leo Lust, Benedict's brother, has generously supplied me with memories and photographs of "Uncle Ben" and other family members. Conversations with Lauren and her sister, Anita Lust Boyd, were inspirational and informative. The historian George Cody has been generous in sharing his thoughts with me. I have benefited from the resources shared with me by Peggy Spranzani, chair of the Butler Museum, in Butler, New Jersey, where Louisa's Bellevue Sanitarium was located and where later the couple's Yungborn cure developed. Support and encouragement also came from Dean Paul Wong, of the College of Arts and Letters at San Diego State University, and assorted grants received by SDSU over the years: Research, Scholarship and Creative Activity, CAL Micro-grants, and two sabbatical leaves. John Weeks and Harry Swope, of the board of directors of the Naturopathic Education and Research Fund; Sheila Quinn, former executive director of the AANP, who was a lively and helpful correspondent in the early years of this project; Clyde Jensen, then president of the Institute for Natural Medicine, National College for Naturopathic Medicine; and Jane Sexton, director of library services at Bastyr University, were also instrumental. I am also grateful to the staff at the Francis A. Countway Library of Medicine at Harvard University and to Amber Dushman, archivist at the AMA Archives.

In the course of this lengthy project I have benefited from the skills and hard work of many others. I am grateful for the skillful work of Danielle Bauer, who spent countless hours on detailed manuscript preparation and indexing. I would also like to thank Teddi Brock and Polly Mason for their work on earlier versions of the manuscript. Friends, family, and many students served as researchers and provided library assistance, including Anna Andrade, Jennifer Baker, Ginger Blackmon, Susan Clark, Andrea Dottolo, Jennifer Higley, Pat Hobin, Roberta Hobson, Deborah Hughes, Claire Jackson, Ann Jette, Kara Jones, Mary Logomasini, Sandra McEvoy, Britton Neubacher, D'Marie Simone, Adam Trey, Elle Vandermark, and Anne Zimmerman.

And to my sister Joani Cayleff, thank you for your support, love, patience, and encouragement.

Introduction

The setting was idyllic at the Whistler Resort in British Columbia in 1991: gorgeous mountains, crisp September air, and energetic, positive people eager for the first-ever North American Convention of Naturopathic Physicians. It was a conference cosponsored by the American Association of Naturopathic Physicians (AANP) and the Canadian Naturopathic Association, the largest gathering of naturopaths in the profession's recent history. I was there because in the spring of 1990 Cathy Rogers, ND, the then president of the AANP, had contacted me. She had read my book on the nineteenth-century cold-water-cure movement, *Wash and Be Healed*,[1] and saw connections between the efforts of that movement and the history of naturopathy. Rogers asked whether I might be interested in researching and writing a history of naturopathic medicine in the United States. Unfamiliar with the word or its meaning, I erroneously assumed that *naturopathy* simply meant an eclectic blend of holistic therapeutics. The convention in British Columbia disabused me of that notion. Naturopathy, I quickly learned, is actually a philosophy for a way of life, demanding that we link body, soul, mind, and daily purpose. Naturopathy is at once a medical system and a way of living in harmony with, rather than conquering, the natural world. I learned about the current vitality of the naturopathic profession. When I researched early naturopathy's medical ethos, my own fascination with natural healing processes led me to undertake this project.

Naturopathy has a compelling, complex past. It developed historically as a system of medicine based on the premise that the body will heal itself when various components are strengthened through the use of nontoxic, natural therapies. It relies on preventive techniques and the interconnections between body, mind, and social and environmental factors that determine the status of health.

Naturopathy brought the nineteenth-century popular health movement into the twentieth and twenty-first centuries. Its leadership and followers resisted the new deference to and reliance upon medical experts as elitist and antidemocratic. Natural healers since the mid-1800s had believed that regular (allopathic) medicine positioned the physician as an irreplaceable sage; it too often dismissed common-sense approaches or used invasive and dangerous techniques. As American Medical Association doctors set out to standardize and professionalize themselves in the nineteenth century, other sects, using tried-and-true methods, saw allopaths who shunned any and all alternatives as autocrats. Naturopaths carried forward the belief in American individualism (albeit as a collective struggle) and self-reliance when it came to health. The natural health movement also offered an empowering and instrumental space for women practitioners.[2]

Late-nineteenth-century naturopaths were an eclectic collection of firebrands who, despite their vast differences, openly loathed many changes taking place in medicine at that time. Their ranks were drawn from Germanic practitioners, philosophical idealists, countless sectarian natural health leaders, self-proclaimed nature doctors with little or no schooling, advocates of Populism and other political radicals, agrarian advocates, those wracked by illnesses that did not respond to regular therapeutics, women barred from regular medical practice, vegetarians, antivivisectionists, religious devotees, spiritual and mental healers, and antivaccinationists. They all shared a self-imposed outsider status that often led them to reject notions of progress and change connected to industrial innovation or to a corporate-style monopoly of power and authority.

At the same time, after 1880 change was evident everywhere, particularly in urban areas, evidenced by massive ethnic immigration; rural migration to northern cities; electrified transportation; electric lights, which skewed the orderliness of day and night; congested streets teeming with loud children, vehicles, animals, vendors, and debris; countless saloons; millions of sporadically employed laborers; and an endless cacophony of tumult, motion, and frequently, chaos. The changing identity of the cities led some Progressive Era reformers, including self-appointed medical experts who used public health authority to define disease and the government measures to control it, to push for social control. Those who believed in the right to decide their own physical fate saw this as the ultimate affront to their way of life and their traditions. From the nature practitioners' perspective, this control stole common people's essential self-determination. It usurped an older, more reassuring approach to illness and health that asserted the superiority, indeed the *purity*, of natural methods.[3]

Naturopaths had an uphill battle because they viscerally rejected some key turn-of-the-century politics and science. They perpetually disdained public health authorities and regularly educated allopathic physicians, whose expertise, they claimed, overruled a person's knowledge of his or her own body, especially during forced vaccination campaigns. Naturopaths' fight against allopaths for medical liberty led to their inability and unwillingness to accept any scientifically based methods that were becoming effective in the twentieth century. In addition, because of the unhealthy pace of life and the poor hygiene of city life, naturopaths thought that improved health had to begin with improvements to the environment. They advocated a rather naïve and unrealistic agrarian lifestyle when millions were urban poor. Initially, they also largely rejected professional quality control by any central power (even of their own kind), and this, combined with the wide spectrum of self-identified healers, meant that some naturopathic patients were unprotected from charlatans. Over time, the leadership continued to waffle concerning their own standards for healing efficacy and claims to expertise. Clearly, unregulated free-market health care carried its own dangers.

The history of naturopathy has been marked by fluid boundaries, diversity, and eclecticism. Over time it came to be characterized by cohesive beliefs, philosophies, and therapeutics. It is not a single methodology. It is a hybrid. Theory and praxis have come from best practices that work with the body to heal itself. Some may see this hybrid approach as a weakness, a refusal to proclaim one absolute truth. But naturopaths' desire to be fluid and integrative has allowed, after much struggle, for the creation of a strong *system* of healing.

The story of naturopathy's journey since the 1890s is one whose time has come. This book is the result of my years of exploring the single most complex, multifaceted, and poorly understood medical system of the late nineteenth and twentieth centuries. As a women's studies medical historian, I was struck by the world-view proffered by the eclectic groups of nature curers. Now, as naturopaths gain increasing legitimacy, the story of their roots—their philosophies, predecessors, proponents, therapeutics, and attempts at institutionalization and professionalism—is told here in their own words whenever possible. Throughout their writings, naturopaths' voices were deliberate and articulate. They regularly capitalized, bolded, and italicized key strategic terms and concepts. It was a strategy to call attention to their arguments and to change American discourse and the language being used about the healing arts. I maintain these unique features for historical accuracy and emphasis.

This is the story of radical visionaries who defied cultural norms and medical demigods. It also details the development of naturopathy as a healing system and the struggle of formally trained naturopaths to professionalize and to be recognized as legal and valid practitioners. I leave to future scholars the history of self-described natural healers, also known at various times as drugless healers, traditional naturopaths, natural hygienists, or straight nature curers. Regional histories also deserve their own studies, as do the countless pathbreakers and personalities who helped create the movement.

Naturopathy's earliest diagnostics and treatments blended methods from nineteenth-century domestic medicine, Thomsonianism, homeopathy, botanical therapeutics, physical and health cultures, hydrotherapy, osteopathy, chiropractic, vegetarianism, electricity, eclecticism, sun and light cures, fasting, iridology, dietetics, hypnotism, neuropathy, and natural medicine. Early naturopaths were also avidly against vivisection and vaccination. These ideologies spanned several centuries and were gradually adopted into—or rejected from—the regimen of naturopathic therapeutics. Naturopathy continues to evolve today with its incorporation of Chinese medicine.

As I learned of naturopathy's rich history, I was drawn to the daring, often oppositional stance practitioners took in the early years at the turn of the twentieth century. Their vision was unapologetically radical. They defended the working classes. They opposed the capitalistic greed of the monopolies and any advances with negative impacts, such as processed foods, pharmaceuticals, environmental toxins, and, later, atomic energy. They also rejected an ethos of self-interest among healers; they had a collective vision for societal well-being through reformed health practices. They rejected the notion of physician authority and omnipotence, as well as the public health movement's forced interventions of vaccination in the life of average citizens, which they saw as a dangerous collusion between the American Medical Association, government, and pharmaceutical companies.

Naturopathy also thrived on women's cocreation and advancement of the system. Women emerged as leaders, financial powerhouses, doctors, organizers, teachers, and authors. Naturopaths' gender ideology was at times essentialist but offered empowerment nonetheless. The leadership courted women and provided opportunities for agency and authority through the care of their own and their families' health, food preparation, caring for the young, living a self-determined existence, and rejecting the cultural reliance on professional expertise and science.

The outsider posture that naturopaths assumed elicited the wrath of organized medicine, public health officials, and the legal system. A formidable foe was the press, which at times lambasted naturopaths accused of medical wrongdoing. The press frequently echoed the accusations of regular physicians who claimed that only they possessed valid health expertise. Constant legal battles consumed decades of work.

The history of naturopathy is also a history of collective action and its attendant difficulties. The most influential national leaders were often involved with progressive politics and allied themselves with natural healers whose visions, therapeutics, political acumen, and education varied considerably from their own. Despite many shared values, this was not a harmonious movement. Both alliances and attacks came from botanical healers, osteopaths, chiropractors, homeopaths, and self-identified nature doctors, many of whom practiced without coherent professional boundaries. Within the ranks of self-identified naturopaths, schisms nearly destroyed the movement several times. The rancorous in-fighting caused numerous organizational splits, competing leadership, and a weakened public identity. Yet these alliances persisted, often shakily, in part because of a profound hope for unity and broad-based reform.

Ironically, within the climate of the movement's inclusivity and liberal democratic values, conservative and exclusionary beliefs also persisted. The largely middle-class Germanic leadership, devoutly Christian, made up the bulk of the late-nineteenth-century popularizers. Despite rhetoric aligning it with the urban poor, the system was too expensive for many who needed it most. Some naturopaths embraced eugenics, as did many other reformers at the time, believing that right living complemented biological and physical superiority. At the same time, however, they championed individuals' ability to improve physically through right living.

We should also look at who was not part of this movement. German immigrants and the middle classes were at the center, but most people of color and those struggling to survive were not part of the leadership. There were some notable Latin American leaders and participants. Cuban naturopaths were prominent, and texts published in Spanish emerged over the years. From the myriad publications it is difficult to ascertain how many nondominant groups were able to visit the naturopathic health facilities that were rural retreats. It was costly to participate in the away-from-home cures, to reject urbanization, and to pay attention to healthful, self-grown foods. Naturopaths encouraged people to adhere to their precepts at home, but that required literacy and access to the books, articles, and food items. Legal cases reveal ethnically diverse patients, such as Italians, Jews, Irish, English, and Northern Europeans. But an unfettered life in harmony with nature was a near impossibility for most of the millions of new immigrants. Likewise, black migrants from the South, drawn to northern urban centers in pursuit of paid work, had little access to trained practitioners and rightly feared allopathic medicine's racial experimentation. Their medical needs were most often addressed by folk traditions, culturally sanctioned healers, and sporadic contact with allopathic physicians.[4]

Naturopathy built upon the ideology of New Thought, the belief that healthy living, emotional equilibrium, and positive mind-spirit strength would lead to personal and global harmony. Naturopaths believed in the healing properties of the sun and clean, fresh air, to which people in the city had little access. They prescribed sun baths, and because the sun could only penetrate the body through full or partial nudity, this added to their controversial reputation.

For all its early activism, eclectic inclusiveness, and eccentricities, naturopathy was pathbreaking. By the 1970s it had been advocating for decades the forms of natural healing that came to be associated with New Age and holistic modalities. But it had distinguished itself from the others. It has been part of the late-twentieth-century national movement in health reform, but a distinct philosophical entity within it.

Many people today look for natural health solutions that do not have harmful side effects. They are learning daily about the dangers of processed foods, overuse of antibiotics, and the debate about the effects of vaccinations. Those individuals also care about local and global environmental devastation, such as depletion of rain forests; global climate change; the effects of hazardous waste; toxins, air pollutants, and overexposure to chemicals and pharmaceuticals; and drinking water.[5] More health-conscious individuals are severely critiquing, even abandoning, dominant medicine altogether after years of questionable results from allergy shots; antibiotics; psychotropic medicating; drugs with dangerous side effects, including even death; and unnecessary surgeries. Even more Americans supplement their treatment by allopathic physicians with complementary and alternative medicine.[6] In addition, today some allopaths admit that naturopathic techniques can augment their own modalities. Chinese herbals, frequent exercise, yoga, massage, acupuncture, probiotics, and methods for ensuring non-toxic home environments are now incorporated into physician referrals or are found within their offices.[7]

This complementarity began when some trained naturopaths embraced insights and methods advanced by medical science in the 1930s. This willingness to borrow from allopathic medicine was largely absent in the first decades of the American movement. Then college-trained naturopaths worked to integrate scientific medicine, convinced that their own methods were truly central to health, not just auxiliary alternatives. In recent decades naturopathic practitioners have published clinical studies and research, and allopaths have studied the effectiveness of combined modalities.

Until recently, naturopathy as a system predicated on mindbodyspirit wellness received more than its share of ridicule and attempts to annihilate it. While some critics still occasionally attack naturopaths, the acrimonious fight to paint them as backward quacks has ebbed since naturopathy's stormy nadir almost a century ago.

Because of the sheer amount of relevant material, I have chosen a hybrid approach that focuses on historical predecessors of natural healing and the major elements of American naturopathic philosophy and therapeutics, the role of women, legal persecutions, immutable antivivisection and antivaccination stances, educational standards and professionalization, and, of course, legal status. I also highlight a few notable leaders and several specific counterculture viewpoints held by naturopaths from 1896 to the present day, along with libertarian stances that existed side by side with moral and eugenic conservatism.

Because this is a history of the naturopathic profession, I do not fully address the popular response to naturopathy. It was massively supported by advocates of medical freedom, antivivisectionists, and antivaccinationists. Its perseverance despite battering in the press and state legislatures speak to its following. Nor do I focus directly on the clientele, although patients' testimonies in print and court are brought in. Naturopaths' licensure successes, albeit embattled, and their ability to maintain practice despite near insurmountable obstacles in the 1950s and 1960s speak to the health needs they continued to meet in America. The scope and multiplicity of sources I encountered are worthy of several volumes. I have not probed the international aspects of naturopathy, except in passing mention. These too deserve a thorough study.

Four themes became clear in this story. First, naturopaths were twentieth-century champions of a health movement that had begun in the early nineteenth century to oppose what Americans saw as invasive and depletive practices of regular medicine. Second, naturopathy was part of a larger, vibrant and contentious resistance movement against corporate and institutional dominance at the turn of the twentieth century. Participation in the natural healing lifestyle and methods was an act of individualism and opposition to the country's growing reliance on authoritative medical expertise and centralized power. Third, what became professional naturopathy moved from a loose collection of eclectic practices to a precise set of philosophies and therapeutics. Finally, naturopathy continued to provide opportunities for women as practitioners and as providers for their families' health.

These themes emerged from the disparate voices that contributed to naturopathy's development on a national level. Multiple definitions of *naturopathy* abounded at any one point in time; distinct philosophies, methods, and credentials (or a lack thereof) of trained practitioners existed and continue to exist today. At any given time in *one* region or state's history, several naturopathic publications, organizations, leaders, brick-and-mortar away-from-home cures, and schools of learning existed simultaneously.

At the turn of the twentieth century, much of naturopathic work emanated from New York and New Jersey, where leadership was centered. But there were also many self-described natural healers with little or no organizational involvement nor clearly defined therapeutics and methods. I focus on those practitioners, publications, and societies with significant followings, membership, and coherent claims. Those who embraced scientific naturopathy, many in the Midwest and in the western United States, had become a significant and prophetic force by the 1950s.

The history of naturopathic medicine helps us to understand what naturopathy has become in the twenty-first century. The AANP was founded in 1985 and is currently based in Washington, DC, after years of headquartering in Seattle, which has long been a stronghold of naturopathic medicine. Because it seeks increased licensure recognition, the cross-country move was engineered to access the nation's capital. The organization defined itself as follows in 2014: "Naturopathic medicine is a distinct primary health care profession, emphasizing prevention, treatment, and optimal health through the use of therapeutic methods and substances that encourage individuals' inherent self-healing process. The practice of naturopathic medicine includes modern and traditional, scientific, and empirical methods."[8]

As for insurance coverage and affordability, increasingly health plans offer some coverage of complementary and alternative medicine, but it tends to be limited and inconsistent. Most people pay for these services and products themselves; however, the Affordable Care Act holds out significant promise for wider US insurance reimbursement and hence the promise of accessibility to trained naturopathic medical care.[9]

What was once a disparate group of practitioners with shared motivations has honed itself into a professionalized, educationally sound, and politically organized healing profession. Over the past decades, after rigorous organized political pressure, laws in several states and two US territories have recognized naturopathy as a legitimate, insurance-reimbursable medical methodology. In these states, naturopathic doctors are required to graduate from an accredited four-year naturopathic medical school and pass an extensive postdoctoral board examination administered by the Naturopathic Physicians Licensing Examinations Board in order to receive a license. Licensed naturopathic physicians must meet annual state-mandated continuing education requirements and have a state-defined scope of practice. As of 2014, naturopathy was legal in Alaska, Arizona, California, Colorado, Connecticut, the District of Columbia, Hawaii, Idaho, Kansas, Maine, Minnesota, Montana, New Hampshire, North Dakota, Oregon, Utah, Vermont, Washington, Puerto Rico, and the US Virgin Islands.[10] In addition, Florida, Illinois, Massachusetts, New Mexico, New York, and Virginia have been debating licensure. State legislatures considering a debate include Maryland, Michigan, Nevada, Ohio, Pennsylvania, Texas, and Wisconsin. Obviously, licensure is a primary aim of the AANP. Its Alliance for State Licensing has set goals for developing campaigns in the states lacking it.[11]

Presently there are five accredited American schools granting the Doctor of Naturopathic Medicine (ND) degree: Bastyr University in Kenmore, Washington, with a new San Diego campus; the National College of Naturopathic Medicine in Portland, Oregon; the Southwest College of Naturopathic Medicine and Health Sciences in Tempe, Arizona; the University of Bridgeport College of Naturopathic Medicine in Bridgeport, Connecticut; and the National University of Health Sciences in Lombard, Illinois. Canadian schools granting ND degrees are the Canadian College of Naturopathic Medicine in Toronto, Ontario, and the Boucher Institute of Naturopathic Medicine in New Westminster, British Columbia.[12]

The ND degree requires four years of postbaccalaureate-level study in medical sciences. The curriculum in the first two years parallels the basic science curriculum of allopathic medical schools. In the third and fourth years students receive hands-on clinical training in botanical medicine, clinical nutrition, counseling, homeopathy, laboratory and clinical diagnosis, minor surgery, naturopathic physical medicine, and nutritional science. True to its roots, naturopathic education places a strong emphasis on preventing disease and optimizing wellness.[13]

There are also dual degree programs in which students study for an additional period to obtain a master's in an area of specialization, such as acupuncture and Chinese medicine, midwifery, or counseling psychology. Chinese medicine and acupuncture training enable naturopaths to do more than treat just the body; it allows them to restore balance to the whole person.[14] Naturopathic obstetrics contrasts sharply with twentieth-century biomedical birthing methods and has influenced the natural birthing trends of the past thirty years. Naturopaths always emphasized natural childbirth in an out-of-hospital setting with pre- and postnatal care. Today this has gained acceptance, with some allopathic medical centers creating comfortable settings for natural births and midwifery on and off hospital sites.[15]

The status of present-day naturopathy is directly connected to its roots. One constant has been to compare naturopathic training with regular AMA schools. The AANP compared curricula to document the hours spent in specific areas of study at the five naturopathic schools compared with the hours spent at Johns Hopkins, Yale, and Stanford.[16] The data are designed to prove that the training for the naturopathic degree is more rigorous, as well as to distinguish the formally trained, credentialed naturopaths from self-described nature healers. In 2007 the AANP was frustrated that "without licensure anyone can 'hang out a shingle' and practice with nothing more than a 'mail-order' degree or a home study course."[17]

Countless self-proclaimed "nature doctors," recipients of certificates from weekend seminars and correspondence schools, also used the term *naturopath*. Some of these were (and are), as the anthropologist Hans Baer characterized them, "partially professionalized." Others are marginal at best. This concern has always been an issue as naturopaths worked for licensure. Online naturopathy programs are not acceptable to the Association of Accredited Naturopathic Medical Colleges and thus are not accredited; they do not offer proper opportunities for clinical training.[18] Naturopaths aim to protect the public from harm and to protect the profession's integrity against unqualified people and charlatans. In 2002 licensed naturopaths were alarmed to learn that only 40 percent (1,306) of the naturopathic physicians practicing nationwide belonged to the AANP. Fully 60 percent of *licensed* naturopaths shunned professional affiliations, and thousands of others who use the title "naturopath" refused to be schooled at the five licensed naturopathic American institutions.[19]

In 2014, consistent with the message one hundred years earlier, the AANP believed that "every American has the right to choose a naturopathic doctor." Today the association has five goals, with strategies to carry out their mission: state licensure; access for all to affordable, cost-effective naturopathic physicians; furthering the profession with practice-based research; involvement in public policy to incorporate naturopathic practices of wellness and disease management; and to "advocate for global health, sustainable human communities, and a healthy planetary ecosystem."[20]

In the current medical culture the origin of naturopathy is particularly relevant. During the decades when naturopathy became a cohesive set of therapeutics, many practitioners were vocal, even belligerent *opponents* to exclusivity and license requirements. What has occurred since then to alter yet empower the work of naturopaths? My hope is that this study of naturopathic medicine and the social critiques and gendered opportunities it generated can answer that question and guide future scholars to explore the countless issues that emerge from this telling.

CHAPTER 1

Following Nature's Path
and Botanic Healing

THE DERIVATION OF *NATUROPATHY* CIRCA 1896

There are stories of a clear and compelling moment, or epiphany, that yielded the name *naturopathy*. This moment was mythologized by a Texas-based hygienic healer in the 1940s and then again in the 1980s. The origin of the term was never as distinct as any one account portrays. In fact, the term arose in response to fierce legal persecutions and a lack of other viable choices. The term is important for the identity and credibility it gave its practitioners. According to one story, the word *naturopathy* was originally coined in 1892 by Drs. John and Sophie Scheel, who combined the terms *nature cure* and *homeopathy*. In 1896 the Scheels, who practiced in New York City and operated a water cure, gave Benedict Lust, the American popularizer of the system who was called by many the father of American naturopathy, the right to use the name in association with his ventures.[1]

A more romantic version of this story claims that one morning in 1896 an enthusiastic young Benedict Lust burst into the office of a colleague in New York City and exclaimed, "We now have a name for our work!" He believed the word *naturopathy*, recently coined by Sophie Scheel, who taught at the nearby Homeopathic College, perfectly encapsulated the healing techniques he advocated. Lust, Scheel's student, developed a strong reputation as a Kneipp-taught water curer and gained legal use of his instructor's copyrighted word upon her death. In two other versions of the word's origin, R. T. Trall had used *naturepathy* in an editorial in the pages of the *Water-Cure Journal* in the 1860s, or it was submitted as part of a naming competition.[2]

Although the origin of the term is murky and still debated, Benedict Lust is credited with helping to form a cohesive naturopathic identity and reducing the number of persecutions of its practitioners. They could not use the titles "doctor," "therapist," or "physician," and they could not use the terms *cure, healing,* or *therapy* to refer to their work. *Naturopath* was the only term they could safely use that denoted nature cure and disease. *Naturopath* was the magic word that set them free. Lust saw the term—and its lasting power—"as a living protest against the autocracy, coercion, imposition, intolerance, and persecution of the New York Medical Society Trust in particular, and the American Medical Association Trust in general."[3]

Despite Lust's promotion of the term *naturopathy* met with immediate (and ongoing) challenge, even from practicing naturopaths. The naming debate mirrored the discordant debates among practitioners over theory and practice. One rejected its syntax, but far more important was the critique that its meaning— sickness or suffering of nature—was objectionable. Yet the term stuck as both a form of protest and a legal tactic to distance the system from allopathic therapeutics and other sectarians. Because only Lust and the American Naturopathic Association could use the word *naturopathy*, other like-minded practitioners who loosely followed Lust but chose to adhere to other principles had to create different names. Before *naturopathy* was chosen, the discipline was termed the *natural method of healing* (1898), *physical culture* (1904), and *neo-naturopathy* (1917). It was also called *natural therapeutics* (1919) and *sanipractic*.[4]

Naturopathy's popularizers were Benedict Lust and Louisa Stroebele Lust. Lust was born in Michelbach, Germany, in 1872, and immigrated to the United States in 1892. Contracting tuberculosis soon thereafter, he returned to Germany, where he recovered under the care of the renowned hydropathic healer Father Sebastian Kneipp. Lust returned to the United States in 1894 a committed zealot commissioned by Father Kneipp to open a Kneippian society, institute, and school and to launch a magazine. Lust earned a medical degree from the New York Homeopathic College in 1902. When Lust's Naturopathic Center opened on East 59th Street on September 15, 1896, several other sectarian schools already existed in New York City. His natural healing methods used water, homeopathy, light, chiropractic adjustments, dietetic advice, exercise, baths, and massage for health restoration and preservation. His practice prompted local authorities to deem the therapeutics medicinal, and hence the practitioners were considered to be in violation of the law for practicing medicine without a license. Lust and his copractitioners adopted the term *naturopathy* not only to reflect their values but also to mask their role as practitioners and thus avoid arrest. So they hoped.

Despite early enthusiasm that the term could provide shelter, more than fifty practitioners were arrested that first year, each of them found guilty of practicing medicine without a license and fined $250. Lust was arrested many times, and heavy fines were levied upon him.[5]

From those practicing natural cures arose a blend of ideas and philosophies that developed into a *system* of naturopathic healing. Naturopaths embraced botanics, often called "herbalism" (Thomsonianism); hydrotherapies; dietetic reform; hygienic living; homeopathy; the restorative use of light, sun, and wind; healing diets; fasting; chiropractic and osteopathy (at times); magnetism; hypnotism; electricity; iridology ("reading" the eye's iris); and myriad other eclectic practices. The historical predecessors of these components created the foundation for the diverse therapeutics and philosophies that came to make up naturopathy. Practitioners of nature cure might embrace some, but not all, of these healing techniques. One strength of early naturopathy was the *fluidity of components*, which allowed practitioners to adopt different versions of naturopathic practice. Ironically, this was also a shortcoming of the system, for it offered no single theory and praxis that promised panacea relief. Its very complexity and numerous components made it at once appealing and intimidating. It demanded of both practitioner and patient a degree of comfort with uncertainty, multiple modalities, inexact expertise, and the acceptance of gradual and subtle progress in the treatment of illness rather than immediate, dramatic results.

CHAMPIONS OF THE CAUSE: BENEDICT
AND LOUISA STROEBELE LUST

In addition to popularizing naturopathy in America, Lust, bolstered financially and philosophically by his spouse, Louisa Stroebele Lust, went on to found the American School of Naturopathy and the American School of Chiropractic, health food stores, and the magazines *Naturopath and Herald of Health* and *Nature's Path*. He produced a stunning array of publications that addressed topics such as an encyclopedia of naturopathic healing, herbs and their uses, water therapies, the dangers of vaccinations, necessity of diet reformation, and specific disease-oriented therapies. To learn the gamut of current health practices, Lust earned numerous degrees. He received a DO from the Universal Osteopathic College of New York in 1898; his MD from the New York Homeopathic Medical College in 1902; his ND from his own American School of Naturopathy in 1905; and a degree from the New York Eclectic Medical College in 1913.[6]

In 1896 Benedict Lust and Louisa Strobele (1865–1925) opened Yungborn, a healing sanatorium outside Butler, New Jersey, where they implemented their eclectic philosophies. In 1913 they opened a second Yungborn in Tangerine, Florida.

When Lust met Louisa Stroebele in 1895, she was already a renowned healer. She had studied curative work in England and for several years had been the personal aide to the radical thinker Lady Cook, Tennessee Celeste Claflin. Stroebele's healing beliefs informed her practices at Bellevue, a healing residence she opened in Butler, New Jersey, in 1892, emphasizing Arnold Rikli's curative air and sun bathing and a healthy diet. Upon meeting Lust, she incorporated the water-cure principles embodied in the Kneipp Cure—which she had long valued—and hired Lust to oversee them.[7]

Stroebele and Lust labored side by side for about thirty years, first as partners, then as spouses. Louisa's intellectual radicalism and insistence on female equality made her a leading force in American naturopathy for many years. She authored the *Practical Naturopathic-Vegetarian Cook Book* (1907), which endorsed a sparse diet. Since naturopaths saw nutrition as a fundamental core of wellness and naturopathic philosophy, her book became the dietary bible for the movement. She urged mothers to use their influence as leaders within the home to ensure their families' healthy diet and habits. Yet her actions were not confined to the private sphere of influence assigned to late-nineteenth-century Euro-American middle-class women. She excelled in public roles and became a leader in the national movement to gain recognition and legitimacy for naturopathy.

Louisa, financed by successful investments with Claflin, provided financial support for Benedict's ventures and for the field of naturopathy in New York, where she served as an instructor of practical naturopathy. She cofounded the two naturopathic journals and paid for the defense in at least seventeen legal actions taken against individual naturopaths and their healing system. She was a rare combination of healer, radical freethinker, and economic powerhouse. The Lusts were the guiding force during the early years of American naturopathy. Theirs was a bond of common goals, mutual respect, and devotion to natural healing.

The Lusts' experiences creating the field of naturopathy paralleled those of many who had advanced new theories and methods before them. Their predecessors were healers from the eclectic school and those who used botanics. But the publications, great public appeal, and self-produced healing remedies of eclectics and botanics had threatened mainstream allopathic medicine and led to legal attempts to annihilate their systems. Like so many other non-AMA schools, they eventually lost funding, were deemed fraudulent, and were forced to close their doors. Despite this opposition, the practices persevered, and botanics and eclecticism were incorporated by the Lusts into the philosophies and methods of their larger movement.

THE ROOTS OF NATUROPATHY: BOTANICS
AND ECLECTICISM

More than any other method, botanics fundamentally informed naturopathic healing. Botanics, extracts of plant derivatives, were in keeping with the naturopathic philosophy that bodies are capable of healing themselves when provided with necessary nurturance and noninvasive support, explained Oregon's William A. Turska, ND, in 1954. Turska, a significant voice in the movement, had graduated from the Seattle College of Chiropractic and Naturopathy and became chair of the International Society of Naturopathic Physicians' Council on Naturopathic Philosophy. He advocated the use of tinctures made of herbs, as well as herbal teas. Used in large quantities, some of them could be toxic, he cautioned, but administered in small doses they were safe and created "synergy," helping the body to build vitality, eliminate toxins, and heal itself.[8]

Many theorists and practitioners had advocated botanical medicine for centuries. Naturopathy continued traditional natural healing practices but went beyond single-cause and single-cure philosophies. In America from the time of the earliest English colonies, in the 1600s, to the 1850s, women and men taught and practiced domestic medicine, which was based largely on botanics, wherein the curative properties of botanics were brought out through preparations such as that of willow bark, which contains salicin, the original aspirin. Domestic medicine was the knowledge possessed by women, who passed it down through generations in written receipt, or recipe, books. Many women shared this vital knowledge verbally.[9]

Available and affordable texts facilitated acceptance of Samuel Thomson (1769–1843), the premier nineteenth-century American botanic physician. He sold his therapeutics in kits to the public. Thomson deplored depletive heroic therapeutics used by allopaths, such as bleeding, cupping, leeching, and the use of emetics, purgatives, and cathartics, aimed at ridding the body of the putrid matter causing illness. Instead, Thomson facilitated health through restoration of a person's vital force. He used simple herbal remedies, enemas, and steam baths. The two herbs he used most commonly were lobelia and cayenne. While his beliefs were foundational for naturopaths, his therapeutics were *not* mild. His approach, like allopaths', sought to restore balance. In Thomson's view, the body's essential elements were earth, fire, air, and water. These composed the body, he believed, and their disarrangement caused illness through diminished body heat; cold influences on the body caused disease and could be cured by heat applications induced by herbs. He later claimed to have sold 100,000 copies of his book, *New Guide to Health* (1825), and to have 3 million followers.[10]

Thomsonianism built on the established tradition of home self-doctoring. The popularity of Thomson's methods is understandable: they gave patients a sense of self-determination and control over the uncertainties associated with illness and produced visible physical results. Thomsonianism also emphasized the patients' ability, even responsibility, to be their own healers. This sense of responsibility complemented mid-nineteenth-century, middle-class Anglo values of self-determination and self-restraint. These traits, they believed, distinguished them from immigrants and the lower sorts. His ideology paved the way for Lust and Stroebele's concept of naturopathy.[11]

Like later naturopaths, Thomson railed mercilessly against allopathic apothecaries and pharmacists and their therapeutics. He called them "purveyor[s] of poisons." His movement included its own infirmaries and retail outlets for the manufacture and sale of his medications. More than one hundred colleges taught his medical system.[12]

Thomson's efforts were continued and modified by his contemporary Dr. Wooster Beach (1794–1868), who created requirements and standards for the practice of Thomsonian medicine. Beach, who apprenticed under the German herbalist Jacob Tidd, hoped to reform allopathic practices from within. Beach, unlike Thomson, advocated the use of medical hospitals, colleges, and botanical medication together as an effective healing system. Thomson, on the other hand, had opposed medical schools because of his belief that herbalism could be self-taught and administered. Yet a decade after he died, independent Thomsonians founded a Cincinnati, Ohio, medical college and came to dominate the Thomsonian movement. These people were later subsumed into Beach's eclectic school. The botanics from Thomsonianism and Beach's hospitals and schools were foundational for the field of naturopathy.[13]

By 1850 Beach's movement was known as eclecticism. Under Beach, the system sought to enrich the vital forces. Eclectics believed that natural medicines were preferable to synthetic ones because the plant bridged the mineral and the animal and made each able to assimilate the other.[14]

Beach, like Lust decades later, advocated standardized education. In 1832 he opened his Eclectic Medical Institute in Worthington, Ohio. The school operated in various forms until 1939. Beach received a regular medical degree from the Barclay Street Medical University in New York. He became a member of the

New York Medical Society, but his professional straddling of allopathy and botanics led his peers in regular medicine to scorn him for his advocacy of botanic remedies. Later, Lust, like Beach, would receive lifelong scorn despite his schooled proficiency in mainstream modalities.[15]

Three others shaped the eclectic school and consequently shaped naturopathy. Two of the three worked in tandem. Dr. John King's (1813–1893) development of botanical concentrates (dubbed resinoids) was also advocated by his colleague Dr. John Scudder. Both were located at Cincinnati's Eclectic Medical Institute. Scudder's contribution was to use small amounts of single botanical medicines according to empirical indications. A third individual, the pharmacist John Uri Lloyd, manufactured their botanical extracts through the Lloyd Brothers Pharmacists Specific Medicines. Publications spread King and Scudder's philosophies. Lust and Stroebele added them to the group of complementary practices within naturopathy.[16] Eclecticism survived and flourished into the twentieth century, until powerful and wealthy philanthropic funding sources commissioned a report about competent versus incompetent healing institutions. In a 1910 report, Abraham Flexner portrayed eclecticism as incompetent and drug mad. By 1920 only the Cincinnati school remained open, and it folded in 1939.[17]

Eclecticism and early naturopathy shared many similar views. When the fledgling Naturopathic Society of America was founded in New York City in 1902, it incorporated philosophies from eclecticism. By 1916, however, when it was renamed the American Naturopathic Association, the affiliation with eclecticism had been dropped, likely because of the Flexner Report. Naturopathic practitioners, however, continued to value Thomson's and the eclectics' early advocacy of botanical medicine, and Benedict Lust had obtained a degree from the New York Eclectic Medical College ca. 1913.[18]

While naturopaths maintained use of botanics, their applications of them evolved beyond the original principles. In the inaugural issue of Lust's *Kneipp Water Cure Monthly* (1900), a discussion of the healing properties of twenty botanics explained that "as a first help in all diseases, herbs should be in every household: and once there, they will be valued very highly." Promoting expanded botanic use, Lust referred to them as an invaluable component of naturopathic therapeutics. Botanics' results, he said, were more quickly felt than those of other treatments—such as air, water, exercise, light, massage, and mental exercises—because they complemented the body's ability to heal itself while expelling diseased substances. He advocated teas and dried herbal concoctions only.[19]

In time, numerous contributing authors, first in the *Kneipp Water Cure Monthly* and then in the *Naturopath and Herald of Health*, extolled the virtues of botanics in all their varied preparations. The latter introduced the regular feature Phytotherapy Department: The American Herb Doctor in 1916 to detail the therapeutic uses of herbs for specific ailments.[20]

Authors' enthusiasm for botanics echoed that of diverse practitioners in the United States and abroad. In Britain, classic texts providing medicinal uses of herbs and dosages were found in nearly every rural cottage. Their authors, like Thomsonians and naturopaths, looked to mothers to prescribe and treat their families' common ailments. The acceptance of such texts fueled early naturopathic praise for the use of plants such as green bean pods, mistle, betony, bear berry, and birth leaves to cure numerous ailments.[21]

DISSENTING VOICES AGAINST BOTANICS

There were self-proclaimed nature-cure practitioners, some lacking institutional credentials, who disagreed vehemently with botanic use. One of the greatest dilemmas in naturopathy's early history was that wildly diverse practitioners claimed the title "naturopath." Many wrote often in the pages of the *Naturopath and Herald of Health* against drugging of any sort, equating drugs with sickness. Professor V. Greenwald, of Kentucky, blamed hospital deaths on narcotic poisons. He recounted a chilling tale of death induced in a healthy male patient who was injected with narcotics and carbolic acid. He did not distinguish between poisonous narcotics and botanics; he saw only drugless treatment at home or at a nature-cure sanitarium as safe.[22]

Most articulate in holding this view was the influential Dr. Henry Lindlahr (1862–1924). In his prolific career as an author and advocate of drugless doctoring, Lindlahr strove to eliminate the use of all medicines, including herbs. He advocated hydrotherapy, manipulative treatments, and dietary renovation. Lindlahr harkened back to Vincent Priessnitz (1799–1852), popularizer of the cold water cure in then Austrian Silesia. His philosophy and therapeutics were successfully transplanted to the United States in the 1830s through American popularizers. Priessnitz's treatment comprised not the use of pills and potions but proper nutrition, plenty of exercise, fresh mountain air, water treatments in cool, sparkling brooks, and simple wholesome country fare. The results of these simple therapies, Lindlahr exclaimed, were magnificent. Little wonder, then, that Lindlahr was a steady publisher of cookbooks that ordained rightful nutrition as the path to health.[23]

The disparate, democratic character of naturopathy's beginnings meant that other naturopathic practitioners, among them some of Lindlahr's own students, disagreed with him and continued to use botanics. One such dissenter, a gradu-ate of the Lindlahr Health Institutes, which comprised the Lindlahr Sanitarium and college in Chicago, was Dr. Anna Abraham Bingesser. She and her hus-band, Carl (both naturopaths), operated the Waconda Springs Sanitarium in central Kansas in 1907. The Drs. Bingesser, like other naturopaths since them, incorporated American Indian botanical knowledge into their own modalities, recognizing the enduring aspects of botanic medicine but cautioning against some poisonous herbs and explaining how to identify them. Naturopaths who accepted botanic remedies warned users about the yew tree, henbane, deadly nightshade, hemlock, and black hellebore. While some potentially fatal herbs possessed healing properties—black hellebore was used for the treatment of epi-lepsy, dropsy, and liver trouble—all could be fatal when taken internally at inap-propriate dosages. Other naturopaths expressed concern that botanics did not offer the appeal of drug medications because some common folk preferred the dramatic aspects of the latter.[24]

On the extreme side of the desire to use botanics as medicinals were some naturopaths who resented their legal exclusion from prescribing drugs. One in 1939 decried Florida's law stating that naturopaths could not prescribe narcotics because they were not legally sanctioned physicians. Thus, the debate surround-ing the use of drugs, and what constituted a drug, continued through the mid-twentieth century.[25]

One author in 1921 countered the accusation that a so-called drugless system that advocated medicinal herbs was hypocritical. He said that the use of botanics was quite unlike "the administration of poisonous and metallic drugs" used by allopaths. Rather, herbs induced cure, as sickness was expelled and vitality in-creased. Among naturopaths who advocated botanics, the controversy over herbs as drugs was frequently answered by this middle-ground claim: herbs were veg-etables. This led to the evolution of an entire branch of naturopathic healing from the early twentieth century on that I refer to as healing foods and dietary renovation. This ideology spawned the lay publication *Nature's Path*. By the late 1940s a definitive shift had occurred, and practitioners' ambivalence or hostility toward botanic medications had virtually disappeared, as demonstrated in the pages of both lay and professional publications. This was owing, in part, to a new generation of practitioners who staunchly reclaimed early naturopathic princi-ples. An author in *Nature's Path* (1947) argued that botanic healers were the first practitioners of nature cure. Several years later, Dr. William Turska went so far as to adapt language from allopaths in his "Catechism of Naturopathy." He likened botanics to chemical remedies—as natural methods of living and treatment. He further argued that nutrients found in botanics profoundly affected tissues that impacted the nervous system as well as the humorals.[26]

Other midcentury prominent naturopaths concurred. A. W. Kuts-Cheraux, ND, MD, argued persuasively in his seminal teaching text, *Naturae Medicina and Naturopathic Dispensatory* (1953), that botanical agents corrected bodily deficiencies and hygienic problems and stimulated the body's innate agents to correct ailments. Several years later, a series of articles in *The Naturopath* addressed folklore and plant drugs. The author claimed that botanics were often trivialized because they were commonly linked with folklore.[27]

Both naturopaths and lay natural healers promoted the vital role of botanics at midcentury. Some acknowledged women's role as administrators of botanics; grandma's medicine chest and her fresh drug infusions or old-fashioned teas with garden herbs were efficacious. One homey article praised grandma's use of ground comfrey root for dental health and a medicine made of five kinds of barks and two medicinal roots for rheumatism. Another explained the benefits of edible and medicinal herbs such as stinging nettle, primrose, milkweed, and bindweed. Authors invoked their mothers as their personal teachers and practitioners. While heralding botanic usage, however, one author added a strong proviso that the authority of a trained doctor was irreplaceable. This injection of authority undermined the self-doctoring that naturopaths commonly linked with botanic applications.[28]

Prominent naturopaths often weighted in. In 1965 Dr. John W. Noble, then president of the National Association of Naturopathic Physicians, investigated and reported on the healing properties of weeds and other plants, primarily St. John's Wort, or *Hypericum perforatum*, used for nervous conditions and depressive states.[29]

By the mid-twentieth century even some allopaths acknowledged the efficacy of herbal medications and delineated the preparatory processes of decoctions, extracts, and infusions, along with their therapeutic effects such as rejuvenative, sedative, and stimulant.[30]

BOTANICS AND THE AMERICAN INDIAN INFLUENCE

In the 1930s, Indian John of the Great Plains was an author whom many referred to as a teacher. His writing contributed to naturopathic knowledge about how to counterbalance toxic botanicals taken internally. Naturopaths readily acknowledged the influence of American Indian plant knowledge on them. One wrote derisively in the 1950s of allopathic doctors who suddenly found a so-called new remedy, such as the Ohio physician who studied seventeenth-century American Indian medicine men and "discovered" an herbal preparation that was effective in clinical tests. Naturopaths and homeopaths learned from Indians' use of squawroot as a blood purifier, ocotillo to overcome diabetes-like conditions, saw palmetto to treat urinary conditions, and Cactus Grandiflora for heart conditions.[31]

Decades later Dr. Jeanne Albin, the first Native American naturopath, was featured in a naturopathic article (1991) for completing her degree at the National College of Naturopathic Medicine at age fifty-one. A member of the Coast Salish tribe and residing in British Columbia, Albin had not been raised with traditional tribal healing beliefs. Yet she and her (non–American Indian) husband, Steve Albin, ND, believed naturopathy was well poised to address the health needs of diverse American Indian communities. The Native American Rehabilitation Association (NARA), a drug treatment program, benefited from Albin's Native American and naturopathic connection and lecturing ability while she was still a student. Her link between the two philosophies resulted in clinical opportunities for naturopathic students and natural health care for NARA clients. Similarly, in the early 1990s a reservation-based hospital in the Midwest negotiated with a naturopathic college to develop a residency program for its graduates. Albin's proposal that the two be linked met with enthusiasm from the students. Albin stated, "The students felt N.D.'s are a bridge between native people and conventional medicine." For Albin herself, her insistence on linking American Indian and naturopathic methods led the East Bank Reservation in Kelowna, British Columbia, to court her to practice medicine there—offering a 2,000-square-foot clinic.[32]

There are other examples of American Indian and naturopathic cooperation. The Indian Health Clinic in Portland, Oregon, uses both traditional Mohawk and naturopathic methods, believing that naturopathy is most like Native American medicine. Likewise, David Paul, ND, of Phoenix, Arizona, investigated how naturopathic doctors could practice on Indian reservations. Reciprocity and acknowledgment of complementary practices continue to characterize both American Indian and naturopathic healing philosophies.[33]

By the 1970s, books published by commercial presses resonated with Native American traditions and signaled growing intercultural acceptance of herbal medicines. One cataloged the historical, therapeutic, and ritualistic traditions individuals should employ. There were several prohibitions, at least some likely well founded, such as never facing into the wind when picking herbs, not letting them touch the ground once harvested, not touching them with cold iron containers or utensils, as this destroyed their healing properties, and not picking herbs with the left hand. Appropriate weights and measures, nutritional elements, and a glossary were offered. The author recommended that herbal self-doctors address the plant to give thanks, following traditional American Indian practices as well as medieval Wiccan and pagan beliefs, all of which instill mindfulness for maintaining an ecological balance while harvesting, as opposed to depleting precious resources.

During a 1989 class on American Indian ethnobotany the ancient ritual of honoring plant use was performed. Jane Dumas, a Kumeyaay tribal elder widely respected as an expert on plants native to the San Diego region and a herbal healer, began the class session on plant usage and healing properties by address- ing the complete plants—with roots—lying on the table at the front of the class- room. She thanked the plants for giving their lives so that humans might learn from them. At the end of the class Dumas gently wrapped the plants and their roots in newspaper in preparation for taking them home to use them fully. Im- plied in these practices is that other species are as valuable as humans and that reciprocity is essential. Plant knowledge is a *gift*, given to people through their herbal teachers, who come to them through the graciousness of Mother Earth and Father Sky.[34]

Late-twentieth-century naturopaths continue to utilize botanics as a central medicinal aide and produce literature to educate the public in ways to link nutri- tion, botanical medicines, and other naturopathic modalities for wellness and healing. Presently, course curricula at accredited naturopathic schools include the medicinal use of botanics.[35]

CHAPTER 2

Spokes of a Wheel

The Healing Systems of Naturopathy

At the turn of the twentieth century, as in our own time, the term *natural* was open to interpretation. Any number of methods could be used, and were, to help the body naturally heal itself. Besides botanics, successful, noninvasive methods that had been used for centuries were hydrotherapy, or water cure, and modes of hygienic living. Naturopaths also began to include variations of ancient and modern applications of blood washing, fasting, massage, eliminative therapeutics, color healing, electricity, homeopathy, hypnotism, iridiagnosis, light baths, magnetism, neuropathy, and zone therapy. While there was never complete agreement about the validity of all these healing systems, they were all spokes of the wheel that became naturopathy.

As their scathing critiques of allopathic medicine increased, naturopaths turned to three well-established remedies: water cure, hygienic living, and homeopathy. Still used today, these modalities have been modified to create a system of naturopathic methods and ideology.

HYDROTHERAPY, OR WATER CURE

One healing agent universally accepted by naturopaths was water. Naturopaths knew their history: water therapies had long been accepted as medically effective, particularly in the treatment of fevers, abnormal skin conditions, childbirth, nervous conditions, and depleted physical energy and as a systemic purifier. In addition, water's healing properties had long been culturally validated by its wide use in religio-magic traditions.[1]

Naturopaths in America aligned with cold water cure, a distinct healing system not derived from the European mineral bath spa experience. It was introduced to the nation in the 1830s by Thomas L. Nichols, Joel Shew, MD, and R. T. Trall, all of whom had observed Vincent Priessnitz's (1799–1852) curing methods and miraculous results in the mountains at Gräfenberg in Austrian Silesia (now the Czech Republic). It was affectionately called the "Water University" by those who knew it well, although its founder did not use the term *hydropathy*. Priessnitz's methods met with immediate success; his patient population increased in number from 49 in 1840 to 1,500–1,700 annually thereafter. Health seekers came from all walks of life, from the privileged and the impoverished, and from numerous nations. Priessnitz's successes stemmed from removal from unhealthy urban or stressful environments, elimination of excesses, a pleasant communal setting, the mystical healing powers attributed to him personally, diet and exercise improvements, ceasing heroic therapeutics, and letting nature right what was reversible. The longevity of his success—and of his techniques— was owing to patients' involvement in their own cures and their adoption of healthy habits.[2]

Although Priessnitz's apostles Nichols, Shew, and Trall introduced his theories and practices in America, medical journals had chronicled his methods and therapeutics for a decade. In the 1840s the three collectively began a prolific publishing trail that made water cure a part of home self-doctoring, most of which was done by women. The *Water-Cure Journal*, published from 1844 to 1913, made the water-cure philosophy and methods accessible to tens of thousands of Americans. Women's leadership was a vital legacy that hydropathists bequeathed to naturopaths.[3]

By 1843 Shew and Trall had opened an establishment in New York City. Their therapeutics depended largely upon cases they had observed Priessnitz treat at Gräfenberg. Trall helped develop hydropathy into a comprehensive healing philosophy dubbed the "Philosophy of Medical Science" and the system of "Hygienic Medication."[4] The philosophy of hydropathy, taught to lay readers and physicians alike in the *Water-Cure Journal*, was to cease drugging, use water as the primary curative agent, employ a skillful water-cure physician and good nursing care, eat simple foods only, partake of fresh air and exercise, stay free of high excitements, and pursue pleasant associations and cheerful companions.[5]

Naturopathic practitioners had embraced hydropathic life theories and therapeutics by the late nineteenth century. Water, pure and simple, not stagnant, was applied according to a patient's "reactive power." The water temperature was determined by the ailment and the patient's constitutional vitality. Specific therapeutic applications included general bathing, pouring baths, shower baths, plunges, dripping sheets, douches (a stream of water falling onto the locally inflamed site), sitz baths, wet-sheet packs, blanket packs, and fomentation (the equivalent of a wet compress), among others. As Mary Gove Nichols, hydropath, author, activist, lecturer, and spouse of T. L. Nichols, wrote in 1850, water processes facilitated the removal of putrid matter from the human system, internally and externally, and fostered healing and rejuvenation.[6]

Publications of the fledgling American naturopathic movement repeatedly pointed to the efficacy of nineteenth-century hydropathic therapeutics. Patient relief was medically observable, with quantifiable results for fever control, reduced inflammation, burn healing, sleep inducement, and treatment of a variety of other common conditions. There was also diminished pain and discomfort when treatment included touch and personal communication.[7]

Benedict Lust built on the rich tradition of hydropathy, entrusted to him by Kneipp, when he brought the *Kneipp Water Cure Monthly* to America in 1900. The front-page subhead proclaimed it "A Magazine devoted to the late Rev. Father Kneipp's Method and Kindred Natural Systems." The page also displayed a large portrait of Sebastian Kneipp (1824–1897). In the premier issue, Lust, who served as the editor and manager, articulated the basic tenets of obedience to nature's laws with his "Return to Nature" system. First was Kneipp's advocacy of water cure, followed by the use of air, light, sun heat, water, rest, exercise, gymnastics, massage, diet, electricity, magnetism, and herbs.[8]

In the same issue, Lust traced the history of medicinal uses of water. Narratives of allopaths' conversions to natural therapeutics created a theme of personal salvation, at times both medical and spiritual. A prime example is the narrative of John Schroth (1798–1856), a schoolmate of Priessnitz's who broke his knee in a livery accident in 1817. Lameness seemed his lot in life until at the urging of a monk he self-administered water-cure therapeutics to his knee. He regained mobility and became a staunch proponent of the system. Schroth, who became a respected healer, also espoused the value of a well-regulated diet to reduce weight in the treatment of sickness. He prescribed curative soups and foods, along with periodic abstinence from food and drink. His dietary treatment combined with water cure became known as the Schroth Cure. In their 1870s texts R. T. Trall and E. P. Miller continued the tradition of combining water and diet as complementary therapeutics. They counseled strict dietary renovation, usually vegetarianism, abstention from alcohol and tobacco, and moderate to no use of spices and no use of grease in cooking. Lust and Stroebele combined strict dietary changes with air and water therapies.[9]

The Lusts also saw sharp social-class distinctions among those practicing water cure versus those seeking allopathic expertise. Lust wrote that water cure was the science of healing for the plain and poor people. The rich, he believed, were perpetually dependent on their doctors to cure their sequential illnesses. Water, the poor man's treatment, relieved him of pain and drugs, improved and cleansed his system, and gave him greater strength when he followed the simple rules of natural healing.[10]

Lust explained his own inspiration to work as an emissary for Father Kneipp in America: He had been a mere shell of a man when he arrived in America. He had endured several operations, and he blamed a half-dozen vaccinations for contributing to his ill health and tuberculosis. He had weighed a skeletal 104 pounds and found no relief through allopathic or homeopathic remedies. He wrote in his memoir that American doctors filled out his death certificate in front of him—possibly a grandiose claim to promote his own beliefs. Devastated, he wrote, he had returned to Germany to expire in his homeland. In a final sputter of self-preservation, he had sought out Father Sebastian Kneipp, who restored him to health in just eight months. Rejuvenated and zealously converted to these new healing methods, he had returned to America in 1894 as Kneipp's sanctioned emissary.[11]

Kneipp's use of hydrotherapy combined with herbalism had earned him great acclaim. Among those singing Kneipp's praises were Theodore Roosevelt, Austria's Archduke Ferdinand, and the pope. Less famous although profoundly influential was Henry Lindlahr, a prominent wealthy businessman who heralded Kneipp's methods after the priest treated his advanced diabetes successfully. After his cure, Lindlahr obtained his DO and MD and became a well-known author and naturopath based in Chicago.[12]

Lust benefited tremendously from his special relationship with Kneipp, who died the year after Lust's return. By all accounts Lust was a skillful practitioner who could apply water a thousand different ways. Yet as a pragmatist Lust began to distance himself from Kneipp's strictest methods and included other modalities. He rejected Kneipp's dictum that his patients walk barefoot in early morning dew and in local streams. Lust believed that this behavior, when carried out in New York's Central Park, made his followers the target of ridicule.[13]

Lust established the magazine *Amerikanishe Kneipp-Blatter*, a sanitarium, and a store in New York. While Lust benefited from the fame accrued to Kneipp the individual, he later went so far as to bar some practitioners of Kneipp's methods from his naturopathic organizations as he developed his own regimens. In 1901 he created the American School of Naturopathy in New York. In 1902 the Lusts established the Naturopathic Society of America. Through these institutions Lust and his colleagues taught water-cure principles. Within six years,

1896–1902, he combined all natural methods under the distinctive term *natu-ropathy*. A defining moment came in 1902, when he changed the name of his English-language magazine from *The Kneipp Water Cure Monthly* to *The Natu-ropath and Herald of Health*.[14]

Rapid momentum solidified Lust's health empire. Water cure and naturopa-thy had become inextricably linked. Under his editorial watch, his journals chronicled tales of successful water-cure treatments. In 1921 Mark Twain ex-plained how he had come to naturopathy: his mother had practiced the Kneipp water cure on him. She had poured a bucket of cold water on him and rubbed him down with flannels, covered him with a sheet wrap, and put him to bed. He said he had perspired so much that his mother had put a life preserver in his bed. Twain, an outspoken ally of common people, detested the growing insistence on credentialed medical expertise and said that he had come to trust a healer based on the healer's ability, not licensure.[15]

Naturopaths wrote constantly on the therapeutic applications of water. They saw water cure as just one highly valued element in a healing regimen. Louisa Lust was an eclectic thinker and innovator who integrated water therapies with plant usage and dietary regimens. She delineated the treatment of fever with rhubarb tea, no solid foods, cool water with a spoonful of lemon or orange juice, lukewarm enemas, hourly water ablutions, and abdominal cold packs. Hydro-therapy, Professor Robert Bieri trumpeted at a 1920 convention, was, and always would be, the backbone of drugless healing. He praised Dr. Lust's Naturopathic College for its thorough instructions in hydrotherapy and the ever more promi-nent position it would hold in future curricula. The naturopathic practitioner Dr. Leo Scott used innovative applications of water cure and electrical stimula-tion. This meshing of hydropathic principles with other therapeutic practices such as the use of herbs, dietary renovations, fasting, hypnotism, and "crisis" (bringing fever to an acute stage) came to be key to natural living and healing.[16]

HYGIENIC LIVING

Hygienic living was another healing ideology widely accessible to a lay populace that preceded naturopathy and was absorbed into it. The principles were pro-moted by Sylvester Graham (1794–1851) and William Alcott (1798–1859), both part of a vibrant network of health reformers who included Thomsonians, eclec-tics, homeopaths, and those opposed to drugs and allopathic doctors. Hygiene, from the perspective of nineteenth-century healers, included all aspects of life that influenced good personal health, including food, exercise, sleep, clothing, and breathing, among others. The word *hygiene* derives from *Hygeia*, the name of the Greek goddess of health, and only after the advent of germ theory in the late nineteenth century did the term refer more to cleanliness than to general health and well-being as in earlier generations.[17]

Graham was a moralistic health crusader who began as a temperance lecturer in Philadelphia in 1830 and then turned his considerable oratory skills to freelancing in 1831. He wrote on hygiene, and in 1837 with David Campbell he founded the *Graham Journal of Health and Longevity*. He also published two volumes of lectures promoting vegetarianism, sunlight, frequent bathing, regular exercise, fresh air, dress reform, and sex health, known then as sex hygiene. Graham believed that any sensual stimulation, including marital intercourse, depleted the body. To guard against debilitating depletion, one should employ simple clothing, a nonstimulating and moderate diet, abstention from tobacco, non-arousing reading, and controlled contact between the sexes. Graham cautioned against overeating because he believed the stomach was the center of bodily systems. Gluttony would overtax the digestive system and unbalance, or derange, other bodily systems as they strove to process excessive, spicy, or fleshy foods. He advocated consumption of whole-grain bread only—unbolted, or unsifted. This led to his invention of the "Graham cracker," which he urged as a staple food in hygienic households.[18]

The educator and reformer William Andrus Alcott, a contemporary of Graham's, also preached physical and moral salvation. Like Graham, he believed in a diet of only water and vegetable matter. A prolific and passionate author, Alcott published numerous singly authored books and countless articles in *The Moral Reformer and Teacher on the Human Constitution*, which he launched in 1835. Among his most impressive works was his *The Young Woman's Book of Health* (1855), in which he argued favorably for women's education and contradicted the widely held belief that women's physiology was debilitating, sickly, and intellectually as well as socially limiting.[19]

From the 1830s to the 1850s Graham and Alcott were powerful advocates of hygienic living and radical dietary reforms. The founding of the American Physiological Society in Boston in 1837 was in part a result of Graham's public lectures. Alcott served as its first president and urged the founding of anticorset societies in keeping with his commitment to dress reform and to women's physiological parity with men. Both activists were contributors to the *Water-Cure Journal*, and both preached the natural intersections of diet and cold-water cure.[20]

At the center of hygienic living was vegetarianism. Graham and Alcott—and their many followers—argued that vegetarianism was morally and physiologically superior to meat eating. A vegetarian diet was more digestible, more nutritious, and more rational given the dangers associated with flesh eating. Meat products not only overtaxed the digestive system, they argued, but also impaired physical health overall, carried contagious diseases, and were therefore unhygienic.[21]

Graham and Alcott's legacy as hygienic advocates was enriched by Dr. John Harvey Kellogg (1852–1943), a noted allopathic physician who had attended the Michigan State Normal School and the New York University Medical College. He was a trained hydropathist, an advocate of vegetarianism, a dress reformer, and a surgeon. In 1901, with his brother Will, he founded the Sanitarium Food Company, which became the Kellogg Food Company in 1908, producing the ubiquitous cornflake cereal. Disease, he wrote, could be lessened by change in intestinal flora, and he advocated vegetarianism, low-protein and high-fiber foods, and laxatives to that end. He was the first to mass-produce health foods such as granola, soy-based products, and coffee and tea substitutes to facilitate hygienic living. He was a Seventh-Day Adventist, and beginning in 1876 he served as the chief medical officer of the Battle Creek Sanitarium in Battle Creek, Michigan. The sanitarium was operated and owned by the Seventh-Day Adventist Church, and vegetarian dietary reform and hygienic living were taught there. Later in life, Kellogg was expelled from the church for beliefs that contradicted its doctrine. He invented the electric light bath. He discovered that oscillating electrical currents applied to the body produced muscle contractions that had the therapeutic effects of passive exercise without pain in disabled patients (this electrical therapy was later termed the *sinusoidal current*). Kellogg denounced alcohol, tobacco, and sexual stimulation, believing that they ruined health and caused social disorder. In 1906 he was one of a growing number of doctors who believed that right living and hereditary traits produced more desirable people. He cofounded the Race Betterment Foundation, which became a major center for the American eugenics movement, and proposed a eugenic registry that would designate proper breeding pairs.[22]

The "rationale for hygienic living," articulated by Graham, Alcott, Kellogg, and like-minded colleagues, became one of naturopathy's foundational beliefs. Dr. Henry Lindlahr argued that health was determined by an individual's hygienic living, and good habits of eating and drinking were central to that life. In 1914 Lindlahr and his coauthor and spouse, Anna Lindlahr, published their extensive compendium of more than nine hundred recipes for right eating.[23]

For naturopaths, hygienic living meant clean bodily systems. William F. Havard, ND, explained in 1920 that according to the naturopathic theory, disease originates "with an accumulation of waste material in the blood, and gives as the cause for this condition heredity plus abuse of the body through unnatural methods of living." The body needed clothing appropriate for one's climate, exercise, sun, and raw foods despite social and governmental pressure to the contrary, and women's role as food preparers was central to hygienic living.[24]

ALLOPATHY VERSUS HOMEOPATHY

Homeopathy posed the largest competitive threat to allopathy in the mid-nineteenth century. Allopathy was distinguished by its college-educated and licensed practitioners, who claimed unique expertise and dramatic, if sometimes harmful, therapeutics. The dramatic, potentially life-threatening results of allo-pathic, or heroic, medicine both attracted and repelled the American populace. By the 1840s, physicians' efforts to professionalize and elevate allopathic methods led practitioners to aim to satisfy the age-old plea of the sick to "do something." But invasive therapeutic results could be dangerous and produce side effects. This conundrum contributed greatly to the popularity of the popular health movement of the 1830s and 1840s and of competing sects offering milder and more self-determined therapeutics. By the time allopaths formed the American Medical Association (AMA) in 1847, they were singularly aggressive in their attempts to control professional health practice. But because of the mixed reception to al-lopathy, many regular allopathic practitioners converted to homeopathic practice, which proved to be more lucrative.[25]

Allopaths, like all competing sects at that time, offered a rationale for disease causality and methods of cure. Prime among these was the notion that the body possessed a finite amount of vital force that must be kept in equilibrium or disease would result. In the early nineteenth century, allopaths employed blood-letting, purgatives, cathartics, and emetics in an attempt to restore balance by ridding the body of putrid matter. In later decades, these methods were replaced by drugs and alcohol-based medications. The goal of these invasive therapeu-tics was to cause discharge via boils, blisters, urination, perspirations, or erup-tions. All of these, it was believed, signaled the release of sickly matter and the possibility of health's return. While their methods varied widely, all of allopaths' competitors—sectarian practitioners of botanics, eclectics, hydropaths, advocates of hygienic living—sought instead to help the body, through natural means, expel toxic matter and prevent and cure disease. At times their therapeutics were also actively interventionist, but the substances used were generally less damaging than the allopathic pharmacopoeias, some of which were metal- or opiate-based, and occasionally toxic.[26]

In this context, hydropathy and homeopathy offered milder therapeutics and signaled a decisive turn away from both the dramatic allopathic treatments and the physician's authoritative posture. These two sects, more so than others, emphasized prevention through right living rather than dramatic episodic interventionism once disease had manifested. Self-control and moderation were at the center of these ideologies. Homeopathy and hydropathy appealed to the common folk as well as to intellectual and social progressives who embraced notions of self-determination and self-control rather than depending upon a credentialed expert. This rejection of professional dependency dovetailed with the leveling of social classes so valued in Jacksonian America. The winds of democratization help explain the popularity of these competing sects at midcentury. But by the late 1800s, as naturopathy was forming, cultural values had shifted dramatically. During the Gilded Age of upper-class extravagance and middle-class desires, medical self-determination became less attractive if one had the ability to employ an expert, a signal of both social status and the success of licensing laws passed by allopaths.

Homeopathy, introduced in 1825 by the German practitioner Samuel Hahnemann (1755–1843), bridged the competing doctrines of patient self-determination, professional education, and credentials. Hahnemann's theory of disease and cure offered three principles whose simplicity appealed to many. First, he asserted the law of similars, like cures like: substances known to produce certain symptoms in healthy individuals ought to be prescribed for individuals suffering from diseases with those same symptoms. Second, he proposed that the effectiveness of medicinal doses was increased when smaller amounts were used. His theory necessitated that practitioners weaken (dilute) preparations up to 30 times their original strength. Third, he posited the law of infinitesimals, the premise that most diseases were caused by a suppressed itch, or psora. As he searched for less invasive ways to treat sickness, Hahnemann self-administered quinine, the drug most used to treat malaria. He noticed that he developed chills and fever, the most common symptoms of malaria. Dilutions increased the potency of the drug when succussions (vigorous shaking) accompanied each dose. These minute doses triggered the body's innate self-healing mechanism.[27]

Homeopathy also appealed to Americans who identified with its links to other reform movements, such as temperance, women's rights, and at times abolition. These views contrasted, often dramatically, with the more conservative political positions taken by regular practitioners, who privileged the elite. Furthermore, the exclusive opportunities in regular medicine were open only to males. Homeopathy, like eclecticism, botanics, and hydropathy, welcomed women into its ranks as practitioners. Allopathic medicine not only categorically shunned women as skilled healers but viewed them as a class in need of constant doctoring. In the

context of these opposing world-views, Hahnemann, like his antiallopathic contemporaries, levied frequent and brutal indictments against the heroic practitioners, something naturopaths would continue to do in the next century.[28]

Hahnemann built on the writings of J. Laurie, MD, an allopath who converted to homeopathy. Laurie said that the selection of appropriate homeopathic
remedies would save many patients from allopathic treatments that compromised even the strongest constitutions and induced permanent maladies. In 1876
Hahnemann stressed the uniqueness of each individual's treatment and the necessity of keeping detailed medical records. He condemned allopaths' use of harsh
evacuative therapeutics and of opium, which another homeopathic physician cited
as part of the "destructive" art of healing.[29]

Homeopathy, like the other hygienic living regimens, advocated self-care, minute doses of tinctures and powders, pure food and water, and a rational and moderate living regimen. Overstimulation was to be constantly avoided. By midcentury,
homeopathy claimed a following of tens of thousands, and the fledgling movement
educated numerous physicians to meet the growing demand. Homeopathy had a
considerable impact because it was an alternative to allopathy, about which the
public had doubts. While homeopathy's scientific approach was attractive to some
allopaths, it also reinforced belief in nature's—and one's personal—curative powers. By the 1880s there were homeopathic medical colleges in major cities, some of
which educated both homeopaths and allopaths—for example, the University of
Michigan or the New York Homeopathic College, where Lust received his MD
degree. There is little evidence that Lust and the naturopathic movement as a
whole, often self-described as drugless throughout the early decades, embraced
homeopathy. In fact, Lust was treated miserably by his fellow students for his advocacy of hydropathy, which is perhaps one reason why he was largely silent on
the topic of homeopathy. But the lifestyle and medical freedom espoused by homeopaths paved the way for naturopaths, and homeopaths were among the many
mixer practitioners who blended therapeutics.[30]

OSTEOPATHY AND CHIROPRACTIC

When the term *naturopathy* was patented in 1896, many practitioners possessed
training in osteopathy, chiropractic, or both. Osteopathy, founded by Andrew
Taylor Still, and chiropractic, founded by D. D. Palmer, shared with naturopathy
similar conceptual frameworks and an assault levied upon them by regular medicine. Lust was profoundly influenced by his instruction in both osteopathy and
chiropractic. Lust created the New York School of Massage in 1896, and in 1901
he began operating both the American School of Naturopathy and the Ameri-

can School of Chiropractic, all housed in his five-story building on 59th Street in New York City. This location sufficed until 1907 as his health empire grew rapidly. During this period, the three sects generally embraced and complemented one another.[31]

All three sects openly challenged the philosophies and methods of allopathic medicine. Each was aided by the charisma and popularity of its original founder, whose death led to chaos both in leadership and in therapeutics. The competition among these sects reveals a tension for gaining legitimacy from the public, allopathic practitioners, and legislative bodies. However, naturopathy, far more so than either osteopathy or chiropractic, avoided the scramble for institutional approval and acceptance at the turn of the century thanks to the antiestablishment, antiauthoritative stance of naturopaths.

While these three sects interacted regularly, none of them embraced the tenets of another rising sect, the Church of Christ, Scientist (Christian Science). Its founder, Mary Baker Eddy (1821–1910), upon recovery from a devastating illness published her healing beliefs. She organized the Church of Christ, Scientist in Boston in 1879, and in 1895 the church issued a manual that gave it structure and outlined its government and missions. It did not appeal to naturopathy, osteopathy, or chiropractic, because it was a Christian-based faith that preached physical healing through prayer. It claimed that matter and illness were nonexistent and ignored all evidence of bodily ruin induced from unhealthful living. Christian Science's rejection of organic disease contradicted all that Lust and other naturopaths observed and sought to rectify. The religion was hardly mentioned in naturopathic texts, in speeches, or by its leadership except to reference legal persecution. Naturopaths and Christian Scientists did share, however, a belief in the importance of mental and spiritual well-being and Christianity, but the degree and rationale of their belief varied so greatly that it did not induce camaraderie.[32]

Osteopathy's originator, Andrew Taylor Still (1828–1917), was born in Virginia and eventually settled in Kansas with his family in the early 1850s. When three of his children were fatally stricken with spinal meningitis in 1864, he questioned and then rejected the methods and principles of the allopaths who had administered to them. He also rejected homeopathy and eclecticism, despite their less toxic regimens, as he became increasingly unconvinced of the efficacy of any medications. One reason for this view was that alcohol was the basis of so many remedies: he had a moral abhorrence of alcohol consumption.[33]

Still then studied magnetic healing. Pioneered by Franz Mesmer (1734–1815), an Austrian physician, its basic premise was that magnetic fluid flowed throughout the body and that its obstruction was a major cause of disease, especially nervous disorders. The fluid was restored to its proper balance by passing hands or magnets over the body. Still was greatly influenced by Mesmer's ideas, and in addition to emphasizing magnetic energy, he stressed the free flow of blood as a solution to disease or deformity. Still valued the laying on of hands, believing that temperature changes in the spinal column could reveal spinal lesions (disturbed nerves). When Still began laying on hands in his community of Baldwin, Kansas, he was summarily scorned and cast out from his Methodist church, which saw his work as the devil's doing. Rejected and ridiculed, Still moved to Kirksville, Missouri, where he set up practice as "A. T. Still, Magnetic Healer." In the late 1870s Still added the manipulative practice of flexion and extension procedures, rubbing the spine and re-placing bones appropriately in the spinal column. He named his new system osteopathy, *osteo* meaning "bone."[34]

With this therapeutic transition, Still offered a mechanistic model. He maintained that the sick human body must be repaired by restoring its parts to their proper relationship, without pills. Still analogized the body to machinery, echoing contemporary language of productivity and mechanization. For women's physiology, osteopathic theory also offered therapeutics of spinal adjustment, massage, and muscular therapeutics. His writing influenced other practitioners looking for methods preferable to orthodox therapeutics.[35]

Still obtained a state charter to open the American School of Osteopathy in Kirksville in 1892. It was popular with patients, readily recruited students, and won legal protection from the Missouri state legislature in 1897. Students recorded their teachers' skills and methods meticulously, and these continue to serve as clear and compelling examples of the system's methods. Still's teachings inspired an impressive body of medical literature. His own *Philosophy of Osteopathy* (1899) announced his theory of natural immunity. This crucial set of ideas was cocreated and echoed by naturopaths at the same historical moment that Lister was developing his first methods of antisepsis and Koch began identifying specific disease-producing organisms. Still's text also outlined bodily systems and explained how to stimulate their proper blood and energy flow through spinal manipulation and circulatory stimulation to induce healing.[36]

From the beginning, Still's students contributed to this body of literature. They proudly identified themselves as graduates of the American School of Osteopathy, some adding, "under the Founder of the Science." Early practitioners and authors portrayed osteopathy as a science to refute impressions that it was a mere art form. They emphasized the need for scientific knowledge to distinguish their skills from the more pedestrian skill of the historically less valued bonesetter.[37]

CHIROPRACTIC, NATUROPATHY, AND MIXER SCHOOLS

Chiropractic shared with osteopathy the belief that spinal alignment was crucial to bodily integrity and health. Like naturopathy and osteopathy, chiropractic met with considerable resistance from the orthodox medical profession; the regulars and the AMA had no need to ally with them and preferred to battle with them for public legitimacy.[38]

The founder of chiropractic, Daniel David Palmer (1845–1913), was, among other things, a Canadian-born former schoolmaster. He practiced magnetic healing for a decade before he performed his first chiropractic adjustment, in September 1895 in Davenport, Iowa. Based on successful patient outcomes, he created the art of adjusting vertebrae to relieve pressure on nerves. He used the spinous and transverse processes as levers.[39]

Palmer incorporated the name of his system in 1896, the same year naturopathy was named, and simultaneously founded Palmer's School of Magnetic Cure in Davenport. His first students entered in 1898, and the classes averaged between two and five students through 1902. In 1907 David's son B.J. purchased the school from his father and renamed it the Palmer School and Infirmary of Chiropractic. Under B.J.'s charismatic and innovative leadership the school flourished, as did a far-reaching radio station he owned. The patients of chiropractors, like those of early osteopaths, came from all socioeconomic levels, especially where medical doctors were scarce and when medical treatments were too expensive.[40]

The response of allopathic medicine to chiropractors was swift and uncompromising. "Doctors," wrote one historian, "by this time organized and better aware of the threat to their livelihood, began to assail them with a systematic virulence unequaled in the annals even of the medical profession." Palmer was jailed repeatedly for practicing medicine without a license, and countless prosecutions were staged, mostly prodded by MDs. Naturopaths would come to know this method well. Dummy patients were sent to chiropractors to trick them into misdiagnosis. These backfired in the court of public opinion, however. In California in the early 1920s, 450 of roughly 600 California chiropractors were thus entrapped. When faced with a fine or jail time, many chose jail, which garnered them massive public sympathy and support. When a chiropractic licensing law came to a vote in 1922, it passed by an overwhelming majority, whereas a similar bill before the mass convictions had failed.[41]

Meanwhile, the philosophy and therapeutics of chiropractic continued to evolve and to be refined in its texts. The basic principle of the system was that disease was a result of violating foundational natural laws. Disease was found in subluxation of the spine, and relief was gained through adjustment of subluxations.[42]

In these early years, graduates of Palmer's school critiqued, challenged, renamed, and differentiated themselves from the founder's view at times. Oakley G. Smith (1880–1967) founded naprapathy in 1907 as a new and distinct school of healing similar to chiropractic, but rather than focusing on irritated nerves, it centered on diseased ligaments. He opened the Chicago National College of Naprapathy. Andrew P. Davis, MD, DO, DC (1835–1915), founder of neuropathy, theorized that pressure induced pain and that removal of that pressure restored health. In 1909 Davis advocated several therapeutic modalities and emphasized harmonizing acid and alkaline secretions in tandem with spinal adjustments.[43]

Years later, in 1946, Thomas T. Lake, a naturopathic chiropractor, argued that neither chiropractic, osteopathy, nor naturopathy could stand alone. "All of them have a missing link per se. None of them is complete." He argued instead for the complementary use of the therapeutics of all three, including some drugs. And he proposed that a blend of all three would be an excellent adjunct to psychiatry and physical therapy.[44]

This constant recasting led to competing national organizations. Dissension among chiropractors culminated in 1905–6, when Solon Massey Langworthy, a 1901 Palmer graduate who operated the American College of Manual Therapeutics in Kansas City, Missouri, founded the American Chiropractic Association (1905), with himself as president. In response, in 1906 B. J. Palmer organized the Universal Chiropractors Association, whose first official resolution scoffed at Langworthy and his claim of modernized chiropractic orthopedics. Within a few years Langworthy, his school, and his association disappeared. But numerous other homebred critics flourished. Amid this fractious environment, chiropractors were particularly vulnerable to outside criticism, attack, and legal censure.[45]

Thus the beginnings of chiropractic medicine were far from smooth. Its history of varied theories and competing organizations is eerily similar to naturopathy's. A new American Chiropractic Association, founded in 1922, hoped to provide a broader alternative to B. J. Palmer's organization. Simultaneously, organized medicine continued to attack chiropractic through legislation requiring basic science education. Yet another group, the International Chiropractic Congress, led the way in bringing consensus to these disparate groups, foreshadowing international naturopaths who bridged competing naturopathic societies decades later. The American Chiropractic Association and the Universal Chiropractors Association merged in 1930 to form the National Chiropractic Association; by 1934 the International Congress had joined to constitute part of its council structure. From this emerged chiropractic's first Committee of Education, which sought to design uniform standards.[46]

Individual chiropractors embraced naturopathy, and at the "mixer schools" chiropractic and naturopathy were often joined, since students could earn both the DC and the ND degree with additional instruction. The schools taught naturopathy, physio-therapeutic modalities, and chiropractic and mirrored individual founders' commitment to both systems. When these two systems were coupled in an educational setting, they borrowed heavily from early osteopathy and emphasized drugless healing and the healing power of the body and nature. Mixer schools proliferated. Some taught applied obstetrics, minor surgery, and the setting of simple fractures. They did not teach major surgery or materia medica. Some chiropractors were not very different from naturopaths, but the former's emphasis on spinal manipulation differentiated the two, as chiropractors always placed it at the center. Other, later innovators recast chiropractic as bloodless surgery.[47]

AND THESE TOO HEAL

For naturopaths, several other healing modalities were clustered with the core therapeutics. The ever-expanding complementary therapeutics included blood washing, eliminative therapeutics, color healing, fasting, electricity, iridiagnosis, massage, light baths, and hypnotism. It is beyond the scope of this book to devote great attention to them, but their use points to the experimentation with techniques that informed the practice of naturopathy. The sheer number and variety of methods, while a potential strength for naturopathy, became a source of diffused identity and internal combativeness. Some of these therapeutics reflect the influence of urban industrial technology upon naturopaths, despite their rejection of it. At the same time they reveal attempts to combat the effects of unhealthy city life.

Blood washing, also known as a water shower, was recommended by Lust and many others. It is clearly a form of hydrotherapy, since the slogan "wash and be clean" appears repeatedly. The masthead of the *Water-Cure Journal* read, "Wash and Be Healed." Blood washing was seen as a distinct treatment, however. Preceded by an enema, ideally the bath, taken sitting or lying down in water at 106 degrees, lasted eight hours or more. The treatment purportedly created feelings of youthfulness, as it energized the person, washed the blood, cleansed the whole system, and relaxed the mind; one's skin, teeth, eyes, and hearing all benefited. Streams of water were aimed at the sides of the head, the spine, the abdomen, and the soles of the feet. Focusing on these body points was also called zone therapy, similar to reflexology or acupressure. The Lusts offered blood-washing treatments at both Yungborn locations and at their Naturopathic Institute in New York City. Typical was a patient who was cured of a chronic sore arm and elbow and permanently relieved of his symptoms with additional massage around the knee and with dental care for badly decayed teeth.[48]

Just as blood washing was thought to cleanse the system, so did therapies that emptied the bowels. These treatments went by various names: colonic irrigation, enemas, natural laxatives, rectal dilation, constipation cures, and internal baths, among others. All aimed to purge the body of obstructed waste materials. Nearly all naturopathic practitioners advocated these practices, and numerous devices, food products, and therapeutic treatments were created to meet the need. As one author wrote in 1923, "By the use of irrigations the retained, concentrated putrefactive material is rapidly gotten rid of. . . . Coupled with a vegetarian diet it . . . results in a most satisfactory change in the intestinal flora, which ends the process of self poisoning." The topic was so important that it was the subject of a keynote address before the thirty-ninth annual congress of the American Naturopathic Association in June 1935, in which the speaker marketed his own device. The therapies continued into the 1940s and beyond.[49]

Another eliminative therapeutic advocated by many naturopaths was fasting, also called the bloodless operation. It expunged lingering poisonous matter in the body. The fast cure shook up the entire organism and removed diseased tissues. This practice ultimately led to legal battles and incriminations. Lust openly remained loyal to those accused of causing patient deaths, since neither he nor the practitioners who used fasting agreed that it was the cause of death.[50]

The concept behind fasting was simple: morbid matter built up in the system as the result of excesses and abuses, and the most efficient way to cleanse the system was through fasting. One 1915 author advised no food or drink except water to restore equilibrium. Animals did this, he wrote, as did people who had studied and practiced it themselves. Not only did it purify the system by increasing the activity of the eliminating organs but it gave the digestive and assimilative organs a complete rest. In short, the body increased its capacity to repair itself. However, fasting alone was not a cure. Moderate exercise, sunshine, air, water, and mental and physical relaxation were also needed.[51]

Bloodless surgery was widely advocated by naturopaths in the 1930s and 1940s. The inaugural issue of the *Journal of the American Naturopathic Association* in 1948 announced the upcoming "Bloodless Surgery Conference," to be held in Long Beach, California. Timed to coincide with the Pasadena Rose Parade and the Rose Bowl, it promised sunshine for northerners. It would bring practitioners together to share experiences and to help form standard procedures. Recently, allopathic research has been rediscovering the impact of fasting on the body's ability to heal; one publication in 2012 stated that "multiple cycles of fasting . . . could potentially replace or augment the efficacy of certain chemotherapy drugs in the treatment of various cancers."[52]

Other complementary therapeutics focused on diagnostic tools, one of which was iridology, the "reading" of the eye's iris. Iridology was popularized by Emanuel Felke (1856–1926), an ardent follower of the Hungarian physician and homeopath Ignatz von Peczely (1826–1911). Together they advocated disease detection by reading facial and iris signs. Felke claimed that he could assess a patient's condition and the presence of mercury, quinine, iodine, opium, morphine, and other drugs from the discolorations in the cavity, or lacuna, (depression) of the iris. Felke used iridology in combination with massage, herbal teas, plants, homeopathic remedies, diet, water applications, outdoor exercise, and air huts (cabins or huts with screened openings at the top of the walls to allow plentiful fresh air). Like so many other natural healers, Felke was hounded by legal authorities. Sixteen lawsuits were brought against him, and in 1909 twenty-four physicians testified as witnesses—both for and against iris diagnosis.[53]

In 1907 full-page ads in the *Naturopath and Herald of Health* explained the benefits of iridology. It allowed self-diagnosis and healing and led to self-revelations of medical value. It also offered evidence of glandular extracts and operations. A few things are notable here: This was a tool for teaching *self*-diagnosis; in these early years the physician was not vital. Iridiagnosis came with such a wide-ranging variety of promises that it was an eclectic panacea unto itself. It took the work of individual naturopaths refining the practice to make its claims less fantastic and more believable. Yet it remained one area in which hyperbole reigned and claims abounded.[54]

Countless prestigious naturopaths heralded iridology. Henry Lindlahr was a powerful and articulate advocate. He said the eye not only mirrored the soul but could reveal abnormal changes and conditions in the organs. By examining "the density of the iris, nerve and scurf rims, itch or psora spots in the iris, signs of inorganic minerals or poisons in the eye . . . excesses in iodine, lead, arsenic, bromids [sic] and coal tar products are seen. Disease in the vital organs are [sic] also seen in the eye as are chronic diseases." Lindlahr was joined by other practitioners who provided instructions for "reading" the eye and employing treatments to complement natural therapeutics.[55]

Naturopaths acknowledged that the claim that a person's physical condition could be determined by examining his or her eyes "sounds like a fairy tale." But one explained that each organ of the body was represented within a well-defined area of the iris. Abnormal bodily processes observable in the iris included circulatory disturbances, lymphatic encumbrance, drug deposits, nerve irritation, and destroyed tissue. Others explained that the iris of a healthy human had a uniform texture; abnormal lines, spots, and discolorations revealed pathological and functional disturbances in the body. They believed that they could use these insights to diagnose a disease and determine its cause, prognosis, and treatment.[56]

In 1926 iridology was deemed a recognized branch of naturopathy, and the meeting of the International Iridology Research Society was promoted in the *Naturopath*, which contained numerous articles on the subject. One speaker questioned the validity of iridiagnosis and scoffed at a machine invented to perform iris analysis that had become popular among drugless healers. He noted that the machine had "dampened our ardor for iridiagnosis to a large extent, [but] we have by no means wholly forsaken this method." The speaker said that many wild and unfounded claims had been made about iridiagnosis, and he concluded that while evidence of sickness might be seen in the eyes, there was no proof that a spot on the iris indicated disease in a certain organ of the body. Yet two months later, in a speech given before the twenty-seventh convention of the American Naturopathic Association in Chicago on the cause and prevention of cancer, F. W. Collins reinforced the more sweeping claims of iridiagnosis. With it, he said, "we can tell many years in advance if cancer is developing and the organs of the body that are affected." While grandiose assertions may have exaggerated the usefulness of iridiagnosis, it was a precursor to the method which ophthalmologists today use to spot diabetes, clogged arteries, hypertension, liver disease, cancer, and autoimmune diseases by "examining the blood vessels, nerves, and structure of the eye."[57]

By the mid-1930s authors linked iridiagnostic insights with the deleterious effects of vaccination. Vaccination, it was said, darkened the scurf rim, and a dark ring or band around the outer edge of the iris was evidence of skin defects. By this time the term *ocular diagnosis* was occasionally used in place of *iridiagnosis*. One author asserted the pupil's ability to show both emotion and disease. The popularity of iridiagnosis led R. M. McLain, a chiropractor and naturopath, to make the claim that it "stands out as the one pre-eminent diagnostic discovery of the age." Since the technique was difficult to master, he advocated the use of the microscope to aid the healer in reading the eye. By the late 1930s his articles were typically outlandish; in 1939 one denounced the need for laboratory corroboration for substantiation.[58]

Naturopaths embraced many more therapies to stimulate the body to heal itself, complicating their claim of a cohesive set of therapeutics. Among them were color and light, electricity, hypnotism, magnetism, and massage. Of these, massage was the most widely used and had the clearest benefits. Unlike other disputed therapeutics, there was complete agreement on the value of physical culture and massage. A national concern arose during industrialization (ca. 1865–1900) that middle- and upper-class Euro-American men were weakened by sedentary labor, which lacked the physical rigor of manual labor. During this time Swedish and German gymnastics grew in popularity, accompanied by sports programs in schools, in the military, and in public venues. Fueled by the model of President Theodore Roosevelt (1858–1919), the emphasis on exercise and physical activity grew.[59]

Most famous and influential in America was Bernarr MacFadden (1868–1955), who began publishing *Physical Culture* magazine in 1899. Its motto was "Weakness is Crime." During its fifty-year run, many of the photos were of MacFadden in tight trunks flexing his impressive muscularity. His numerous books espoused sexual virility, the perfection of motherhood through his lessons, and racial strength. He would tour the country, sometimes with his entire family of six, and put on displays of physical fitness, then also called physical culture. MacFadden and his family would do handstands, lift weights, flex, and demonstrate their strength in a variety of other ways before appreciative audiences.[60]

MacFadden's first school of physcultopathy opened ca. 1903, and Benedict Lust taught in it. All the students were also enrolled in Lust's American School of Naturopathy. Lust took pride in his own robust, strong physique and spoke glowingly of MacFadden. He "was the strongest man in the health field in America. He awakened the people to the idea of personal health. He did more than any other man to make the people health conscious." Naturopathic leaders and practitioners heralded MacFadden's work, no doubt fueling his fitness empire, and they extended that appreciation and endorsement to other exercise and strengthening regimens as well. MacFadden likewise advertised naturopathy in most issues of *Physical Culture*.[61]

Massage, another component of naturopathy, was classed by the AMA and leg-islative bodies as medicine. The classification was used repeatedly to arrest nature curists and naturopaths for practicing medicine without a license. Naturopaths advocated the *manual* application of massage rather than the mechanical. In 1923 Lust wrote that the disdain many medical men had for the therapy came from their fear that it would lower the status of their profession to that of a barber or a person who gives rubdowns. Only the soft human hands could be ideally effective, he said. They could not be replaced by lifeless automatic contrivances. Here Lust ridiculed physicians who refused to touch their patients to bring them relief. Throughout the 1920s the column The Massage Operator appeared regu-larly in the *Naturopath*. Written by Dr. P. Puderbach of Brooklyn, it detailed skeletal structure, bones, joints, and soft tissues that produced the movement of the spinal column. His early articles read like medical texts, predominately offer-ing information and definitions. At other times he focused on prostate-gland massage and Swedish gymnastics. The massage was illustrated with a photograph of a Burdick Radio Vital Light Bath administered in the patient's bed concur-rent with massage.[62] Naturopaths saw massage as working the muscles, which was particularly valuable for an invalid unable to exercise. It stimulated circu-lation for the heart and lungs and all body parts. It was also recommended for athletes, whose accumulated fatigue waste products, compounds accumu-lated in the body as a result of exercise, would be removed through quickened circulation.[63]

Light cures may have been most connected to industrialization. They were *not* based on the natural sun. They were developed for those who could not ac-cess the sun for its curative powers. As a 1918 author wrote, solar light's impact on human well-being was observable by contrasting the robust farmer or outside worker with the pale and bloodless factory worker. The outdoor sun could be rep-licated for those indoors with electric therapeutic lamps. Yet many naturopaths spoke out against light cabinets manufactured for profit that made one-sided claims. One naturopath wrote in 1918 of the three groups of therapeutic lamps for sale on the market: the arc lamp, which yielded the same spectral composition as the sun; the Finsen lamp, which produced almost pure ultraviolet rays; and the incandescent, or Leucodescent, lamp, which produced mostly thermal and lumi-nous rays. He claimed these were indeed true elixirs of life. These indoor lights signified an important naturopathic adaptation, since they no longer presumed that patients could leave paid work, journey to the countryside, and utilize natu-ral therapeutics. As the twentieth century evolved, with an oppressive factory system and cramped and squalid urban conditions, these devices demonstrated a clear recognition that adjustments must be made to accommodate the exegeses of modern life. The question who could afford these devices went unaddressed, as did the fact that practitioners hawking their products were profiting immensely. Later discussion of products differentiated between wattage, filaments, and infrared rays. In the 1920s one innovator reported that he had treated twenty-five patients at once, each on a couch or table. Interestingly, the patients were referred to as female—recognition that middle- and upper-middle class women were largely confined to the home and targeted as consumers of this therapy. These treatments, administered by trained personnel, were eventually called electric light baths, or radiant energy. They were integral auxiliary naturopathic therapeutics and harkened back to Arnold Rikli's sun baths. They supplemented the income of many practitioners.[64]

Color was a healing agent often linked with light therapy and at times with iridiagnosis. Colors were used in combination to control and alter psychophysiological responses. From the late 1930s to the 1950s a spate of articles emerged on this topic. One author identified seven colors in the spectrum: red, orange, yellow, green, blue, indigo, and violet. Like sounds, colors were produced by vibration oscillation, thereby producing energy, which was measured in octaves. The author scolded physicians for not utilizing light and color, encouraged patients to self-administer them (he did not say how), and asserted their universal importance as curative agents. The *Naturopath and Herald of Health* advertised publications on the topic and devices to generate color.[65]

As often happened, nature curists enamored with one particular method would market their own device or publication as offering the surefire way to diagnose and treat disease. A book written by the self-proclaimed "World's leading authority on colors for many years," Dr. Ernest J. "Rainbow" Stevens, described colors in nature, their basic principles, and their uses in psychology and health. This modality's popularity was evidenced in the ten book titles listed under the heading "Color Therapy" in *Nature's Path* in 1950. Each sold for a dollar, and six of them were by Stevens.[66]

Some therapies, like light and color, were minimally invasive and focused on psychological and mental well-being. Naturopaths saw electricity, magnetism, and hypnotism as triggers of the body's immune responses through external stimulus. These methods overlapped and were often used in combination. The possibilities seemed endless. In 1908 one enthusiastic author asserted that nuts, fruits, and seeds stored vital electricity. Proof that uncooked food provided much-needed electricity for the body could be found with a highly sensitive recording instrument, a Kelvin Astatic Galvanometer, with a resistance of 80,000 ohms. He invoked the popular motto "Electricity is life" and detailed the charge of the electricity in the body. This correlation between the charge stored in food, fruit in particular, and the electricity in the body led one author to make exaggerated claims of improved living and moral conditions. It allowed Christians to live purely, it could solve the problem of Ireland's decreasing population by introducing nut culture, and it could eliminate poverty in Europe and America. Drunkenness would cease, as would violent crimes. Disease, war, and insanity would be reduced, surgeries would become rare, and positive traits such as kindness, along with plenty and happiness, would be restored. The author's unbridled enthusiasm for keeping the body well charged through diet is palpable on the page, but again, extremist claims did not help naturopathy's case for legitimacy. Another author claimed that high-frequency electricity could retard old age with a new apparatus. He wrote glowingly of treatment with the Hyfrex coil. A patient with arteriosclerosis was placed in a solenoid and connected to the apparatus. He was subjected to a bombardment of millions of oscillations per second, which caused reduced blood pressure and a slight elevation in temperature. He said this "séance"—an interesting word choice that implied a supernatural presence— could be repeated three to four times per week. Through it bodily systems were energized, blood flow increased, general nutrition and waste functions improved, and functional activity stimulated, all of which led to the shedding of poisonous materials. Over time nature would take on this role, with no need of electricity.[67]

Several years later Dr. J. B. Bean asserted that electricity was life in three ways, through chemical action, friction, and the magnetic. It manifested in light, heat, and mechanical power. Like others before him, he argued that electricity stimulated the body's ability to process food most efficiently. But later that year Bean shifted his argument significantly, linking healing with the laying on of hands, prayers, charms, and incantations, with the electrical currents generated between the healer and the patient. He cited Christian Science, mental and divine healing, and magnetic healing as possessing this ability. These methods provided a harmonizing power that was "the One Great Life Force." While Bean was an outlier in his beliefs, he articulated a key concept of naturopathic healing: the spirit and body must function together holistically. Electricity practitioners lambasted their critics, saying that anyone who had unsatisfactory results must have misapplied the electricity. The treatment could only be safely performed by a skilled electropath. In the 1930s authors continued to assert the value of electrotherapy and devices for its application. Ernest's "HW-100 Portable 6 Meter Short Wave Apparatus" sold for a whopping $129. Drawings showed electricity being applied to the leg, arm, neck, waist, eyes, shoulder, stomach, or foot. Dozens of additional articles over the decades used the term *short wave therapy* to describe electrotherapeutics. Countless device ads bore the names of their practitioner-advocates, and their claims were often fantastic, improbable, and self-aggrandizing. By the 1940s and 1950s many devices had been banned by the FDA.[68]

Hypnotists, who relied on personality and suggestion to alter human behavior, were not universally embraced by naturopaths. They were known to use magnetism, and their activities and popularity were duly noted. But Lust, writing in 1908, said that their ability to induce certain behaviors was a dangerous substitute for truly reformed ways of living. Being told to sleep was not the same as falling asleep naturally, which came from natural tiredness, exercise, and calming of the nerves. Hypnotism was detrimental to soul and body. Many naturopaths believed that conscious behaviors and mental self-discipline were worthy ways to change oneself. Self-control should be learned; autonomous actions must outweigh psychological chicanery. Conversely, thirty years later an article in the *Naturopath and Herald of Health* asserted that hypnotism was a scientific psychological procedure of value for babies. Four months later another naturopath called it a "psychological infection," decrying abuses of the practice and calling it a fad, if not a racket. Hypnotism increased suggestibility in general (even when a person was not in a hypnotic state) and could be used for the commission of crimes. It was typical of Lust's editorship to fuel debate by printing all these views. In this case, part of the disdain for hypnotism can be attributed to its widespread use among medical professionals such as psychologists and by untrained individuals. Naturopaths were leery of both. This ambivalence is notable: despite criticisms to the contrary, naturopaths did *not* embrace all alternative therapeutics wholeheartedly.[69]

These auxiliary therapies, with their attendant gadgetry and the panacea claims of their popularizers, combined with the major components of naturopathic therapeutics. Their inclusion helped enlarge the community of alternative healers. This accounts for the thousands of natural healers Lust included in his *Naturopathic Encyclopedia* in the second decade of the twentieth century. Yet they continued to dilute the credibility—and sometimes undermined the medicinal value—of the key elements of naturopathy.

THE POWER OF NAMING: JUST WHO IS A NATUROPATH?

Over the decades, the naturopathic leadership struggled mightily with the fact of its being inclusive and collectively attacking the medical monopoly, yet including too many theories and practices, bringing condemnation and charges of quackery upon the profession. Who, then, was entitled to use the title "naturopath"? Clearly, Benedict and Louisa Lust were naturopaths; Benedict had been trained and received his credentials from multiple schools, and Louisa had been trained in multiple venues and at some point must have earned the ND degree through their American School of Naturopathy. What *exactly* were the differences between NDs, unlicensed nature-cure advocates, and practitioners holding degrees other than the ND, many theoretically complementary to early naturopathic degrees? Reciprocal admiration was often accompanied by confusion and internal hostilities among those claiming the label. The terms *nature cure* and *naturopathy* were often used interchangeably in the literature between 1896 and the 1920s. Adolph Just's text *Return to Nature!* (1896) was translated from the German in 1903 by Benedict Lust and embraced as a cornerstone text by the fledgling profession. The friendly relations between those using the two appellations was demonstrated in an advertisement in Just's book for Lust's Yungborn Sanatorium in New Jersey. Their mutual respect for each other was obvious.[70]

F. E. Bilz's *The Natural Method of Healing* (1898) was also widely included as a foundational naturopathic text. It was in large part a water-cure text but also advocated the use of fresh air, light, and massage. It also embraced hypnotism, Kneippian thought, and magnetism. The frontispiece contained a large print advertisement for the First Brooklyn Light and Water Cure Institute, founded in 1895 under the direction of Mrs. Carola Staden.[71]

Nature. Natural. Naturopathy. Given the plethora of therapeutics and range of training of those using these words, what *were* they? In his lead column, "Editorial Drift," Lust stated, "Naturopathy is a hybrid word. It is purposely so. No single tongue could distinguish a system whose origin, scope and purpose is universal." The descriptor above the editorial column illuminates what naturopathy was in 1902 according to Lust. The *Naturopath and Herald of Health* was a monthly magazine devoted to "Natural Healing and Living Methods, on the basis of Self-Reform and Popular Hygiene, Hydrotherapy (Priessnitz, Kneipp, and Just Systems), Osteopathy, Heliotherapy (Sun-Light and Air Cure), Diet, Physical and Mental Culture, to the exclusion of Drugs and Non-accidental Surgery." Broadly speaking, Lust proclaimed, "naturopathy stands for the reconciling, harmonizing and unifying of nature, humanity and God."[72]

It is little wonder, given these inclusive definitions, that a mixed cadre of practitioners laid claim to the system *regardless of their own credentials*. J. H. Tilden,
MD, argued that nature could aid herself, and he aligned his therapeutics and
hygienic living methods with nature cure. Similarly, Henry Lindlahr, allopath
and naturopath, felt quite comfortable expanding upon the principles of nature
cure. Dubbed by recent biographers "the Founder of Scientific Naturopathy," he
established a sanitarium for nature cure and osteopathy under his name. His

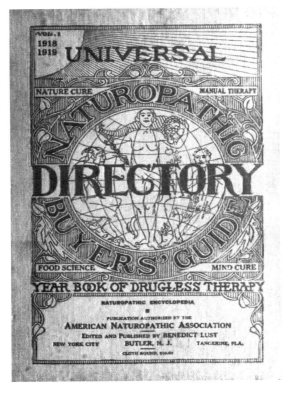

Cover of the *Universal Naturopathic Encyclopedia, Directory and Buyers' Guide*,
showing the cornerstones of what Lust hoped would be a unified definition of
naturopathy and its therapeutic universalism: nature cure, manual therapy, food
science, and mind cure. Lust wanted this reference for practitioners and patients
to help solidify the professional standing of naturopaths. At 1,432 pages, the book
provided Lust and the American Naturopathic Association evidence of a united
front of qualified practitioners, and he distributed it nationwide. The most
respected leaders in naturopathy's beginnings contributed articles on philosophy
and therapeutics. Benedict Lust, ed., *Universal Naturopathic Encyclopedia, Directory and Buyers'*
Guide: Year Book of Drugless Therapy, vol. 1, for 1918–19 (New York: privately printed, 1918).

writings, especially *Nature Cure* (with at least twenty editions by 1922) thoroughly described naturopathic philosophy.[73]

In the early years, self-definition and self-identification as a naturopath went a long way toward forming a professional identity. Intellectually, medical kinship was formed by some shared historical healing methodologies and therapeutics, as well as by political harassment. Kinship was also created by way of shared *philosophical agreements*, varied as they were, about the healing power of nature.

By 1918 Lust felt the need to reiterate and embrace the professions of natural healing. Noting that there were at least forty thousand naturopathic practitioners in the United States, he compiled the *Universal Naturopathic Encyclopedia, Directory and Buyers' Guide: Year Book of Drugless Therapy* for 1918–19. He willingly included every practitioner of drugless therapy, but as the subtitle implies, he intentionally excluded homeopaths, whose healing premise revolved around drug use, however natural they might be. Lust's vision of universalism included "electromedicine, neuropathy, dietology, chiropractic, mechanotherapy, osteopathy, phytotherapy, apyrtrophy, physiculture, natural divine healing, astrocopy, and phrenology." His inclusion of all these methods prompted two historians to comment that he embraced "the strangest healing systems [and] every Naturopathic practitioner, from the Arctic circle to the furthest limits of Patagonia." Lust firmly believed that "In Unity There Is Strength" (in full capital letters in the original), and he thus compiled his encyclopedia as a veritable who's who in the realm of drugless healing. The inclusion of countless methods allowed for a rich cross-fertilization, but it also meant that a concrete definition delineated by an educational core was slow to emerge. This did not mean that Lust opposed educational standards. It did signal, however, the complexity and diversity of what came to constitute naturopathic practitioners and philosophy.[74]

CHAPTER 3

"Nature Takes the Right Road"

Naturopathic Philosophy

The philosophy of naturopathic medicine was at once simple in its message but complex in its methods and porous in its boundaries. It was the antithesis of one of its key components, water cure, in which followers had merely to live hygienically and use only water in its varied applications as a healing agent. Water cure had been considered a panacea system in its heyday, providing answers for all of life's uncertainties. Users of naturopathy, in contrast, had to be secure about their place in the world and comfortable with constantly changing health definitions and modalities. They had to have acute skepticism about, if not abhorrence for, new science. Naturopathic practitioners and patients went against the grain in the late nineteenth century as new applications of scientific advancements and methods became an organizing principle in America. Science was the new religion by 1890.

Naturopaths and their patients were also part of a greater cultural movement that had been in the making for generations. As naturopathy was coming into its own in the second decade of the twentieth century, it was influenced by New Thought, a movement that had gained steam since the 1840s. Individual naturopaths mentioned that they were believers. When Dr. Henry Lindlahr wrote in 1912 that he was "a thorough and consistent optimist and New Thought enthusiast," he was referring to an ideology that had been given this title in 1894. A set of ideas that was not easy to define at that time, much less today, it was "a theory and method of mental life with special reference to healing, and the fostering of attitudes, modes of conduct and beliefs which make for health and general welfare." It had a spiritual component—largely but not solely Christian—and behavioral and philosophical components. The lifestyle promoted by New Thought paralleled early naturopathic life strategies and positive right thinking: "Man leads an essentially mental life, influenced, shaped and controlled by anticipations, hopes and suggestions. . . . Life is largely what we make of it, what we bring to and call out of it." Early naturopaths advocated the individual's need for self-control of both mental and physical behaviors. Passions, moods, and inclinations had to be focused and subject to self-control.[1]

Definitions of natural medicine and its philosophical underpinnings proliferated from a variety of early-twentieth-century advocates. They had in common a healing ideology based on the premise that the body would heal itself with the help of natural methods, and they shared a loathing of invasive allopathic therapeutics. Pioneer naturopaths, such as Germany's F. E. Bilz and the American Louis Kuhne, argued that the natural healing method was based on the new art of healing, which did not include drugs and operations. One eclectic argued that nature alone could rid the body of accumulations of abscesses if no complications had been induced by poor allopathic treatments and bad care. One also had to cease all bad habits.[2]

The more naturopathy became defined, the louder the rejection of allopathy became. Lust's enormous *Universal Naturopathic Encyclopedia, Directory and Buyers' Guide*, published in 1918, revealed the contributors' collective scorn of allopathy and its inability to eliminate the root causes of disease. He called allopaths professors of the irrational theories of life, health, and disease who treated their victims with dangerous drugs and animal vaccines and serums. Their magic pills, potions, and poisons, he wrote, attacked the ailment and suppressed the symptoms instead of addressing the ailment's real causes.[3]

THE "CONSTRUCTIVE PRINCIPLE OF NATURE" AND PERSONAL RESPONSIBILITY

Both Lust's and Henry Lindlahr's definitions of nature cure emphasized building, or constructing, bodily elements, compared with what they saw as the allopathic method of destroying them. Nature cure harmoniously built up, repaired, and improved the body mentally, physically, morally, and spiritually. It opposed the destructive principle of fighting symptoms in a way that destroyed existing forms of life. William Turska's and Henry Lindlahr's seminal discussions of "Catechism of Naturopathy" outlined six components of the constructive principle: establishing normal surroundings and habits of life in accordance with nature's laws; boosting vitality; building up the blood with its natural components in right proportions; promoting the elimination of waste materials and poisons without injuring the human body; correcting mechanical lesions; and finally, arousing the individual's consciousness to the highest degree. Patients had to take personal responsibility and actively help themselves to heal.[4]

The goal of all naturopaths was to increase patients' vitality so that they could work with nature to attain the highest level of health. Naturopaths and other sectarians discussed vitality or vital force much in the same way that we discuss holistic well-being today. Vitality meant overall health and strong body energy, immunity, and functionality. Vitality was created by normal, healthy nerve, circulatory, and organ functions. Boosting vitality was analogous to what we mean today when we talk of boosting immunity, strengthening the core, building endurance, or reducing one's risk of disease through healthy activities and choices—all rolled into one.

Naturopathic approaches to disease prevention and disease remedies were designed to balance the body's function and thus increase vitality. In the words of Samuel Bloch in 1906, "The blood and nerves are the vehicles of this vital power, therefore the vitality of any organ is due to the amount and quality of blood supplied to that organ and the condition of the nerves leading to that organ." Lust's *Universal Naturopathic Encyclopedia* listed the methods for doing this: regulate the diet, learn proper breathing, exercise, bathe, and eliminate poisons. Dr. Louis Kuhne's essay in the book described how his own ill health had been turned around with water therapies, diet, and what he called the science of facial expression, or personifying—expressing—spiritual well-being. He believed that his philosophy was distinct enough to warrant its own name. He called it "neo-naturopathy," and this terminology was embraced by other early naturopaths. Kuhne also originated the theory of the Unity of Disease, also referred to as the singleness of disease. According to Kuhne, there was only one cause of disease and hence only one disease that showed itself in various ways. Disease was simply the presence of foreign matter in the system. He wrote that one should not differentiate between different diseases—only between different forms of disease. Lindlahr emphasized that the skin, the kidneys, the bowels, and the lungs had to work harmoniously and efficiently, and regulating food intake reduced the organs' burden and increased vitality.[5]

The responsibility for health was ultimately up to each individual, who should not resort to the invasive intrusion of an allopath and his methods. This fact could not be emphasized enough. When Lust presided over the American Naturopathic Association in 1922, he referred to the necessity of living healthfully rather than relying on periodic crisis intervention. American naturopathy, he said, was a revolutionary movement that combined rational medicine, a system of disease prevention, and medical freedom worth fighting for. It would "usher in the new day of better health, more happiness and liberty for our bodies and souls." Lust had urged readers in the *Naturopath and Herald of Health* to think for themselves in order to become healthy. Each individual was personally responsible for balancing, harmonizing, and unifying nature, humanity, and God. It was a demanding and unsympathetic approach that called for abandoning the self-indulgent abuse of one's body, taking personal responsibility for one's behavior, and constantly reevaluating one's place within the larger cosmology. In short, followers were obligated to ask the fundamental question, Am I living in harmony with myself and with the larger world around me? It was not an easy exercise, given that by 1890, for the first time in history, more Americans were living in cities, where they faced the chaos of a loud, crowded, and often filthy environment.[6]

Natural health required three key behaviors: natural habits, self-regulation, and positive spiritual well-being. Exactly what were natural habits? Arnold Rikli (1823–1906), affectionately called the Sun Doctor, considered light, air, and sun baths central to natural healing. Rikli also advocated variation, believing that constant repetition in diet and daily routine weakened muscles and caused stagnation and death. In addition to light, air and sun, the renowned Adolph Just, in *Return to Nature!* (1896), advocated water, reform clothing made of natural, loose, breathable materials, and a dietary regimen. The pioneer naturopath F. E. Bilz believed that of all these core components, the most crucial were fresh air and light. These three men provided the core principles of the naturopathic lifestyle for decades to come. These fundamentals were the only healing modalities that were uncontested among naturopaths. The countless other methods each had vehement detractors.[7]

The naturopathic follower, then, had to make *choices* that were at once both appealing and disconcerting in a healing philosophy. It was easy to just be told what to do; it was another matter to become a right-living person, constantly making decisions about the proper course of action in daily health, which was what naturopathy demanded from its followers.

DISEASE CAUSALITY

Henry Lindlahr, MD, a leader in naturopathic philosophy, explained the five specific conditions that caused disease: lowered vitality; abnormal composition of blood and lymph, resulting mainly from wrong eating and drinking; accumulation of waste, producing morbid matter and poisons in one's system; mechanical lesions, that is, pressure, tension, or strain on nerves and nerve centers caused by luxations (dislocations) of bony structures or straining of muscles and ligaments; and discordant or destructive mental and emotional attitude. These conditions more or less remained the core of naturopathy for decades.[8]

Others articulated causes a bit differently. California's J. T. Work said that the nervous system controlled the body, whose condition was determined by heredity, learned behavior, and habits of living. Nature's rule was moderation in all things. Disease was caused by excesses and weakness that made the elimination of wastes and toxins impossible. Louisa Lust attributed ill health to one great cause: accumulated waste material or foreign substances in the body. She advocated allowing the body's vital force to throw off the poisons by fever. Suppressing fever thwarted this natural process.[9]

THE UPAS TREE OF DISEASE

EVIL IS NOT AN ACCIDENT, NOT AN ARBITRARY PUNISHMENT, NOT ALWAYS
AN "ERROR OF MORTAL MIND." IT IS THE NATURAL AND INEVITABLE RESULT
OF VIOLATIONS OF NATURE'S LAWS. IT IS INSTRUCTIVE AND CORRECTIVE IN
PURPOSE, AND WILL REMAIN WITH US ONLY AS LONG AS WE NEED ITS SALU-
TARY LESSONS.

"Violation of Nature's Laws." The pioneer naturopath Henry Lindlahr used this "Upas Tree of Disease" to illustrate the consequences of violating nature's laws. "Evil is not an accident," he wrote; evil, or disease, "is the natural and inevitable result of violations of Nature's laws. It is instructive and corrective in purpose." Ignorance, indifference, lack of self-control, and self–indulgence led to lowered vitality, abnormal blood, accumulation of morbid matter, and the ultimate outcome: diseases. Henry Lindlahr, MD, *Philosophy of Natural Therapeutics* (Chicago: Lindlahr, 1918), 8.

Early naturopaths agreed that sickness was often the result of people's violating nature's laws—deliberately or inadvertently—and introducing impurities into their bodies. Louis Kuhne said in 1917 that women's ailments "are to be traced to women's wrong matter of living, to neglect of bodily health, to want of regular exercise in the open air, to inattention to the natural and prompt satisfaction of the bodily needs, and an exaggerated quest after pleasure," as well as other deviations from nature's path. Kuhne was articulating his era's belief that some women were inherently frivolous, but as he pointed to women's personal irresponsibility, he was also revealing women's behaviors owing to cultural constraints imposed upon them. Middle- and upper-class women in cities were considered society's chief consumers and were targeted by businesses as permanently leisured clientele; they were confined to indoor living and received little encouragement to engage in physical activity. Conforming to naturopathic ways meant deviating from a class-based gender norm and required a monumental effort. Naturopaths' main point, however, was that both females and males became sick when they deviated from nature's path because of lack of self-control or ignoring the body's signals.[10]

TOXEMIA THEORY

Naturopaths argued that nature provided cures by producing toxemia, which some believed to be the one basic cause of disease. One naturopathic chiropractor wrote that disease was nothing more than an abundance of toxic matter in the blood. The greater the amount of toxins, the worse the disease. Three phases of abnormality indicated toxins' progress: retention, invasion, and enervation. If the toxins were removed, the disease disappeared. By 1936 this theory was disseminated widely by the naturopath Harry Benjamin, who authored several books and was a member of the British Naturopathic Association. His *Everybody's Guide to Nature Cure* went through thirteen editions, with fifty-three thousand copies in print by 1958. Waste accumulated in the body through wrong habits but also through enervating habits, such as overwork and worry. Since his fundamental principle was that eliminating toxins allowed the body to heal itself, sickness was *not* to be suppressed. Disease showed the body's desire to continue living, and nothing was gained by voiding, suppressing, or aborting symptoms. As one practitioner put it, this was "another way of saying that disease quite often *is* the treatment."[11]

Leading naturopaths continued to rely on ideologies related to toxemia throughout the twentieth century. In 1969 James Hewlett-Parsons, former vice president of the Guild of Naturopaths and Osteopaths, wrote that lost vital force resulted from violation of natural laws, accumulated waste matter, and highly toxic substances. By the 1980s toxemia theories were more sophisticated and conceptualized at the cellular level, but they continued the intent of naturopathic predecessors. Waste products and chemical toxins found in food and drugs accumulated in the body's tissues and caused cellular damage, which in turn obstructed vital functions. Cells need adequate oxygen and nutrients and the ability to effectively eliminate waste products.[12]

VITAL FORCE AND THE LAW OF CRISES

Naturopaths believed that the vitality, or vital force, within the body was awakened through a healing crisis, which then stimulated the body's own ability to heal. Long before the concept of autoimmunity existed, the naturopathic concept of vital force posited its existence.

The role of vital force, according to naturopaths, was to prompt healing through expulsion of morbid matter and then support the body's innate ability to heal itself. F. E. Bilz wrote that remedies did not heal disease; vital force within each person allowed nature to take the right road to heal a disease. The healer's job was to aid the body's ability to heal itself, not place obstacles in the way of vital force. Self-healing resulted from applying the principal elements fresh air, light, and water and avoiding medications. The life force, or the energy of vitality, was activated by, and relieved disease through, healing crises, defined as the expulsion of toxins by natural forces. The process promoted recovery and was not to be suppressed through medical means. Vital organs and fluids were not to be destroyed or disorganized. However, the body required sufficient vital force and reactive powers to integrate treatment and any change of habits. Naturopaths acknowledged that the concept of vital force was difficult to explain as a phenomenon beyond physicality. It had confounded philosophers and physicians throughout the ages. The best they could do was state that a healthy vital force allowed bodily responses to occur. Conversely, vitality was abnormal when the state of tension or the bodily materials were insufficient or improperly balanced.[13]

Vital force was a component of toxemia theory. The body tries to throw off and expel built up toxins, uneliminated wastes, bacterial and environmental poisons, and so on. A healing crisis, Lindlahr said, was marked by "acute eliminative activity." That could be "fevers, inflammations, skin eruptions, diarrhea, boils, abscesses, perspirations, hemorrhages and mucopurulant discharges." Nature did not undertake a healing crisis until the body was sufficiently cleaned and strong enough to benefit from the crisis. If a person was too weak, then he or she needed treatment to help the body complete the process. While allopaths provided medications or surgeries that affected the organs, early naturopaths be-lieved that bodily evacuations were key; by the 1950s some asserted that elimina-tion had to start at the cellular level, not in the bowels or kidneys.[14]

GERM THEORY

Naturopaths believed that germ theory, which posited that outside invaders caused disease, was deficient in rationale and analysis. Medicine's acceptance of germ theory came after Louis Pasteur (1822–1895), a French professor of chemis-try at the University of Lille, sought solutions for local industries' practical prob-lems. Focusing particularly on manufactured alcoholic drinks, he demonstrated that bacteria soured beer and wine and that boiling and then cooling the liquid would remove the bacteria. He later extended this process to milk, the removal of bacteria from which is now called pasteurization. He posited that bacteria were environmentally introduced. His work convinced him of the germ theory of disease, according to which germs attacked the body from outside. This ex-plained the origins of certain diseases, in particular rabies, cholera, tuberculosis, and smallpox, which could be prevented by vaccination.[15]

Ironically, as urban overcrowding, filth, and poor nutrition increased among immigrant communities in the United States at the turn of the twentieth cen-tury, Pasteur's claim that the environment introduced bacteria to the body was overshadowed by his advocacy of vaccinations. Naturopaths argued that un-hygienic living, poor nutrition, lack of exercise, and bodily degradation caused disease. They advocated preventive measures aimed at strengthening the body to ward off disease without introducing toxic matter into the blood. They knew of germ theory and accepted the concept of germs, but they rejected germs as a sole or primary cause of disease and especially rejected vaccination as a cure-all. Pub-lic health officials, operating through the auspices of the American Medical As-sociation, focused almost exclusively on vaccinations, which introduced small doses of toxins into the body to stimulate the body's ability to fight off the virulent disease strain. The environmental conditions that spawned outbreaks of disease in tenements were not addressed sufficiently by public health officials, said naturo-paths. The positions taken by both medical systems created a false dichotomy in which the two sides saw what they viewed as the cause—on the one hand, germs; on the other, environment plus individual vitality—as mutually exclusive.

Naturopaths were limited in their thinking in that they rejected public health reformers outright, viewing them as merely the enforcement arm of the AMA. They ridiculed campaigns for clean milk because they considered breast milk far preferable. Naturopaths discussed wretched living conditions but did not actively address that issue as the Progressive Era tenement house reformers did. They rarely addressed the quality of water supplies, although they later opposed fluoridation. They opposed anything involving governmental AMA-backed efforts, which meant that they could not acknowledge any scientifically derived advances. One naturopath disdainfully declared that only morons believed in the germ theory.[16]

Naturopaths' rhetoric of resistance to germ theory was often displayed in their opposition to vaccination. One 1918 article did not hold back, mentioning a "twentieth century superstition . . . that disease can be cured by taking poisonous drugs, or by injecting germ excreta of filthy animal matter into the blood." Naturopaths also rejected quarantine and hospitalization, which they believed went against the public good. Instead, they advocated common-sense solutions that would also improve living conditions, such as the simple use of antiseptics and disinfectants. The decrease in typhoid fever, they said, was owing to control of the mosquito populations by protecting homes with screens, doors, and windows and to mothers learning better methods of feeding and caring for their small children.[17]

Prevention was key for all major diseases. Naturopaths ridiculed scientific medicine's reliance on surgery for cancer, saying that allopaths' explanation of germ theory had failed, so that they were forced to adopt a policy of cutting and then waiting to see if it worked. One hundred years ahead of their time, naturopaths promoted nutrition and physical activity to prevent cancer, just as the American Cancer Society does today. They said that if more people embraced a diet high in vegetables, air, and exercise and abandoned self-indulgent behavior, fewer would be living with weakened bodily resistance to disease.[18]

One way that naturopaths worked to discredit germ theory was to repeat the reservations, or reversed positions, of former germ-theory advocates. The most high-profile of these was none other than Rudolph Virchow, dubbed the father of the microbe theory or the father of pathology. Virchow was thought to have said upon his deathbed that "if I could live my life over again, I would devote it to proving that germs seek natural habitat, diseased tissue, rather than being the cause of diseased tissue." It is a quote still repeated today. It was a powerful example, since Virchow was best known in his lifetime as a social reform crusader who critiqued the social and environmental conditions that fueled the spread of disease rather than focusing on compromised bodily systems. While he was a pathologist, he would come to be identified with social medicine.[19]

Critiques of germ theory continued in the twentieth century, but most naturopaths recognized the role of germs in disease early on, and some even proclaimed their usefulness. Decay, already under way in the body from dietary mistakes and blocked elimination, bred germs that facilitated waste elimination. Naturopaths believed that science interfered with these functions, and bacteriologists wrongly claimed that this natural elimination process was unhealthy. The naturopathic leader Dr. Jesse Mercer Gehman was one of those who acknowledged that germs were valuable. "They are the result of disease and are present everywhere. Breeding bacteria and germs is a necessary part of investigative science to determine their characteristics." Gehman credited Pasteur with introducing the concept but ridiculed the hero worship he had since received. After all, he had likened "the human body to a barrel of beer and went so far as to pronounce it, like that substance, at the mercy of, or subject to invasion from extraneous organisms, that is germs." Gehman said that focusing so closely on germs outside of the body invalidated the need to study the body as a breeding ground for them. A colleague of his called germs "friendly little scavengers" that were not detrimental to mankind. When extensive stagnation and decay of body tissues were eliminated, normal bodily functions would alleviate chronic disease.[20]

Ultimately, many more naturopaths came to understand the benefit of analyzing allopathic germ theory and naturopathic toxemia theory together. In 1951 one naturopath argued that the "toxemia hypothesis" (that disease results from toxic matter in the blood), long a bulwark of naturopathic philosophy and practice, was finally proven valid by scientific medicine. He stated that "modern research in chemistry, bacteriology, physiology and pathology has resulted in overwhelming data" showing that toxemia was a prime cause of disease. He was convinced of the hypothesis's legitimacy because of evidence in empirical data and responses to antibiotics. He went so far as to claim that there had been a simultaneous acceptance of toxemia theory by allopathic science and of germ theory by scientific naturopaths. He thought that members of both groups had abandoned their single-cause-of-disease theories to embrace each other's ideas. At best, his view was an exaggeration; at worst, he misrepresented the facts. His optimism ignored decades of traditional naturopaths' rejection of germ theory. However, if this acceptance had been true in his own circle of scientifically based practitioners, that would have foreshadowed the mutual recognition at the turn of the twenty-first century.[21]

Early naturopaths' rejection of antibiotics went hand in hand with their arguments about germs and the destructive qualities of manufactured drug use. Instead they used natural methods that destroyed harmful germs and stimulated natural immunity without disturbing the body's natural flora. They employed specific, effective antimicrobial herbs such as horseradish and fresh watercress or garden cress. Allicin from garlic was an antibacterial, and extracts from sagebrush, juniper, buttercups, and California Spanish moss could also be used when antibiotics were insufficient or a therapeutic change was required. Finally, naturopaths relied upon echinacea for its value in treating infections and septic conditions locally and systematically—it inhibited staphylococcus and stimulated the glandular system.[22]

However, by late 1950 the influential leader John B. Bastyr, ND, had embraced the use of antibiotics. Bastyr is called by two biographers "the most important link between the old-fashioned nature doctor, diagnosing by close observation and collecting his own plant medicines, and the modern naturopathic physician, who uses the latest laboratory techniques and employs standardized extracts of phyto-medicines." Bastyr did not criticize or oppose the use of antibiotics but noted that the term itself literally means "opposed to life." Explaining that they were derived from fermentation of lower plant life, he laid out penicillin indications (pneumonia, ear and many other infections), contraindications (none except those sensitive to it), physiological and toxicological effects (apparently none unless due to sensitivity), and research on dosage and absorption. This was a turning point for naturopathy because Bastyr's stature in the profession was immense. To coin a phrase: when Bastyr spoke, people listened. His appreciation for antibiotics signaled that the scientific branch of naturopathy had cautiously accepted their use.[23]

If early naturopaths believed that antibiotics masked wrong living, then healthy living through diet was the primary method they used to stimulate vitality and natural immunity to germs and disease. In this they followed the path-breaking, influential texts of Louisa Lust, and it continues to be a foundational principle to this day. The healthful, healing properties of food were also promoted in Benedict Lust's lay journal, *Nature's Path*. By the 1920s it contained a series written by Florence Daniel, "Food Remedies," which discussed a variety of fruits, vegetables, spices, and herbs that healed. Frequently discussed were burdock, catnip, cowslip, dandelion, dock, horseradish, leek mustard, peppermint, sarsaparilla, wild ginger, wild turnip, and wintergreen. Naturopaths were forced to assert that herbs were vegetables to avoid the then rampant legal persecutions that accompanied *any* medical advice dispensed by naturopaths.[24]

Quite simply, there were benefits to be had from right eating, and wrong eating could be dangerous. In the words of one author, "The only way to restore normal conditions of health is to live on natural foods." Naturopaths cautioned against overeating and bolting (gulping down) one's food and warned of the dangerous side effects of fad diets. A 1908 article warned mothers against using candy to quiet children. Naturopathic advice ranged from preparations for stimulating yet chemical-free beverages to the reasons for eating only natural foods and the necessity of avoiding all meats, excessive protein, and certain remedies suggested by regular MDs. Vegetarianism was one of the most crucial aspects of right eating. Naturopaths believed that resulted in a more sound physical constitution and a sense of contentment through humanitarian principles. They insisted that vegetarianism was the healthiest method for avoiding the buildup of morbid matter in the system.[25]

The very essence of health—vitality—was strengthened or weakened through food consumption. To understand the importance of diet in naturopathy, one need only turn to the most influential early texts. In addition to Louisa Lust's cookbook, there was Henry and Anna Lindlahr's 1914 *Lindlahr Vegetarian Cook Book*. Henry devoted entire sections of his pathbreaking *Nature Cure*, of 1913, to "Natural Dietetics," "Mixing Fruits and Vegetables," and "Mixing Starches and Acid Fruits."[26] Lust's 1918 *Universal Naturopathic Encyclopedia* likewise contained several essays devoted to diet. Kuhne argued that overnutrition in stout persons was caused by poor food choices that the body could not utilize. The least suitable foods and beverages were flesh meats, eggs, extracts, wine, beer, cocoa, coffee, and tea, among others. Instead he advocated rapidly and easily digestible food, because vitality depended upon food digestibility. Kuhne particularly condemned denatured foods, that is, foods changed from cooking, smoking, spicing, salting, pickling, and placement in vinegar. They were less digestible and thus lowered vitality. He was also in favor of periodic fasting, which he modeled after behaviors observed among some animals.[27]

The bottom line for naturopaths was that nature protected the body's naturally healthy and balanced state: the body was a self-regulating organism, and food was central. If "nature is the healer of all disease," then "let foods be your medicine and medicine be your foods."[28]

CONSCIOUS LIVING

Body and spirit, naturopaths argued, were the two columns supporting life. Destroy one, and life ceases, because a well body cannot exist in the face of mental distress, worry, fear, uncertainty, and anxiety. Simply put, naturopaths saw that the mind's influence on disease was unquestionable. Dr. Henry Lindlahr said that naturopathy embodied the best in science, philosophy, and religion. Central to naturopathic philosophy was recognition of the individual's well-being and his or her place within the cosmos. This metaphysical approach was quite distinct from biotechnical medicine, which historically had viewed the physical and spiritual realms as separate and distinct. Medicine healed, while the philosopher pondered and administered to the soul. According to naturopathic philosophers, however, bodily well-being was inseparable from one's worldly relationships.[29]

In the late nineteenth century, Adolph Just emphasized that spiritual well-being was a precursor to physical health. Then, in 1902 Benedict Lust advocated the philosophy of human perfection achieved through Christianity. Naturopaths were not the only Americans to connect healing with a state of mind and spirituality. Others included hydrotherapists, Seventh Day Adventists, Christian Scientists, and those in the New Thought movement. Lust, influenced by his Catholic background and New Thought, referred to mental healing as set by the example of Jesus Christ. Leading naturopaths such as Lust, Lindlahr, and Edward Purinton embraced Christian underpinnings within New Thought, but scriptural Christianity was not the only spiritual guide. Lust believed that each individual could replicate mental healing. He believed that individuals could make their own theory and method by letting the New Testament be a guide to their spiritual life. Christ could become more of an inner or universal principle, accessible to everyone.[30]

New Thought had become an international movement with annual conventions. In attempting to define the movement in 1917, Horatio Dresser, one of its leaders and the author of its history, laid out its goals, which also appeared in naturopathic articles. New Thought promoted a spiritual philosophy that affirmed life and happiness. It held that right living and thinking made the highest ideals attainable. Treatment of disease would benefit from systematic intellectual and spiritual methods. In short, the theory advanced the power of positive thought, producing bodily well-being—not unlike the effective results of meditation or guided imagery that scientists have recently found among cancer patients.[31]

Naturopaths regularly made connections between mental and physical well-being and Christianity and New Thought. Health depended on one's ability to practice focused purposeful thinking for a positive outcome. In this way, faith and prayer could cure nervous diseases caused by conditions of the mind. Religious faith could uniquely "heal moral maladies which neither drugs nor hygiene nor massage can cure." Edward Earle Purinton, a fervent proponent of New Thought, believed that spiritual alertness would protect one from the decay of health and the weaknesses of age. In 1908 he said that spiritual symptoms involved feeling, thought, desire, and action. He provided a list of actions to "grow" the soul that mixed individualism and communal responsibility. They encapsulated the dominant naturopathic philosophy of the time, with New Thought ideology and some unique twists: consciousness must be earned for oneself through instruction and hard work; one must not repeat mistakes; adherence to religious methods was meaningless unless it was accompanied by true introspection, so one should embrace all forms of religions because each had unique offerings; one should hold one's character to the highest standards and maintain a positive attitude in the face of diversity; one should embrace hardship and allow a healer to bring light (spiritual healing) before physical healing; one should work among the poor and oppressed; one should expose oneself to things one finds repellant, for from them one can learn the most; happiness comes from balancing anguish and ecstasy; one should keep alive one's "child-heart" and maintain hopeful innocence for human good; and most interesting, "our attitude toward the sexes is reversed": there should be a blend of traditional notions of male and female interests, skills, and virtues within each woman and man so that they may "acquire balance and a sane vision."[32]

Naturopaths seemed to agree with New Thought's practical approach to Christianity. They believed the adage "Do unto others as you would have others do unto you" would inspire devotion to one's fellow men, improve one's character, and result in better service and civilization. This emphasis on Christian living was also part of the larger Social Gospel reform movement in the early twentieth century. Ministers, women, and child welfare reformers critiqued the harsh realities of cities, industry, and the changes brought on by modernity, hoping to restore the nation to the true spirit of Christian community. One naturopath believed the Bible could become the lost fountain of health and happiness. Those suffering from nervous afflictions, mental illness, or sexual diseases could turn to the Bible for relief.[33]

Other naturopaths were more likely to refer to spiritual, as opposed to Christian, beliefs. Both mental and physical well-being were inextricably linked to spiritual well-being and faith. One practitioner felt that not only spiritual health but physical health should be preached in temples. Henry Lindlahr, on the other hand, was pragmatic about the relationship between maintaining good health and religious practice: "If there is not self control enough to resist a cup of coffee or a cigar, whence shall come the willpower to resist greater temptation?" Integral to achieving mental well-being was the concept of *vis*, or the patient's will. *Vis* could be awakened through water applications, electricity, vibration, light, massage, or diet. These methods, all physical, *produced* mental health—what the naturopaths called "an invigorated sensibility," or a sense of purpose and an ability to be in the moment.[34]

The overarching question was whether metaphysical beliefs were components of the constructive principle in nature. Under what conditions did mental healing aid and complement physical healing? Metaphysical systems of healing could work if they complemented nature's efforts, awakened hope, and facilitated vital force in the body. They *contradicted* constructive principles in nature if they induced blind faith in metaphysical formulas and prayer or weakened the ethos of personal responsibility. Leaders like Lindlahr embraced nonexcessive mental and spiritual remedies "such as scientific relaxation, normal suggestion, constructive thought, the prayer of faith, etc." Others reflected their social-class bias. One wrote that the thinking man, through concentration enabled by physical health, could accomplish infinitely more than the laborer, revealing contemporary beliefs in Social Darwinism—that educated Americans had greater potential than those consigned to factory and agricultural work.[35]

Lindlahr and Lust derived this language and theory from what some called mental science, the predecessor of New Thought. A mind-healing movement founded on both metaphysical and religious ideas, it was popularized in the last quarter of the nineteenth century. Lust was particularly supportive of Helen Wilmans (1831–1907), of Seabreeze, Florida, who had been an influential pioneer in mental science. The *Herald of Health and Naturopath* in 1918–19 ran a set of Wilmans's serialized lectures on mental science posthumously, offered as a home course. Lust then published her entire lecture series himself in 1921 and wrote the introduction. Lust admired Wilmans's rags-to-riches story. She had founded a profitable publishing house in San Francisco. She had published a paper called *A Woman's World* and a weekly magazine, *Freedom*. She then had begun teaching about mental science, facilitated healings, and written several books, all of which were very lucrative for her. Lust wrote that her teachings had influenced all New Thought ideas. She had been a pioneer who broke away from orthodoxy and its limitations. She had rejected the idea of a personal God who watched over the destinies of human beings and demanded flattery and praise.[36]

Wilmans's course was a primer on New Thought, therapeutic suggestion, psychotherapy, and spiritual and metaphysical research. Her basic philosophy was that "we are as we think"—the ultimate American belief in self-determination. She made the heretical claim that God had been distorted by man to frighten humanity into better behavior. She boldly rejected this idea and instead argued for a new life principle, sometimes called the Universal Spirit of Life. This concept was profound; it meant that each person had the ability, through conscious intelligence, to rule over all things, most importantly oneself. New Thought advocates did not subscribe to Christian Science and other beliefs that faith would make one well. Instead, Wilmans taught the link between mental exercises to instill self-determination and one's physical well-being. Negative, nervous energy, sometimes called nerve force, or neurasthenia, was a debilitating condition. To combat it Wilmans and other New Thought authors advocated the use of affirmations. Repeated often, they strengthened the good and crowded out harmful thoughts. In the mid-twentieth century Dr. John Benedict Lust, son of Benedict Lust's brother Louis, wrote that in addition to physical health, heredity, and harmony, the formula for health included life power based on the soul's will and thought and the spiritual power that formed mind and body. At the turn of the twenty-first century this became "mindbodyspirit" wellness.[37]

Lust republished Wilmans's work in 1929. Her coauthor and editor was now Edward Earle Purinton. She had been deceased for many years, but Purinton and Lust renamed her work *The New Psychology and Brain Building*, a nod to the ascendency of psychological practice and theory. A year later Purinton hailed "diagnosis of temperament" as a new field of scientific study. Its study, he wrote, was both mentally enriching and fascinating. Purinton argued that once human nature was truly understood, accurate diagnosis of temperament would be standard fare at all educational and employment institutions. Naturopathy had embraced its own version of psychology and personality studies.[38]

Needless to say, the naturopathic views of mind-body-spirit connections were not widely held in the allopathic Western healing traditions. Biomedicine has had a compartmentalized view of the human body, going back to Greek thought but popularized by Descartes. In this theory the human form is comprised of distinct entities of the mind and body. Physicians came to believe that treatment could focus exclusively on an affected area of the body, without requiring attention to the rest of the person. Naturopaths disagreed with this model and critiqued it at every opportunity, saying that it was mechanistic in that it did not consider the person's mind-set, determined by his or her place within a community, life factors impacting health, or spiritual beliefs that impacted healing. They echoed old American values of the Puritans, as well as the nineteenth-century conviction that work in a community environment, religious fidelity, sound morals, and personal initiative determined one's destiny.[39]

In 1915 Benedict Lust coached, "You are born to be master. . . . Do everything from your own free will and decision, under full consciousness of all consequences." He criticized the semiawareness of people who ordinarily moved through life "slumbering" and counseled readers to practice consciousness constantly—much as mindfulness is practiced in the twenty-first century. This meant abandoning blind habits that controlled behavior and replacing them with deliberate, chosen action. Naturopaths argued that adults would not have to work so hard at this consciousness if they had been trained as children. The formation of good habits during childhood "does away with the necessity for conscious regulation of many details of life."[40]

INDIVIDUAL TRAITS AND MENTAL HEALTH

Naturopaths believed that bringing out patients' best traits would restore them to health. One argued in 1918 that nature cure—or any other system—failed if the healers did not win patients' confidence, cheer them up, and encourage them. Optimism allowed a "person [to] see sunshine through clouds a mile thick," and this trait brought "friendship, success in business, and self-acceptance." Laughter, of course, was a trait sure to foster mental well-being. A 1915 author called it an art, noting that a full and open laugh purified, enriched the soul, and sustained one through life's sorrows. Enthusiasm was invaluable—like a diamond drill, which can cut through the hardest rock. An enthusiastic person became persuasive and hence more powerful and would not succumb to dull routine.[41]

Play, as described by Purinton, meant keeping the "child-heart" nurtured; it allowed one to avoid overwork, morbid thoughts, and pretentious actions. In 1949 a form of play recommended to foster health was development of a hobby, such as building miniature ships, photography, or collecting. Having a hobby was considered so central to well-being that Hobbies for Health was a recurring feature in Nature's Path.[42]

Detrimental to health were psychopathic influences, and Wilbur Prosser, ND, theorized in 1923 that heredity, fetal life, and early childhood were primary determinants for manifesting them. One's sensation of well-being or ill-being was passed through one's lineage and combined with elementary educational influences. This argument explains the early naturopaths' constant attention to childrearing and mothers' crucial role in it. The irreplaceable teachings of Mother, when true to nature's laws, gave her offspring the proper habits and "the moral and intellectual elements which make for national greatness." By the 1920s, in the wake of World War I, this emphasis on producing national greatness was used as an argument akin to eugenics: only those higher-evolved individuals were suited for reproduction, leadership, and governance.[43]

There was a long list of negative traits that people must eliminate. Melancholy was the "disease of over civilization," and hypochondria or low spirits combined with dyspepsia to create melancholy. One argument was that only highly civilized, sophisticated people were aware of their nerves and stomachs. Espousing a racialized, romanticized view of nonindustrialized life, author Lillian Russell in 1915 wrote of melancholy, "The savage and primitive nations have no such consciousness or knowledge." Another trait to avoid was sulking, which naturopaths associated with lost fun and money, misery caused to others, chronic fear, and overworking. Mental fatigue, or "brain fag," was described by homeopaths and naturopaths as mental confusion, indecision, and chronic weariness. It was attributed to improper elimination by the organs.[44]

Worry, wrote one 1915 commentator, was "a constant thorn in the flesh" that destroyed mind and body, unsteadied the nerves, wrecked health, and led to a life with imaginary ills. This pathetic condition, he concluded, was brought about by lack of faith in God and oneself. A naturopath at Yungborn once spoke about a man in Switzerland who was so overcome by the belief that he had been bitten by a rabid dog that he consulted many doctors despite proof that the dog was free of disease. Without right thinking, his mind was poisoned by crippling worry, not rabies. Naturopaths by 1920 discussed anxiety, or fear thought, observing that thoughts about possible tragedy so immobilized people that they ceased to enjoy life. Negative emotions turned automatically into self-destruction. As Benedict Lust put it, "Fear, malice, jealousy, hatred will squander vitality to such an extent as to be quickly seen in bilious attacks, with accompanying headaches and coated tongue." One could avoid these woes by learning habits of healthful thinking and practicing them daily. In a lively commentary one naturopath observed that "the Devil Worry spoils more digestions than Whiskey . . . and Hate is a low-down, degrading over-civilized detestable thing." Anxiety did not appear to be a gendered condition. Mid-twentieth-century articles about mental distress showed images of men and women in equal number.[45]

In general, naturopaths were clearly interested in psychology. The *Naturopath and Herald of Health* reported lectures on practical psychology and how and why psychology helped drugless practitioners. In the second decade of the twentieth century, naturopathic definitions of psychology sometimes included hypnotism and mental suggestion. But scientific psychology only went so far with naturopaths. "Cults galore are arising," cautioned one author in 1920. They were based on the assumption "that health once lost can be regained by thought alone." There was a place in natural healing for mental and psychological measures, he wrote, but one should not rely on them alone.[46]

Those involved in the medical freedom movement (the National League of Medical Freedom and the American Medical Liberty League, as well as naturopaths) tended to lump psychotherapeutics with allopathic medicine and state power. Purinton noted the limits of psychology, saying that alone it was never an adequate means of cure. What was needed was a sensible doctor who practiced correct diagnosis and taught right habits of eating, drinking, breathing, bathing, exercising, sleeping, clothing, resting, working, and playing. He told his colleagues that "mental control of and by the sick is fundamental. A cheerful, calm, active mind often helps recovery more than medicine does." In 1939 naturopaths opposed psychotherapeutic philosophy because it stopped short of including the role of poverty (among other factors) and focused too much on the sexual difficulties of the sufferer. It was a direct critique of Freudian psychoanalysis. The naturopathic healer instead utilized constructive suggestion, affirmations, and other mental or spiritual healing methods that involved the active cooperation of the sick person. This approach minimized the expert-patient divide and shifted causality to social rather than personal history-based problems.[47]

There were also more blanket critiques of Freudian theory and methods. One was that Freud devoted too much of his time in the study of abnormal individuals, whose behavior differed strikingly from usual behavior. No wonder he conceptualized personality as a battleground. From a naturopathic view, mental conflict was actually normal and could be ameliorated by balanced and healthful ways of living.[48]

LOCATION AND THE SIMPLE LIFE

Location and climate were key to attaining physical and mental well-being, but naturopaths were not always reasonable in their analysis of the connection. Wellness might depend on climate, but some had impractical notions that health also depended on rejecting civilized ways of life. One naturopath, observing where the great religions had been founded, said that "the true Christian, the true adherer to Buddah [sic] and Mohamed can only live in the tropics with absence of wants. . . . Moving away from the equator compels men to care for themselves—makes them selfish and unreligious." He distinguished between Euro-Americans and indigenous people and asserted that tropical dwellers were more innately natural. Temperate climates allowed for nudity, useful for the sun baths favored by naturopaths. Naturopaths tried to acknowledge the harm caused by Eurocentric imperialism. One faulted European missionaries for insisting that natives adopt clothing, since temperate climates were conducive to nudity. Others blamed the white man's diet for depleting the health of Africans they encountered; also ridiculed were the British attempts to Europeanize the clothing of natives in the Gilbert and Ellice Islands. Even as the naturopaths sympathetically applied their theories to other cultural experiences, they came to naïve and utopian conclusions: one should live in a climate that demanded little, reject wealth (which was unnecessary under the tropical sun), and choose the palm trees or the virgin forest. This romantic view insinuated that overcivilized urban Americans could find respite in the so-called less civilized tropical regions.

Conversely, urban, colder climates were more likely to induce physical and mental distress.[49]

Nudity—or even partial nudity—while desirable as a natural state, was not welcome in most locations. That did not stop enthusiasts who felt the benefits of vitamin D synthesis before it was fully understood. Claus Hansen, writing from Iowa, said that he had thrown off hat, shirt, and shoes when he worked in the field on hot summer days, believing this was in accordance with God and nature. He was met by the sheriff and taken before an insanity commission, and four neighbors testified that he was daft. In self-defense he had to secure a certificate attesting to his sanity. Needless to say, the experience was devastating to his mental well-being.[50]

Southern California provided health benefits. A resident at the Coronado Health Home and School bragged in near reverent tones of its year-round balmy, soothing atmosphere and the relief it could provide for the "overstrung, enervated inhabitants of our over-civilized cities." Referencing the weakened condition of urban white Americans, he extoled the virtues of tropical climate. Of course, not everyone could live in the tropics, but naturopaths suggested which materials to wear in each season. Carl Strueh lamented humans' inability to adjust to climate changes as animals did and the need to resort to clothing to compensate for this defect.[51]

While there was a romantic view of natural living, the bottom line was that life in the city was debilitating. When temperatures hovered at 100 degrees, a person on the street was depleted in all ways. Some described the off-putting colors, the vile smells, and the general atmosphere in office buildings, inhabited mostly by businessmen and doctors. In fact, the romanticization of tropical life and nudity accompanied displeasure with urban life, industrialization, consumerism, and life's fast pace and its attendant worries. Proponents of "the simple life" asked people to eschew excesses, focus on meeting their needs, and abandon many aspects of modern life. Yungborn and other sanitariums were designed to remove people from urban life and provide a simple life for renewal.[52]

Simple living also meant abandoning the cultural demigod of materialism. One could reduce the high cost of living, Edythe Stoddard Seymour told her readers, by reducing expenditures, learning to live well on a small salary, staying out of debt and saving when possible, buying sturdy clothing and furniture and repairing rather than replacing them. In this way, one could avoid overwork and maintain mental calmness.[53]

Yet Benedict Lust recognized in the wake of World War I and again in the midst of the cataclysmic Great Depression that getting away from the city or

avoiding mental distress was difficult. In 1931 he conceded that all modern people, regardless of age and social class, were nervous. The condition was so pervasive that it was considered normal. It appeared to be caused by the hurry and
haste of the times and the fierce competition for success. There was little difference, Lust said, between nervousness, neurasthenia, and hysteria. They were all
caused by negative mental inclinations, as well as inborn, abnormal conditions,
he said, and his colleagues concurred wholeheartedly. Thus, the dictate to live
the simple life had been subsumed by the reality of industrial urban life.[54]

Overwork was linked to the dangers and illnesses synonymous with urban
life. Overwork was constantly deemed a cause of poisonous thoughts, but naturopaths failed to point out that the *ability* to work less and still survive materi-
ally was based on class. Edward Earle Purinton offered a series of case studies
in 1914 about overwork and mental frames of thought that induced either distress or optimism. In one case, a healer, lecturer, and health reformer was so
immersed in his work that he never rested. He aged prematurely, suffered from
nervousness, and could not find peace. His devotion to reform ruined him.
Purinton pointed to the irony of devotion to health reform that led to poor
mental well-being. Speaking to Social Gospel reformers—or perhaps to New
Thought followers—he argued that it was not God's will to have reformers labor
to their own detriment. The balance between meaningful work and overwork
was difficult to achieve; fulfillment and success were linked to being in harmony
with other people and achieving a balance between one's work and one's personal life.[55]

In the following decades, naturopaths' counsel to avoid overwork remained
constant, but the rigors of the business world made this unrealistic. Articles were
aimed at middle-class Anglo men who were pictured in business attire. By 1948,
during the postwar boom, workers in intellectual (versus manual) jobs were urged
to exercise. Gone was the exhortation to avoid overwork—its prevalence was
accepted. Instead, one should participate in outdoor sports, stretch while standing, and lie on one's back or face down to achieve physical relief. Purinton
linked prosperity with faith and healthful vitality. His use of the word *faith* was
curious; he invoked Buddha as saying that "faith is the best wealth" but then
described famous millionaires who had achieved because they could envision
their goals and because they strongly believed they would accomplish them.
These successful men found faith in their ability, their work system, the loyalty
of their workers, the satisfaction of their customers, their economy, their quality output, and their plans for the future. The link with spiritual faith had all
but vanished.[56]

Perhaps the strongest signal of a move away from early urban critiques came in a 1950 article arguing that cities could be soothing. The author of this article, aimed at laborers, was oblivious of workers' reality. Discord, he said, was abnormal, unnatural, and evil. But it was not environmental; it was caused by one's reaction to one's environment. The feverish bedlam of the steel mill was not itself discordant, but one's unhealthy reaction to it could be. "Mental awareness of the joy of doing, and spiritual comprehension of the whole magic of existence" could overcome environmental obstacles. In this case, the author's own social standing rendered him simplistic at best.[57]

What had begun as a cohesive and consistent mental health philosophy for practicing meaningful, controlled consciousness and avoiding poisonous thoughts shifted over the decades from a largely pragmatic yet Christian philosophy to a spiritual and psychological one and then to a materialist one. By the mid-twentieth century only remnants of the original ideas about mental well-being remained, to resurface at the end of the century. But the core naturopathic ideologies surrounding vitality, constructive principles, personal responsibility, toxicity, environment, and diet continued to be foundational.

CHAPTER 4

Louisa Stroebele Lust, Benedict Lust, and Their Yungborn Sanatorium

Louisa Lust was a visionary, a devoted laborer and economic powerhouse who drove the naturopathic movement and embodied its core ideals. Born Aloysia Stroebele in Sigmaringen, Germany, in 1868, she was the daughter of the mayor. While written records of her early life are not available, her devotion to the cause, her associates as an adult, and her collaborative relationship with Benedict Lust reveal her to have been a compelling historical force.[1]

Before immigrating to America, Stroebele went to England, where she studied Rikli's and Kuhne's nature-cure methods. While in London she also served as the personal assistant to the American Tennessee Claflin on three world tours. On these tours Claflin and her sister, Victoria Woodhull, promoted their political, social, and sexual ideologies. In their personal lives and in their writings and speeches Claflin and Woodhull were two of the most notorious radical women's-rights thinkers in the 1870s through the 1890s. Stroebele accompanied Claflin from approximately 1888 to 1891 and became a forceful and competent young woman as a result of her exposure to Claflin's activities, beliefs, and economic savvy. The combination of her nature-cure studies and her sociopolitical activities explains why and how Stroebele ultimately became a formidable cocreator of naturopathy in America.[2]

THE TENNESSEE CLAFLIN INFLUENCE

Tennessee, the younger of the two Claflin sisters, was born in Ohio in 1856. She was twelve years older than Stroebele. Claflin's father, Buck, capitalized on his daughters' beauty and dramatic flair. He marketed his daughters and himself as fortunetelling healers to eke out a living. He was somewhat successful because of the rise in nineteenth-century America of a widespread quasi-religious spiritual-ist movement, wherein people tried to contact the deceased through rituals. There were others, like the Fox sisters, Kate and Maggie, in rural upstate New York, who commanded handsome prices, as they supposedly contacted the dead. Within a few years, tens of thousands of believers flocked to séances.[3]

Tennie first showed signs of having second sight at age five, when she told a farmer where to find a lost calf; and she predicted a seminary fire so precisely that she was actually suspected until she was proven innocent. She was shrewd and learned some clever behaviors from her father. By the time she was eleven, Tennie was advertised in Columbus, Ohio, as being endowed with a supernatu-ral gift and able to advise which medicines would cure diseases. Buck Claflin promoted himself as Dr. R. B. Claflin, the American king of cancer. His thera-peutics were guaranteed to cure if patients followed Tennie's revelations and his medicinal apothecary; years later Tennie admitted that she had been forced to humbug people for money. In ensuing years Tennessee and Victoria marketed themselves as clairvoyants and mediums, and they were charged with fraud more than once.[4] Louisa Stroebele knew about this, which preceded her affiliation with Claflin, and it may account for her abhorrence of nature-cure charlatans.

By the Claflin sisters' adult years, accusations against them were constant, in-cluding accusations of illicit sexual activity. Tennie had a brief, ill-fated marriage to John Bartels (in which she refused to take his name). It dissolved after an un-successful trip to sell séances in Missouri and Arkansas. Plagued by rumors of fraud, named in an adultery case, and implicated in blackmail, the sisters left the Midwest and traveled to New York City. Eventually, they arranged to meet the seventy-six-year-old shipping and railroad tycoon Cornelius Vanderbilt, who was an unabashed devotee to spiritualism and faith healing and a notorious woman-izer. He was charmed by Tennie. Thus began a reciprocal relationship in which Tennie received investment and financial advice and Vanderbilt received spiri-tual and, presumably, physical favors from Tennie.[5]

All of this preceded the conservatively raised Louisa Stroebele's working relationship with Tennie. By the time Stroebele became an assistant to Claflin, Tennessee was a charismatic, convincing, and notorious personality. More im-portantly, Tennie and her sister Victoria had gained legitimacy among some pro-gressive women and men as intelligent critical thinkers and activists on behalf of women's rights. Among those rights was the freedom to have sexual relationships as long as they resulted in mutual happiness for a couple. They saw loveless mar-riages in which a woman was forced to stay married to a man for economic sup-port as little more than prostitution. At that time, a woman was obligated to submit to her husband sexually and could be forced to do so; there was no such concept as marital rape. Tennie's intimate relationship with Vanderbilt was an example of her belief in free love. Free love, as advocated by Claflin and Wood-hull, meant that consenting adults had the right to pursue romantic and sexual fulfillment outside the confines of monogamous marriage—that serial monog-amy *with* love was more logical than a marriage in which love had died. In their definition, free love was a single standard of morality for both women and men, in contrast to a sexual double standard in which the community pretended not to notice men's sexual indiscretions. The sisters eschewed traditional views of women's passivity, economic dependence, and physical and sexual timidity. While Louisa Stroebele copied little of the behavioral exploits of Tennie and Victoria, she did model herself after Tennessee with respect to her business acu-men, her critique and distrust of marriage, and her feminist thinking.[6]

Tennie's ideas about sexual equality and women's health influenced Stroe-bele. In the early 1870s Claflin gave a lecture at the Academy of Music in New York City on the ethics of sexual equality. She argued that there was a connec-tion between women's physical health and morality, since men had absolute sexual freedom and dominance, and women were forced into sexual submission or exploitation. At the same time, she argued against regular physicians' methods that professed to cure disease, because they did not produce genuine physical health, leaving women vulnerable.[7]

Claflin then lambasted regular physicians, "with their vile stuffs [who] profess to, and in some instances think they do, cure disease," but they did not cure, be-cause the causes were not addressed. Another, even more controversial claim of hers was that women's health was deteriorating overall because of their legal ob-ligation to be completely obedient to their husbands. Feminist free lovers and female moral reformers were usually on opposite sides of arguments, but when it came to voluntary motherhood, they agreed: women were entitled to resist and even refuse their husbands' sexual advances. Women's bodies suffered from both excessive demands of intercourse and the dangers of serial childbirth.[8]

Stroebele absorbed this ideology and its implications for an individual wom-an's life. She shared Claflin's loathing of regular medicine and of its physicians, who were silent about "the class of diseases resulting from sexual abuses" and their "unfortunate results, which legitimately flow to women from our present marriage system, and to men, from its attendant fact of prostitution."[9]

Stroebele's mentor argued that sexual inequality was exacerbated when middle-class men, presuming their wives to be passionless and sexually over-taxed, sought sexual release with working-class and ethnic prostitutes. This idea that middle- and upper-middle-class European American women were innately asexual, or passionless, was fueled by ministers and regular physicians and led to a rise in prostitution by midcentury. As more men frequented prostitutes, women activists decried the results: increased pregnancies and physical ruin of prostitutes and increased transmission of sexual diseases between husbands and wives.[10]

After three world tours with Claflin and thoroughly indoctrinated with feminist ideologies and sound investment strategies, Louisa Stroebele visited her only brother, Father Albert Stroebele, in Butler, New Jersey. She found the town to her liking and immigrated to the United States in the late 1880s. In 1892 Stroebele used funds from investments with Claflin to open and become the doctor in charge of Bellevue Health Resort, a sanatorium in Butler. She used therapeutics she had studied at eclectic European nature-cure institutions and advocated moderate, reasonable ways of living with stellar results. She used cold water cure à la Priessnitz, contemporary hydropathy, air baths (Rikli's air cure), light baths, and rational eating—vegetarianism and a balanced diet. When Benedict Lust met Stroebele two years later, he described her as efficient, capable, and successful.[11]

Louisa's brother, Albert, was the priest at St. Anthony's Catholic Church and led pilgrimages called the Procession through Grace Valley to the shrine in his town. Benedict Lust, a devout Catholic, belonged to the German singing society and joined these pilgrimages in 1893. As Lust recalled in his memoir, the singing pilgrimages served as an antidote to the criticism and oppression he faced in the city.[12]

When he first saw Louisa Stroebele in 1894, he found her to be a vibrant and attractive woman. He admired her from afar, "a strong, independent, yet warm-hearted girl with a queenly carriage and forceful personality." Before long, he knew he wanted to marry her. However, he was immediately confronted with Louisa's views of gender roles and marriage and did not imagine that she would accept him. Louisa was focused on her work. Lust was reluctant to press the issue of marriage because "from the standpoint of offering her material security I could offer her no more than she had. Financially she was independent. Not only was Bellevue successful which assured her an income but she owned free and clear another property in Butler."[13]

Stroebele planned to expand her business. She was familiar with the Kneipp water-cure treatments and offered Benedict the job of Kneippian physician at Bellevue, since he was an expert in its applications. His work was in New York City, thirty miles north, so he could not personally be in residence, but they struck a deal that he would train and engage an operator for her. Benedict visited

Bellevue each weekend and was pleased that their reciprocal referrals increased the clientele of both the sanatorium and Lust's city-based practice. She also helped him with his English-language skills.[14]

"As I reminisce now," Benedict wrote decades later, "I can discern the influence of Lady Cook's [Tennessee Claflin's] theories of feminine independence—the emancipation of woman in Louisa's reluctant attitude toward marriage." Claflin, after all, had refused to take her first husband's name when they married and likened the marital bond to sexual servitude.[15]

Lust was well aware of Claflin's political actions. Her crusading was internationally known. In 1871 Tennessee had written the political pamphlet *Constitutional Equality*, in which she discussed how the political system debased women through its exclusion of them, the need for rights and privileges for children, the necessity of a strong physical constitution among political citizens, the community's role in caretaking, and the mother-child bond. This pamphlet was published one year prior to her sister's women's-rights-based presidential campaign, with Frederick Douglass, the renowned abolitionist, named as her running mate. Claflin had a short-lived affiliation with the women's suffrage movement; she and Victoria were categorically shunned by the movement when they proclaimed the ideology of free love. Claflin's pamphlet also argued, in what probably struck some as an ironic note, for the positive influence of women's superior moral purity: "To whatever depths of degradation some of the sex [women] have fallen, women, as a whole, is [sic] possessed of a healthful, saving, purifying power that is needed everywhere."[16]

These were the messages of her talks as she traveled with the young Louisa Stroebele. As radical as some of these claims were, others were actually quite moderate. Claflin took an antiabortion stance, arguing that birth control should prevent conception. She stood by the essentialist argument that women were naturally maternal caretakers and that the mother-child bond was sacred.[17]

Another reason Lust had heard of Claflin and Woodhull's influence was that the sisters applied their theories publicly, in their own lives, and in the pages of their radical newspaper *Woodhull and Claflin's Weekly*, published intermittently from 1870 to 1876. The paper advocated for women's rights, less rigid divorce laws, protest of domestic violence, socialist views of housing and transportation for workers, and spiritualism, among other causes. The paper was the first in the United States to publish Marx and Engel's *Communist Manifesto*. Then, to unmask the sexual double standard among men of high standing, they went public with the hypocritical and secret sexual affair of the Reverend Henry Ward Beecher in 1872. His sisters, the nationally known educators and authors Catharine and Harriet Beecher Stowe, had been vocal critics of Woodhull's activism, making Henry a likely target for Claflin and Woodhull. A famous Brooklyn minister, Henry was a social justice advocate who regularly preached marital sanctity and monogamy while pursuing a lusty tryst with his best friend's wife. Elizabeth Tilton was a staid and proper moral Victorian who was Beecher's parishioner and the wife of the well-known editor Theodore Tilton. Woodhull demanded that Beecher admit his behavior and endorse free love. Beecher denied the charge, labeling the accusing sisters "two prostitutes." His two famous sisters came to his aid publicly.[18]

Woodhull revealed the tawdry affair in a speech and published it in *Woodhull and Claflin's Weekly* in 1872. Within hours of Woodhull's speech, Tennie and Victoria were arrested and jailed on obscenity charges. Woodhull and Claflin had done their damage, if only temporarily. In 1875 Beecher stood trial for adultery in one of the most sensational trials of the century. The free-love doctrine and Claflin and Woodhull once again captured national headlines. The trial resulted in a hung jury, and Beecher's wife stood by him. In a perfect example of the sexual double standard fought against by Woodhull and Claflin, Elizabeth Tilton confessed to adultery with Beecher two years later and was excommunicated from the church because of it. Henry Beecher, however, remained a popular church and cultural leader. Tennie and Victoria fell from public grace and fled to England, and ultimately each married well.[19]

This was the stunning political past that Lust associated with Louisa. Yet Benedict Lust, of proper and, one can even argue, rigid Germanic upbringing, found himself attracted to and working for, and with, Louisa Stroebele. Knowing that Stroebele had been Claflin's protégée, Benedict admitted that their traditions were antithetical: "Lady Cook's influence was at great variance with the German conception of marriage. I must admit at that time . . . masculine superiority was stressed in that institution and . . . the offer of security [marriage] was tantamount to acceptance [of male authority] by the bride." Benedict, a realist, smartly summarized, "So I had everything against me, at least so it appeared. For not only had Louise's conception of the institution of marriage and the position of women in the family undergone a decided change through Lady Cook, but during this change and with it, Louise had found personal [economic] security that few girls of the day enjoyed." Benedict mused that "today [1923] we would term Louise a professional or business girl with a career, and therefore in a position to defer marriage until she desired it."[20]

CLAFLIN'S LEGACY TO STROEBELE:
FINANCIAL INDEPENDENCE

Claflin and Woodhull preached that married women's financial dependence on men enslaved women and denied them autonomy, dignity, and adult status. Not surprisingly, they took a series of daring, even shocking steps that allowed them to be free of this servitude. At one time they enjoyed independent wealth.

Tennie was twenty-three when she met the millionaire Cornelius Vanderbilt in 1868. Vanderbilt had already embraced spiritualism and faith healing and entertained anyone who came to help him with his attendant aches and pains. He was also recently widowed and seeking companionship. The Commodore, as he was called, was spritely, aging, and lonely. The charismatic and mystic attentions of Tennie C. (as she began to call herself) appealed to him, and her magnetic hands-on treatments led to a physical, curative intimacy that soon became sexual. The sisters all but moved in with him.[21]

Vanderbilt, smitten by Tennie and spiritually seduced by Woodhull's visions, offered to set them up in a career as financial speculators. They began with real estate ventures and stock market investments, initially with his capital. Theirs was the first women-owned brokerage business when they opened a storefront in Manhattan in 1870 and dubbed themselves the Woodhull & Claflin Company. The sisters' notoriety preceded them, and business flourished for what some called the "Bewitching Brokers." They relocated to fancier surroundings in the center of the financial district, and with Vanderbilt's guidance and their own connections they invested in stocks, his New York and Harlem Railroad, and real estate. Much of what they amassed was the result of their own diligence, intelligence, boldness, and willingness to take risks. However, their successful tips from the Commodore helped them long before insider trading was illegal. As one scholar noted, "Their quarters swarmed with the curious and skeptical, and while the men of Wall Street scoffed over their brazen stratagems, the sisters unquestionably prospered at their new calling." They all made a fortune. Claflin claimed that Vanderbilt had promised her marriage, but his kin intervened and maneuvered him to someone more suitable.[22]

These earnings made possible the production of *Woodhull & Claflin's Weekly*, in which the sisters pioneered muckraking journalism a full decade before the practice was commonplace. When Louisa immigrated to the United States, she brought with her a sizable stash of money that did *not* come from familial wealth. It appeared to have come from wages earned from Claflin and fruitful investments.[23]

A CURIOUS UNION: LOUISA STROEBELE
AND BENEDICT LUST

For the remainder of her life, Louisa Stroebele lived thriftily so that she could fund innovative and foundational naturopathic work. She was empowered as a financially independent entrepreneur; she had found her passion in nature cure, specializing in women, and she was in no rush to marry.

Benedict Lust acknowledged that Louisa gave them their start, as she owned the Bellevue Health Resort long before they were married. When Stroebele opened Bellevue in 1892, she promoted the establishment's proximity to New York City (it was just 30 miles away) and dubbed it "an ideal summer resort for lovers of nature." She and Lust together expanded the Bellevue clientele and therapeutics after she studied Kneippian water-cure methods at the Columbia Institute, taught by Lust.[24] During these years they were successful with Bellevue, and they made plans to open a larger institution on Louisa's other piece of property in Butler.

Because of Louisa's financial independence and beliefs, the courtship of Louisa and Benedict was labored, long, and filled with uncertainty for him. He understood the source of her abhorrence for marriage but was puzzled by her determination to avoid it. He recalled that despite his persistence, he did not seem to be getting anywhere with his hopes of marriage. Chances for broaching the subject dwindled as he felt her disinterest in discussing it. When he raised the topic, he felt rebuffed and silenced by her, because "she was so logical and practical that my arguments for marriage melted under the influence of her conversation."[25]

Louisa's brother, Father Albert Stroebele, emboldened Lust to propose to her. As the two men stood in a field at Butler, Albert picked a flower, placed it in Benedict's hand, and metaphorically counseled: "Benedict you are 'the master of all you survey' of Bellevue. He said nothing about marriage but I understood his meaning." At Father Stroebele's prodding, Benedict mustered courage one final time and proposed marriage. She accepted.[26]

There was further trouble between the couple when a meddling relative of Louisa's implied that Benedict was outclassed by her. Lust had heard that he could not expect Louisa to do the physical things that he did, such as help in the health food store he ran. She was described as "a lady not a servant." So displeased with this pronouncement was Benedict that he cut communication with Louisa before he even broached the issue with her; he wrote no letters, made no weekend visits. When she finally contacted him, they worked things out and agreed, according to Benedict's terse memoir, "that Louise would have to be satisfied to share my life and come down to my level if that is what she considered it." They also agreed to reject interfering kin. They did not marry until 1901, after a six-year courtship and working relationship.[27]

Louisa's subsequent role in establishing the field of naturopathy built on this negotiation with Benedict for power and control in both their relationship and their business dealings. Her progressive background and his traditional upbringing were fated to elicit flash points on key issues. Yet theirs was not a union peppered with discord. Louisa, once she agreed to marriage, took Benedict's surname and was willing in all ways to be a partner, and in some ways, a traditional wife. All the while, she remained an intellectual, economic, and philosophical force in her own right. One can speculate that the change in Louisa was a direct result of her infinite shared passions with Benedict, their work, and her early familial teachings, which had emphasized duty and marriage.

They married on June 11, 1901, at St. Patrick's Cathedral in a ceremony performed by Reverend Joseph Dailey, who presided over the diocese. Dailey's brother, John, operated another Kneipp water cure and the Berkley Gymnasium at Forty-fifth Street and Fifth Avenue. Benedict had bonded with the priest, who was a frequent visitor in Butler.[28]

Their union was greeted as a near royal match. Both were well respected in their fields, and the series of legal harassments and prosecutions he had suffered from the New York State medical authorities had brought him considerable notoriety and sympathy. The couple's wedding was celebrated by their peers with a sumptuous wedding breakfast at Benedict's institute in New York City. Lust wryly admitted that not all the food was in accord with Kneipp's doctrines.[29]

Following the indulgent breakfast, there was another celebration at the Terrace Garden Opera House, the hub of German culture. This was attended by church dignitaries and the singing societies. The evening's elegant dinner and wedding festivities were held at the Arion Club, a place of fine dining. The venue was the result of Benedict's widespread connections: he had worked intermittently as a waiter for the Arion Club in Geneva, Switzerland, in his youth. Later they took the train to their Niagara Falls getaway. En route they visited the Pan-American Exposition in Buffalo, New York, and made a brief stopover in Poughkeepsie to visit a mutual friend, a woman who operated the Meyer Kneipp Cure. The Niagara Falls honeymoon was less than ideal, as both Louisa and Benedict contracted food poisoning at an elegant restaurant on the Canadian side of the Falls; they became violently ill and were unable to leave the town for several days.[30]

The food poisoning left a lasting impression on Benedict, who became permanently suspicious of all restaurants and hotel eateries. He strongly preferred

$1.00 A YEAR   SEPTEMBER, 1907   10c. A COPY

THE NATUROPATH

AND HERALD OF HEALTH

BENEDICT LUST, N. D., Editor and Proprietor, 124 E. 59th St., New York

THE NATUROPATH IS ALSO PUBLISHED IN GERMAN

BENEDICT LUST, N.D.   LOUISA LUST, N.D.
DIRECTORS OF THE AMERICAN "YUNGBORN," BUTLER, N. J.

Photographs of young Benedict and Louisa Lust, the directors of the American Yungborn. Cover page, *Naturopath and Herald of Health* 8, whole no. 92 (Sept. 1907). Courtesy of the National College of Natural Medicine Archives, Portland, Oregon.

food prepared at home, even though their travel schedules frequently disallowed it. This helps explain why he so vehemently extolled Louisa's food philosophies and preparations in both theory and practice.

On their honeymoon they planned for the future of naturopathy, which was realistic, given their mutual business successes and Louisa's financial backing. "We would open a larger institution," Benedict recalled. "We would together spread the message of not only the Kneipp Water Cure [whose applications he continued to oversee at Bellevue] but the air cure, of Rikli and Louisa's own dietary theories and all that was known of Nature Cure to all America. We were in ecstasy with our plans for the future."[31]

Louisa's Bellevue Sanatorium had opened in 1892, but in 1901, after their much-celebrated marriage, they began expanding their interests in Butler. They had in their favor Louisa's credibility, entrepreneurial experience and savvy, the skills she had brought to Bellevue, her successful operation of it, and her savings. Benedict recalled her vital role: "Her hard earned savings she gave generously and cheerfully to help spread further propaganda for the Cause." His monetary contributions came from the two Kneipp magazines, the *Kneipp Water Cure Monthly* and the *Kneipp Blotter* (1896); lectures; the Kneipp Society, the treatments he had offered at the Kneipp Water-Cure Institute since 1895, and his health food store.[32]

BENEDICT'S EARLIER CONVERSION TO THE CAUSE

Once married, Benedict enjoyed the stability and support that had been lacking in his single days. Emboldened, he became determined to earn an MD degree, which he hoped would shield him from the constant legal harassment and public scorn he regularly suffered. His determination was also fueled by an unsuccessful encounter with allopathic medicine a full decade earlier. Upon arrival at Ellis Island in 1892, he had worked as a waiter at the Savoy Hotel in New York City and kept fit through frequent bicycle riding. When he traveled to the Chicago World's Fair in 1893, he had a terrible accident. While on a streetcar, he realized that he had passed his stop and let go of the grip strap. He was thrown from the car into the gutter, where he lay for some time. His left temple, eye, and lower jaw were severely injured. After convalescing in Chicago at the home of a friend, he returned to New York and resumed his work as a waiter. He grew increasingly weak from operations and six mandatory vaccinations. He contracted a severe case of tuberculosis. He recalled that as his health declined he became depressed. He performed the simple water ablution procedures he had learned at Neuchatel in New York City, but his body did not respond.[33]

He decided to return to Worishofen, Germany, in 1893 to consult personally with Father Sebastian Kneipp. This decision, Lust notes, "was to influence my life and give birth to an integrated Nature Cure movement in America." Lust was Kneipp's patient for four months, which he paid for with the money he had earned as a top-notch waiter. These months transformed his health, his priorities, and his goals. "I saw delicate people like myself," he recalled, "restored to health and vigor so that they could go home with the admonition from Father Kneipp 'to go and live more naturally' ringing in their ears." Many got better under Kneipp's cure, but equally important, no one was harmed. The combination of hip and lower-leg baths, vigorous exercise, peasant food, and attending and applying Kneipp's lectures on water cure gradually restored his vigor. Months into his stay, Father Kneipp took Lust aside and counseled him to learn water-cure methods, become a teacher of the Kneipp Cure, and assume the mantle as its emissary.[34]

Within two weeks of his return to New York in 1894 Lust decided to abandon waiting tables and pursue Kneipp water-cure practices as a business. Staying in Saratoga, New York, for the summer season, he became acquainted with Louis Kuhne's vapor baths, skin frictions, and diet reform and Arnold Rikli's aerotherapy. Kuhne's theory of the unity of disease greatly informed Lust's philosophical and therapeutic approaches. According to Kuhne, there was only one cause of disease—foreign matter in the body—and therefore there was only one disease that took different forms. As Lust put it, "It was he who pointed out that one special treatment centered on the part of the body where the deposits mainly occur, the abdomen . . . [and] this treatment was designed to act on the chief secretive [sic] organs of the body, the kidneys and bowels." The fact that both Louisa and Benedict were familiar with these approaches helped fuel their partnership.[35]

The conversion experiences were profound for Lust. When shortly thereafter his brother Louis, a baker, became delirious with fever, Lust treated him successfully with methods learned under Kneipp at Worishofen. Louis was cured. This too confirmed Benedict's determination and his purpose.

THE LUSTS' EMERGING HEALTH EMPIRE

Thus inspired, Benedict opened a singularly unique health food and Kneipp supply store on September 15, 1896. On the first page of the ledger he wrote, "God is with me." By the turn of the century the *Kneipp Water Cure Monthly* was being issued from this establishment.[36]

The early years of his business pursuits reveal Lust's conservatism and stalwart work ethic. In the early 1890s he had openly scorned waiters who did not give their best efforts, and he left a gymnasium when it became "a hot bed of rabid socialists." He loathed anyone who questioned his integrity, and he self-righteously so severely reprimanded his brother Louis in the store for inappropriate conduct that they were temporarily estranged. After operating a catering business in Tampa, Florida, Louis, quite broke, wired Benedict for a loan so that he could return to New York City. "He came home meek as a lamb," Benedict recalled with obvious satisfaction, "and inclined to heed suggestions." Amends made, Benedict loaned Louis money to open the Lust Health Food Restaurant and Bakery on East Fifty-ninth Street. It became an intellectual hub where luminaries gathered, such as the so-called Swami of India, the chief advocates of progressive movements, and adherents of the Kneipp water-cure and physical-culture movements.[37]

By 1896 Lust had amassed a miniconglomerate comprising publications, a school, the restaurant-bakery, and the store. But he was shocked to find that his ambition to develop a natural health culture was opposed by both medical authorities and Kneipp purists. While Lust credited Kneipp with influencing him, as early as 1902 Lust distanced himself from the exclusivity of Kneipp's system. He saw the Kneipp Cure as basically prescriptive. Naturopathy involved instruction and inspired people to grow. His adoration of naturopathy in all its complexities was already entrenched.[38]

During his career, Lust was arrested nineteen times for practicing medicine and performing surgery without a license. Lust's fines, amounting to thousands of dollars, were often paid from Louisa's funds. The Lusts also bailed out other naturopaths similarly charged. As one colleague noted decades later, "She [Louisa] lived for others . . . her unostentatious benefactions provided succor for thousands who never heard mention of her magic name. . . . Naturopaths were . . . wholly unaware of the influence she wielded [so that they could] practice naturopathy through the propaganda and protection wrought by her hands, heart and brain."[39]

Desperate for credibility early on, Benedict convinced himself that if he possessed the MD degree, the legal bludgeoning would cease. He decided to enter the New York Homeopathic Medical College and Hospital, which was accredited to confer the MD degree. He anticipated the worst (which came to pass) but was willing to learn regular medicine. He had been studying nights in preparation and knew well the institution's high ranking. The Medical College required the same basic science entrance criteria as the other eleven schools in New York State: anatomy, pathology, bacteriology, and physiology.[40]

Louisa, who was committed to the utopian grass-roots vision of naturopathy, disagreed with his approach, but Benedict proceeded with a characteristic stubbornness. As he recalled, she believed that nature cure would survive without the help of a medical degree. But, he wrote, he was "much against Mrs. Lust's wishes as I was sure that if I had an M.D. Degree the medics would have to let me alone. But time proved how wrong I was."[41]

The reception he received was ghastly. He was taunted and harassed both physically and psychologically. Fellow students as well as professors worked hard to drive him out. He was a pariah. They treated him cruelly because of his promotion of water cure and naturopathy. They once rigged up an elaborate suspended contraption that dumped a deluge of cold water on him in class. Benedict, clearly demoralized by this assault, recalled that it felt like a devastating thunderbolt, delivered with the complicity of the professor, who just happened to be gone at that moment. When a devastating fire destroyed his printing press, they asked why he had not doused it with hydrotherapy. An article in the college's journal, the *Chromium*, jeered, "It serves Lust right for publishing such an organ and such should happen to all [publications like it]. Lust stood there with a hose, but it was a case where hydropathy utterly failed." They called him "quack," "Hydropathy Ignoramus," "Cold Water Doctor," and "Old H_2O."[42]

After six weeks of torment, when he thought things could not possibly get worse, Lust was called before the faculty and accused of spreading radical propaganda to students. He had distributed copies of the *Kneipp Water Cure Monthly*. All but one professor denounced him. It was a turning point for Benedict; his resolve doubled, and he determined to earn his degree and "fight to bring the freedom of choice of treatment I believed the American Constitution guaranteed."[43]

What he was taught distressed him as much as the abusive treatment. Benedict's outsider status was heightened by his visceral loathing of vivisection. He passed out during more than one demonstration; once he was taken to a hospital. The students were taught about vaccination, drugs, and serumization, but in his view they were not taught to heal. Vigorously opposed at every turn were the benefits of sun baths and the physical culture found in gymnasia and on playgrounds.[44]

Outside of school, he continued to administer Kneippian treatments and was upbraided by his faculty for doing so. In fact, he created the American School of Naturopathy in 1901, while he was studying at the Medical College. He later wrote that the true lessons he learned at the Medical College were that medical power and authority corrupt their possessors and that one did not necessarily find support from other alternative (nonallopathic) practitioners such as homeopaths or chiropractors. In short, he became convinced, as Louisa was, that naturopathy must stand alone as a healing system in order to survive. Amazingly, despite these trials, Lust persevered and earned his MD degree in 1902.[45]

Benedict Lust earned other medical degrees. While at the Medical College, he simultaneously entered the New York Eclectic Medical College, where he benefited from the classes and escaped ridicule, and over time he received a degree there. The eclectics did *not* view drugs as a cure-all, and nature cures were studied, as well as the uses of organic medicines. He had also earned an osteopathy degree. Now armed with degrees in eclectics, osteopathy, and allopathic medicine, he felt ready to branch out. His education from all these schools, even the allopathic, increased his commitment to natural healing. When he and Louisa established a second Yungborn in Tangerine, Florida, he took the Florida State

Board Examination and passed with a grade of 94. Now the question was, where and how could he best apply his knowledge?[46]

THREE YUNGBORNS

When Louisa Lust became the proprietor and doctor in charge of Bellevue Sanatorium in 1892, she promoted it as an idyllic summer resort closer to New York City than to the Catskills. After their marriage, however, the Lusts let Bellevue go. Louisa had a parcel of land, and with the help of her brother in 1896 the Lusts established the Butler Yungborn (sometimes spelled Youngborn), where she headed the Ladies Department. They were gradually able to expand, with adjoining land purchased over time from the Catholic Church. It was promoted as Benedict's enterprise; however, the cover of the *Naturopath and Herald of Health* featured photographs of both Benedict and Louisa, identifying them as directors of the American "Yungborn." Yungborn was a true outdoor enclave conceptualized, managed, and nurtured by them both. Louisa's character traits were evident in the atmosphere created at the Yungborns. Unwaveringly unselfish, sacrificing, cheerful, kind, loving, and unostentatious, she ministered to people's ills. Louisa's energy, talent, and finances allowed her to assert her battle "against stifling conditions of overcivilization, and for the full use of 'God's given earth and power' for life, health and happiness." Louisa was not as high-profile as Benedict, but she was the leading financial partner at Yungborn, and he was the consummate promoter of their work. He also excelled at using his trials and tribulations for publicity. Louisa received public acknowledgment for her publications and for her financial backing of practitioners' legal defense, and she was rewarded with appreciation and respect at gatherings of naturopaths. Louisa also transcended traditional sex roles through her professional expertise, her leadership, and her ownership of sanitariums.[47]

In a 1907 full-page ad, Yungborn Sanitarium is pictured nestled amidst trees, with several pristine tents in two tidy rows. Another ad from the same year touts it as a "First-Class Naturopathic Institute, also for Convalescents, Vacation Seekers and as a General Resort." A decade later an image showed a hut made from tree branches roofed with boxes from Lust's store. It was a true woodlands retreat. Its therapeutic open-air structures, outdoor therapies, routines of daily living, and dietary rules counterbalanced the chaotic pace and unhealthy activities and environs of urban America. The Butler woodlands offered a healthy stillness for patients. Birds and flowers fostered a sense of spirituality, as did the proximity to Father Albert Stroebele and the Sisters of Mercy. Once well established, Yungborn was described as the "Natural Life and Rational Cure Health Home for

Dietetic—Physical—Atmospheric Regeneration Treatment. Fount of Youth, and New Life School for those in need of Cure and Rest, for the physically and spiritually weakened, for those overworked, and the convalescent."[48]

Juxtaposed to the crowded city, filthy tenement life, animal waste from carts, the heat, noise, and dangers, Yungborn stood as an idyllic, rustic, preindustrial enclave. Idyllic, yes, but Yungborn was a wildly encapsulated utopian vision only for those wealthy enough to afford it. It was out of reach for the poor ethnic immigrants, for the struggling working poor living in squalor a mere thirty miles away, and for most blacks emigrating from the South.

By 1911 one neighbor described it as situated on sixty acres of beautiful scenery. It could hold eleven hundred patient-clients and was often full. It was also controversial to some locals, as nude or seminude sun bathing (called "sun baths") occurred behind fenced areas. One enthusiastic author called Yungborn the "Eden Spot." A typical day entailed the following regimen: up at 5:00 a.m. and begin water therapies, followed by mud baths, naps, sun bathing, walks, organized sports play and informational talks, and vegetarian nourishment—most or all done in the nude; the sexes were separated. Bedtime was 9:00 p.m. Disallowed were alcohol, tobacco, meat—and people with contagious diseases.[49]

Benedict and Louisa were billed as one regular physician and one lady naturopath, signifying Benedict's formal training (although "ND" sometimes appeared after Louisa's name in advertisements, so she may have obtained a degree from their American School of Naturopathy). They used physiological and dietetic remedies, including the Kneipp Cure, gymnastics, air, sun, and light baths, and a wide variety of mild water applications. Each individual had a diet designed according to her or his needs. The location itself was heralded as part of the nature cure. It offered a panoramic view of the Ramapo Mountains, a healthy climate, magnificent large private parks, and paths for romantic walks. A reference to "all comforts" in an ad likely refers to sanitary facilities, access to bathing water, and sleeping quarters. Its appealing location and temperate seasons made Yungborn particularly suited to spring and winter cures.[50]

In 1908 the reputation of Yungborn received a boost from Wu Ting Fang, the Chinese imperial ambassador to Washington, DC. He had cured himself of his sciatica via natural therapeutics and chose to stay at Yungborn. A Brooklyn patient lectured before the Franklin Literary Society after his stay, praising Yungborn's beauty and the higher class of patients there. Among the elite residents, besides Fang, were "a Spanish consult, a wealthy New England shoe merchant, a German artist, a count, a genius, a poet, a mental scientist, a number of

"Part View of B. Lust's 'Yungborn,' Butler, New Jersey." Postcard advertising the wooded setting of Yungborn in the early twentieth century. The large building on the right is the main house, and the two buildings on the left are resident quarters with wrap-around porches and roofs that allow a full circumference of therapeutic, fresh, circulating air. From *Collected Works of Dr. Benedict Lust, Founder of Naturopathic Medicine*, ed. Anita Lust Boyd and Eric Yarnell, ND, RH (AHG) (Seattle: Healing Mountain, 2006). Reprinted with the permission of Healing Mountain Press and the North Jersey Highlands Historical Society, Ringwood.

seasoned athletes, a Canadian philosopher, a charming octogenarian from San Francisco [and] a score of ladies." His observations revealed who had the ability to pay the steep fees. Superlative praise was duly recorded in the press, giving Yungborn and the Lusts much-needed credibility at a difficult time. By 1918 the New Jersey Yungborn was pictured as a large three-story building amidst trees and shrubbery on a well-traveled dirt road. A tended walkway led to the front door.[51]

Other sanitariums emerged, influenced by the Lusts' work. One, the Syracuse (New York) Naturopathic Institute and Sanitarium, run by Herman C. Schwarz, ND, DC, DO, offered treatments in water cure, light, air, electricity, mechanical manipulations, diet, and mental and spiritual healing. The institute's advertisement noted that Lust methods were used there.[52]

Emboldened by their success at the New Jersey sanitarium, the Lusts opened a second Yungborn in 1913, called Qui-si-sana And Recreation Home Yungborn, in Tangerine, Florida. One naturopath writing about Tangerine's Yungborn in

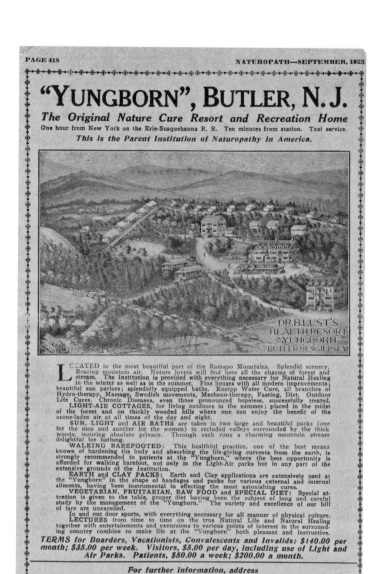
Advertisement for Yungborn, Butler, NJ. The full-page 1923 ad provides a vision and a description of "the Parent Institution of Naturopathy in America." Splendid scenery and "bracing mountain air"; the light-air cottages; sun, light, and air baths; walking barefoot; earth and clay packs; vegetarian, fruitarian, raw food, and special diets; and lectures are promised to successfully treat "even those pronounced hopeless." From *Naturopath* 28, no. 9 (Sept. 1923): 418.

November 1918 waxed effusive. He praised it as a home of health excelling in diet therapeutics, beautiful surroundings, simplicity of life, drugless healing, walking, and boating. He recommended it to all readers. Patients wrote praising testimonials after their stays at one of the two Yungborns, which Lust reprinted in the journal. Qui-si-sana (which the Lusts translated as "where you get well") was opened at an idyllic location on a terrace 90 feet above Lake Ola, 400 feet above sea level, with a dry climate, untroubled by fogs, malaria, mosquitoes, and temperature extremes. Shortly after it opened, a large sunroom was built for patients taking air and sun baths. Later, Tangerine offered two sun parlors segregated by sex. Qui-si-sana offered a full range of natural therapeutics, including all forms of hydropathy, Swedish movements, massage, mechanotherapy, chiropractic, osteopathy, and sun, light, air, clay, and lohtannin oak bark baths. The Tangerine Yungborn also offered a diet of fruits and nuts, as well as fasting. The resort was open year-round for nature cure, boarders, and visitors. Fire destroyed the Tangerine Yungborn in 1943, marking a turning point in Lust's life and leaving him with an injury that contributed to his death in 1945. The Lusts had also established a third cure facility, called the Natural Life Colony, at Palm City, near Nuevitas, in Cuba.[53]

Benedict and Louisa flourished at Yungborn. Both published pamphlets and articles, and Louisa published a detailed cookbook about the methods of healthful living. Louisa learned some of what she knew about food from the writings of Luigi Cornaro, who improved his own health by eating little. Her recipes and counsel were shared at Yungborn, in their restaurants, and in the pages of the *Naturopath and Herald of Health*. In one Yungborn ad called "Planning a Vacation," Louisa praised the beauty of nature, all living and growing organisms, and radiant human faces. One photo showed a brook with a footbridge over it. In addition to the dietary fare and location, "here, then is what we offer you: A camping expedition, a tour of the Alps, a sojourn at a health resort, and a family picnic-party all combined." Guests were expected to be active participants in their own health care, without staff or private nurses.[54]

The rates were inaccessible to all but the upper middle class and the wealthy—$2.50 per day in 1908, equivalent to about $66 in 2014. The weekly rate of $16 compares to $425 in 2014; the monthly fee of $60 to about $1,590. The true meaning of these rates can only be appreciated by comparing the fees with the salaries of workers. Wages in all industries, including those of farm workers, averaged $516 annually, while the average public-school teacher's salary was $455. In 1914 the Florida sanatorium charged the exceedingly steep sum of $100 per month ($2,440 in 2014)—and upwards—which included room, board, and treatments. One could be a boarder, a convalescent, or an invalid for $60 ($1,470) per month or $16 ($391) per week. The daily fee was $2.50 ($61).[55]

The lack of social-class and ethnic awareness resonates throughout the Lusts' brick-and-mortar cures. It is possible that reduced rates or fee waivers were available to those who applied. But this was not explicitly stated in print, memoir, or oral accounts. The Lusts knew, and emphasized, that the system could work for anyone if specific dietary laws were followed at home. But for urban dwellers struggling to feed their families, *any* food was scarce, making adherence to naturopathic right eating highly unlikely. Living at a Lust sanatorium was only possible for the well-to-do.

REJECTING THE "FLESH POTS"

In 1907, in response to requests from her patients, Louisa penned the *Practical Naturopathic-Vegetarian Cook Book*, the first exclusively naturopathic guide to food choices, preparation, and consumption. Naturopathic cooking meant no red meat, pork, poultry, or fish; instead one was to eat fresh vegetables, whole-wheat baked goods, milk, eggs, some sugar, and simple toppings. "The flesh-pots simply cease to attract under the savory spell and tasty appeal of this kind of cuisine," reads one ad. "It's more than vegetarian, it's sane." The recipes in the book, which sold for one dollar, were simple, wholesome, and scientific. Louisa wrote that how one ate lengthened or shortened one's life span.[56]

Louisa's counsel in the book was simple, straightforward, and widely used. The frontispiece chronicled her impressive credentials: she was the naturopathic director of the Ladies Department of Yungborn in Butler, NJ, and for the Naturopathic Institute and Health Home in New York City and an instructor of practical naturopathy at the American School of Naturopathy. In the text's early pages, she writes: "The preparation of food is a science as well as an art. As a practicing Naturopath and instructor of dietetics for about fifteen years, I have found the need of a simple, wholesome vegetarian cookery book, with reference to dietetics in health and in disease." Later in the book she states that "despite 1,700 plus dietary and cookbooks . . . dyspepsia prevails." Women in the home, she counsels, can cook inexpensive food by baking and stewing without destroying its value. The ideal diet comprised nuts and fruits. Besides abstaining from meat, critical for good health were raw foods. She suggested alternatives for animal meat such as pea, lentil, and macaroni cutlets. The recipe categories include rice and wheat; vegetable pies, stews and dumplings; soups and gruels; vegetables; and sauces, among others.[57]

In the *Naturopath and Herald of Health* Louisa wrote a series of weekly menus entitled "Naturopath Health Kitchen." Based on German cookery, they included the following: Monday's lunch, potato salad with poached eggs; dinner, morel soup (with young carrots and green peas), Bavarian vegetable balls with butter, lentils with poached eggs, filled zwieback fritters with jam or fruit sauce, farina with berries. Another day's offering included a lunch of macaroni with golden egg dip and a dinner of rhubarb salad with Bermuda potatoes, rice fritters with jam, whole-wheat bread pudding with fruit sauce, buttered tarts with strength broth, and hulled grain with raspberries. Foods tended to be boiled and browned.[58]

Louisa's influence was profound through dietetics. On the one hand, this work complied with the traditional female role. But less commonly acknowledged is the fact that food values and practices had powerful cultural impact and influenced behavior. Precisely *because* food preparation was women's work and unpaid, until recently it had been minimized and too rarely analyzed in social, cultural, and scientific terms. Louisa, as well as naturopaths in general, encouraged women to embrace this power and control over health through their preparation of food. Rural communities had always understood and valued women's role as food producers, preparers, and preservers. But in the cities, too often this status had been lost in part because of industrialized food production. Louisa's expertise and leadership in naturopathy reemphasized women's role in health at the same time that the nascent naturopathic system emphasized food's importance for healing and building vitality.[59]

Writing in *Naturopath*, Louisa also guided her readers in the ways of healthful living. She advocated hydrotherapy, the circulation of air and sunshine within the home, the importance of cleanliness and teaching children this healthy living. "Do not mind if the furnitures, curtains, draperies, etc. will fade [from exposure to sunshine and air]. It is better to let them fade than should you." Good housekeeping, she wrote, combined absolute cleanliness with frugality and comfort, ensuring healthy bodies through healthy food and careful attention to sanitation and hygiene. At times she spoke directly to mothers' power from the pages of her columns. "If mothers would learn how . . . to prepare simple healthy food and by example teach their children the all-important lessons of how and what to feed the body, sickness would be out-grown and forgotten in one generation."[60]

Louisa's approach was realistic only for those *able* to follow the system. She realized the limited dietary reform that the average reader could adopt. She believed that the fruit-and-nut diet was ideal, but she knew that most people found it difficult. Thus, she believed that an incremental approach was wisest.[61]

The Lusts shared their ideas and generated revenue with a pamphlet series that was a multifaceted marketing tool. In addition to promoting their three cures, the pamphlets also detailed principles of naturopathy and rightful living that could be learned at the two locales in the United States. They also advertised the *Naturopath*, which sold for $2 a year in 1907 (equivalent to $52 in 2014), single copies for 20¢. Pamphlets reprinted articles that appeared in the *Naturopath* in 1915 and sold for 25¢ each. Pamphlet number 17 expounded on the arts of nourishment, bathing, and training. Dr. Benedict Lust's School of Diets is credited as the author, but given Louisa's expertise in diet and cookery, it is likely that she was the actual author. The next in the series urged readers to take control of their own lives by changing their habits. There is considerable repetition in the pamphlets, yielding one cohesive, relentless message: one should live right, be well, and take responsibility for one's way of life.[62]

For Louisa and Benedict Lust, the first fifteen years of the twentieth century were years of professional growth, mutual excitement over their burgeoning health enterprises, financial success, and personal well-being. Their experiences were the springboard for Benedict's nationwide activism and influence. Louisa's leadership made possible multiple realms of power and authority for women in all phases of the movement.

CHAPTER 5

Women, Naturopathy, and Power

Naturopathic philosophy and therapeutics offered an authoritative space for female practitioners and followers. It spoke directly to American women in their traditional roles as caretakers and food preparers for their families. The advice was written for women of all social classes, although the costs and the themes of moderation, self-restraint, and rejection of excesses resonated more with the middle and upper middle classes. Women's leadership in naturopathy—at home and as practitioners—was particularly important given the dominant prescriptive behaviors for women. The combination of conservative gender ideology and progressive ideas led naturopaths to view women as guardians of home and of society at large. Women were encouraged, indeed recruited, to assert their authority as mothers, wives, and progressive thinkers—and oftentimes to lead the rejection of consumer culture. Women were naturopathic physicians, leaders, and advocates of women's rights. Many male naturopathic leaders denounced the limitation of roles for women. Both women's traditional and their new progressive roles placed them at the center of naturopathic right living and societal transformation because of their vibrant health and well-being. Louisa Lust personified the passion, devotion, and unswerving commitment that drove the cause. Both her practice and her writings addressed dietetics, mothers as guardians of family health, sexual relations between married couples, and therapeutic applications for specific diseases. She was an inspiration for other female healers.

Early naturopathic writings reveal a complex tension around women's issues. Naturopaths held progressive views on female physicians and leadership, women's physical strength, nude sun and air baths, rejection of commoditized beauty culture, and women's right to control the frequency of sex—and hence

"The Winner: Easy to Guess." In the battle over female clientele, the ANA as Ana Naturopath is healthy and content. The AMA, as Ama Hippocrate$, is aged and pinched with life's worries. From Benedict Lust, ed., *Universal Naturopathic Encyclopedia, Directory and Buyers' Guide: Year Book of Drugless Therapy*, vol. 1, for 1918–19 (New York: privately printed, 1918), 61.

pregnancy—within marriage. Yet at the core of these views was a basic belief in sexual biological essentialism: women's primary functions were reproduction, nursing, caretaking, and marriage. Naturopaths' understandings of women's bodies came from viewing them in terms of those functions.

The common naturopathic belief was that women's constitution made them compassionate, understanding, and nurturing. The desire to be a mother was in keeping with the natural order of the universe. In this way, naturopaths were no different from their allopathic counterparts. The nationally known naturopath Edward Earle Purinton wrote in 1915, "Women are so made that they must work for their lovers, and play with their babies." Silence and spirituality were ideals for women, and they were more capable of achieving them than men. Yet Purinton also asserted that women could be both feminine and masculine: "The ideal woman is the one who can play a man's game yet keep a woman's heart."[1]

Male and female practitioners alike saw female traits as powerful and necessary for the success of society. But the capabilities that empowered women also supposedly made them subject to nervousness—a belief naturopaths shared with allopaths, although they were less vehement. Over time, naturopathic articles and advertisements distinguished between women's and men's nervous reactions and how to treat them. One advertisement proclaimed in 1929 that sick nerves were a result of overactive emotions and the constant turmoil in women's domestic and marital relations. For men their causes were worries, intense concentration, excesses, and vices. In later years product advertisements invariably pictured distraught women, not men. These ads were created by manufacturers, not necessarily naturopaths, but the images perpetuated strong beliefs in women's vulnerability.[2]

By midcentury the roles of women and men were more bifurcated. Benedict Lust wrote that World War II culture was causing women to fall into decadence, smoking, drinking, and deserting the ideals of womanhood and motherhood. A Massachusetts naturopath wrote that childbirth was the crowning fulfillment of a woman's life. These comments reflected tension resulting from unprecedented numbers of women flooding the industrial sector, performing men's work, and serving in the military—disrupting traditional gender roles. At the war's end there was an unceremonious push for women to return to domesticity and the traditional roles of wife and mother.[3]

Early on, expectations about women's morality reflected these essentialist beliefs. From 1902 to 1917 the Victorian belief in female morality existed side by side with notions of the New Woman. Naturopaths reflected the dominant middle-class assertion that women's morality was innate, valuable and necessary to society. In 1907 Purinton, a self-appointed guardian of virtue, argued that morality was the duty of both women and men. He praised Anthony Comstock, the US postal inspector and author of repressive legislation that curtailed any spoken or written word he deemed obscene, including any public literature that taught about women's reproduction. Purinton ridiculed the new trends in women's clothing and women's sensual adornment to attract men, which compromised their character. No similar caveats were directed toward men. He advocated against women showing their bodies except in marriage. While some in the movement advocated nude sun, air, and light baths, it was within the privacy of home or sanatoriums. Purinton derided the (new) beauty-show posters for their licentiousness. He outlined the mutual duties of husband and wife, yet emphasized women's role more strongly: they were to be polite, dress well, converse at meals, clean house, save money, extend hospitality, attend funerals, save sinners, and respect authority.[4]

Gender norms relaxed during World War I, and after women won the vote in 1920 the urban youth-driven flapper emerged. Birth control was discussed and advocated, and some American working-class ethnic girls "traded" sexual favors in return for treats in dance halls and amusement parks. In naturopathic literature (and mainstream magazines) a spate of morally based articles proliferated. Naturopathic articles equated diet with morals; one associated character with social purity. By 1915 some naturopaths linked moral decline with liquor consumption, jewelry, tobacco, cars, soft drinks, tea and coffee, patent medicines, millinery, and chewing gum. The fear was that these excesses all gave rise to women who "trafficked in sin"—prostitutes.[5]

THE DOMESTIC PRACTICE OF NATURE CURE

Naturopaths exalted mothers and expected them to lead and preserve the sanctity of familial life. Maternal instinct was a given, and the joys of motherhood were emphasized regularly. Articles from 1907 to midcentury discussed the value of breast milk, how to feed a baby, how to care for a child with a finicky appetite, and the proper methods of cooking for and feeding children. They advocated teaching children healthy habits, enjoying the outdoors with them, cleanliness within the home, and treatments for childhood ailments. Advertisements touted helpful, complementary products and the New York Parent Health Center.[6]

Among leading female authors, naturopathic doctors, and lay writers was Martha B. Opland, ND, who wrote the popular column For Mothers and Children. She wrote in 1920: "If we would eat simpler, dress plainer and throw out half the furniture, draperies, rugs, knick-knacks, especially out of bedrooms, then there will be time for a little out of doors with the children." With typical naturopathic philosophy, she recommended that mothers spend less time on housekeeping and devote more time to right-living strategies, walk barefoot outside as weather permitted, and prepare foods that averted tooth decay. Healthy children should not have sugar, white bread, corn sugar and malt, powered buttermilk, animal fat, or salt and yeast. Exercise was imperative. She counseled mothers as household physicians on the proper feeding of babies and between-meal snacks for older children (fruit, milk, and raw vegetables only) and discouraged the removal of tonsils and adenoids and school-mandated vaccinations. She valued raising children in the country instead of the city; she blamed the trappings of urban consumerism for causing ill health. At midcentury Grace L. Dudley, ND, continued these themes in her series "Motherhood," as did John B. Lust (Benedict's nephew) in his serialized column Domestic Practice of Nature Cure.[7]

Naturopaths spread the word about the central role of mothers in natural living via the airwaves. At the 1925 meeting of the American Naturopathic Association, Dr. G. W. Haas, of Los Angeles, gave a radio talk aimed at mothers as physicians. His advice was simple and straightforward: "Be a student of home remedies, but above all, learn how to prevent sickness in the home." Sickness could be prevented through cleanliness, the study of nature cure, thorough elimination, and careful diet. The responsibility of trained mothers was crucial. "The destiny of our nation," Haas wrote, "is determined by the quality and stability of the American home."[8]

Early naturopaths also told women to care for themselves so that they could better care for their loved ones. The Women's Column began in 1902. Its topics included female education, familial caretaking, marital rights and harmony, romantic love, right living, strength, and clothing. The inaugural column argued for the education of women so that they could care for their families' health and discussed how a woman could persuade or influence her husband. Another asserted that "so-called 'female weakness' is the outcome of a women's corset, or a man's passion, or both." Ideology in this early period, during a resurgence of the suffrage movement and a new feminist movement, was more progressive than in the 1930s and beyond. In March 1902 the column stated that a husband ought to let his wife be her own person.[9]

The naturopath Louis Kuhne claimed that the causes of women's diseases were traceable to women's wrong manner of living. This ranged from bodily neglect and lack of regular exercise in the open air to inattention to bodily needs and quests for pleasure. In each case, following the natural laws of living would correct problems. Women's so-called self-sacrificing nature also damaged their health. Advice to women in 1916 pointed out that the housewife puts the needs of her husband and children and then housework before her own needs. This left her too weary to care for herself. Naturopaths warned that putting herself last could have disastrous results; overlooking pain and not consulting a physician jeopardized the health of the whole family. The housewife should learn what was normal female health, such as during menstruation. Healthful measures included water treatments, proper diet, avoidance of sexual excesses in married life, and proper hygienic washing.[10]

POWER FROM THE KITCHEN

A primary means of authority for women was through the preparation of healthy and healing foods for their families' consumption. Wives and mothers should radically alter their families' unhealthy eating habits, especially by rejecting processed foods. Naturopaths were not alone in this belief; crusades against food agribusiness had begun prior to Upton Sinclair's *The Jungle* in 1906. In 1900 the Illinois senator William E. Mason had decried the level of food adulteration in the United States, drawing attention to the flour industry's use of fillers in products. As one historian noted, nearly every manufactured food in the early twentieth century contained one or more potentially harmful chemical additives. Before World War I, thirty-five major corporations that processed perishable foods used chemical additives to enhance their products' smell, flavoring, or coloring. Many adulterants, including acids used to mask spoiled beef, turned out to be dangerous. Naturopaths decried the use of any additives, not just the clearly poisonous ones, because artificial products did not sustain life. They also felt solidarity with working people. Working-class families spent up to half their income on food, and of course the foods available varied according to the season. Layoffs, unemployment, injury, or illness dictated that food budgets decrease, so what little food they consumed had to be healthy.[11]

This cultural setting drove naturopath efforts to reform national dietary habits. In 1902 Louisa Lust penned columns for the *Naturopath and Herald of Health* that contained information useful to women about food, hygiene, and choices. The topics were as varied as how to clean silverware with borax, preparations using cold hard-boiled eggs, the value of raw tomatoes, and the necessity of cleaning a refrigerator weekly.[12] Articles by other naturopaths echoed and went beyond Louisa's teachings. M. C. Goettler, ND, was among those who warned that women needed to learn the combination of foods required to sustain a moral life. "If, with too much starch matter, sugars and fats, you are continually kindling the bowels what can be expected but a fire in the sexual organs?" he wrote. It was believed that constipation would cause pressure on glands and blood vessels and hence improper thoughts. Naturopaths also advised housewives that food chosen well and eaten in a natural state had the power to heal. They discussed the value of raw foods and mucousless diets, what to eat when pregnant and breastfeeding, foods' nutritional content, when dieting was necessary or damaging, and where to find health food stores and vitamin products.[13]

FEMALE PHYSICIANS AND AUTHORS

Female physicians and authors contributed to all the naturopathic journals from their inception, and women identified with them. Readers wrote in with questions, shared stories, and praised the advice they received. Several women had recurring columns; there was Louisa Lust, of course, but other high-profile naturopaths like The New Thought pioneer Helen Wilmans and Minna Beyer Madsen, ND, also had columns. In 1927 Madsen began a column that ran for years, Occasional Intimate Little Chats to Women by a Woman. She was introduced as a twenty-year practitioner specializing in women's health. As a female entrepreneur who operated a cure, she catered to female readers. She addressed them as "My dear women friends" and "Dear mother," and she counseled them on a wide range of subjects, from diet to frequency of bathing and the need to study nature cure. Her style was both empowering and marketable: "Be sane, be natural, be fearless to a degree." She taught women how to present themselves to others, telling them not to talk about food in terms of personal likes and dislikes or weight gain and loss. She contradicted cultural trends: "Beauty parlor, beauty shop, beauty, beauty, youth, youth, that school girl complexion and figure, that is what we hear, see and talk about everywhere. If women would simply bathe inside and outside they would have no need for the weekly facial and manicure." Rejecting fears of aging, she wrote that it was foolish for mothers to want to look like their daughters. Like early naturopathy itself, she positioned herself as antimainstream and reformist. Unfortunately, this gutsy approach to female self-confidence disappeared decades later when postwar naturopaths placed value on women's self-beautification and attractiveness. The early naturopathic leaders Martha Opland and Alice Reinhold also provided advice on a range of topics, not just women's issues. They discussed the importance of urinalysis and shared their expertise on herbs combined with nature cure, natural treatment of pneumonia, the ill effects of nervous energy on the body, and the general precepts and efficacy of naturopathy.[14]

In the ensuing decades two other columns for women were introduced: Your Health Problems, by Dr. Alice Chase, and Virginia S. Lust's Health Menus.

Chase's column had a question-and-answer format, and the vast majority of correspondents were female. Topics included arthritis, diabetes, vertigo, weight reduction, eyesight, ultraviolet radiation, blood clots, menopause, dry nose, and a plethora of other health problems. Chase answered readers' questions bluntly, explaining each disease or condition in clear and simple terms, describing its symptoms, suggesting regimens, and providing examples of cases she had treated. At first no photographs accompanied the column. Then, in the ensuing months the images went from a coquettish nurse in full uniform with lips pursed; to a man and a woman in business clothes, the man behind a desk and the woman taking dictation; to, in 1948, a man in a suit and tie on the telephone rifling through the drawer of a file cabinet. Why these transitions? Despite more women working for wages outside the home after World War II, popular culture mirrored the admonitions of psychiatrists and psychologists: women who veered away from childrearing and domesticity—toward careers or as authority figures—suffered from penis envy, were lost, or were unlovely. With increased societal ambivalence toward professional females, the imagery in the publication was designed to counterbalance the female authority in the text.[15]

Virginia S. Lust and her husband, John (Benedict's nephew), were the second-generation naturopaths in the family. Virginia's column, similar to Louisa's decades earlier, provided weekly menus for three meals a day. She explained her recommendations of fruits, vegetables, tea, dry cereals, simple soups, nuts, soybeans, coffee substitutes, buttermilk, and cheese. Charts were followed by cooking instructions, and again, images played a marketing role. By 1947 the images alternated between a middle-class housewife in an apron using an electric mixer—turned away from the camera and focusing on her task—and a male chef in cooking whites and chef's cap assertively facing the camera and making the "OK" sign with his fingers. The September 1948 photo shows a middle-class woman, hair permed, canning vegetables. Dressed in a full apron and smiling winsomely, she has both hands raised in greeting, one holding a spoon, attempting to appeal to the American postwar back-to-the-kitchen culture and to middle-class housewives. The images valorize male culinary expertise—even though the author was female. However, these visual messages were offset by the many articles written by women. One 1948 issue of Nature's Path contained five articles written by women on diverse topics. Accompanying images showed women dressed in the styles of the day, smiling contentedly. These articles shared the common middle-class postwar themes of food preparation, self-beautification, gardening, leisure, and moral commitment.[16]

Throughout the decades, women's ideal roles and character traits were constantly defined in these writings. Louisa signed her earliest articles "Mrs. B. Lust" to appeal to other married women. By 1923 her pieces were credited in the New Woman style, "Louisa Lust, ND." Gone was the self-negating Victorian signature. Echoing New Thought ideology, she counseled her female readers about emotional well-being. When she advocated cheerfulness in 1902, she detailed the health-giving merits of approaching life with a positive attitude to enhance home life and combat the stress from the public sphere. She admonished her readers not to provoke people, and she identified various irksome personality types encountered in life. She joked about whether hers might be among them. While Louisa and others were espousing turn-of-the-century New Thought and naturopathic ideology, the emphasis on women's cheerfulness in the home was a carryover from Victorian middle-class womanhood. It was a culturally defined feminine trait that remained consistent through the mid-twentieth century.[17]

Naturopaths used first-person testimonies and doctor-reported case studies to capture the interest of readers, particularly with regard to women's issues. Mrs. Krueger, a self-proclaimed magnetopath, described successful cures she had employed. She said that a bodily exchange occurred as the healer sacrificed her own health for the sake of her patient. Her description fit the Victorian notion of the self-sacrificing woman, despite naturopaths' admonition to be mindful of one's own health. Other practitioners similarly related their patients' stories, some helping to legitimize the cause of naturopathy more than Krueger.[18]

Female readers shared their conversion narratives and praised practitioners. Dr. Julie La Salle Stevens in 1923 told a lively tale of six eminent physicians who had decided that she needed surgery for cancer. She had lost thirty-eight pounds, and there was no other way to save her. She began reading about how nature cured, and "Behold a miracle!," she wrote, "I began to recover! No operation was necessary!" After this profound turnaround, she dedicated her life to helping others use diet to achieve health and happiness. She saw the dangers of drugs and all unnatural devices and said that she felt a bountiful spirit thanks to healthful living. Female readers also made use of the Correspondence section; in 1927 many women expressed their satisfaction with, and great benefit from, the valuable advice in *Nature's Path*.[19]

Throughout the first half of the twentieth century, female naturopaths influenced the creation of naturopathic science and culture. By the 1950s the *Journal of Naturopathic Medicine*, a publication of the International Society of Naturopathic Physicians, ran a regular column written by the prominent Dr. Ellen Schramm, vice president of the society's Council on Obstetrics, Maternal Health, and Child

Care. The journal's pages also recruited female practitioners for the American School of Naturopathy, profiled notable practitioners, ran obituaries of female luminaries, chronicled openings and relocations of women's offices, reported on talks by prominent leading women naturopaths, and advertised books by female authors. The market was there, as Dr. Nora E. Thrash noted, writing from her Health Sanitarium in Fort Worth, Texas: "Women are especially eager to listen to lectures [by women] on interesting subjects, such as diet, exercise, reducing and a host of others." The index of *Nature's Path* in February 1950 contained fifteen authored articles, six of them written by women.[20]

SOCIAL CLASS, OBSTETRICS, AND GYNECOLOGY

Because women were expected to be the moral and health core of the family, middle- and upper-class leisured women (whose ways of living contradicted naturopathic directives) were blamed for ill health. Yet naturopathic critics failed to address how middle- and upper-class women's socially and medically constructed roles framed their approach to life. For decades, allopaths and popular literature defined women of these classes as both physically weak and asexual by nature. *Ovarian determinism* was the allopathic theory that women's bodies were complicated by their reproductive systems, thus their normal state was sickly. This theory influenced the theory of women's passionlessness, or asexuality, which had been around since the late eighteenth century and was the basis for middle-class women's claim to moral authority. Women who believed these theories worked to uphold what they thought was their moral superiority against the supposedly wanton sexuality of men, working-class women, and women of color. These theories did not apply to women of the lower sorts, who—so the thinking went—with their more base instincts, were inclined to unrestrained pursuits of pleasure or even prostitution. By 1900 these class-based beliefs came from century-old assumptions about women who worked for wages outside the home and about poor working women's unchaperoned activities in urban public entertainment venues, such as amusement parks and dance halls. These race- and class-based beliefs, generated by men as well as women, also reflected widespread cultural anxiety over massive influxes of immigrants. The wealthier leisured women believed they needed to conserve their rather minute amounts of energy, become involved in urban reform, and depend upon the lower sorts to perform manual labor.[21]

Naturopaths ignored these complexities, and they directed their criticism at privileged women. In 1908 one practitioner blamed the lifestyle of the middle and upper classes on women, believing that daughters were doomed to be sickly if, like their mothers, they were given drugs, wore corsets, stayed indoors in their bedrooms and parlors, avoided domestic physical labor, got insufficient rest, and ate and drank the wrong things. This wrong living caused anemia, menstruation and blood cramp, and abnormal uterine hemorrhages. The practitioner suggested remedies and preventative measures, including a nonstimulating diet, water treatments (sitz baths), drinking water and milk only, bed rest, and avoidance of quick, strenuous exercises in the midst of an acute episode. This was similar to the rest cure prescribed by allopaths for weakness and nervousness. In addition, in an era when women's class-constructed ideals of pale beauty and tight clothing caused ill health, naturopaths advocated dress reform and short nude sun and air baths. In the early literature, naturopaths, like allopaths, also commonly referred to hysteria, also known as nervousness or neurasthenia. The disease was portrayed as more prevalent among women than among men. By the mid-1920s, however (after women received the vote), naturopaths believed that it was equally prevalent among women and men and that it could be corrected through a comforting physician-patient relationship and naturopathic living.[22]

Some middle-class women were convinced of their weakness, believing it a sign of femininity and class status and making them natural opponents to naturopathy. Lengthy articles blamed members of women's clubs, wives of the medical trust, and some nurses for their complicity in subverting naturopathy and for girls' lifelong ill health. One naturopath called leisured women "slackers." He described them as "loafers and luxurious parasites" who were obsessed with thinness and shirked all responsibility and effort. He praised Camp Fire girls for their out-of-doors activities and praised tomboys, who jumped, dug, and yelled. Benedict Lust added his derisive voice, saying that nervous women at the turn of the century manipulated others for clothing, hats, and seaside stays. These women tormented other people because of their own weariness; they were disgraceful. While a critique of the affluent lifestyle and its effects upon women's health was understandable, naturopaths failed to realize that they were blaming women for much larger materialist-culture and gender ideologies.[23]

Yet even as these class-based critiques of mainstream women continued, naturopaths worked to reform gendered behavior and treatments of female health issues. The fitness and natural health pioneer Bernarr MacFadden proclaimed in 1936 that motherhood was the loftiest and holiest of ambitions; it was the crowning glory of a woman's life, and women's health should reach its height during pregnancy. His glorification of motherhood was not unique; its valuation was perpetuated throughout American culture. But this degree of sentimentality was particular to the middle and upper classes. In this case, the best and most admirable women were those whose pregnancies and proper practices yielded healthy naturopathic children. Naturopaths also believed, as so many cultures had for centuries, that women might facilitate sex preselection through the timing of intercourse or even by following astrological signs.[24]

Young girls' health was particularly important because they would become tomorrow's mothers. Their food habits were crucial because healthy choices while young produced a strong constitution in adulthood. But self-indulgent and poor choices led to chronic ill health. To maintain mental and moral health, girls were directed to marry for love, not money; to be industrious; to be content and home-loving; and to be wary of prospective spouses. Miss J. Rachel Walker told girls in 1915 to avoid promiscuous flirting and excess emotion. Girls needed to identify their ideal mate and practice absolute fidelity. An anonymous poem told girls that they should ask a prospective mate if he was free from whiskey, tobacco, and sin. And of course they were encouraged to understand the facts concerning motherhood.[25]

From naturopathy's beginnings, the care of pregnant women was an important endeavor. In the first decades, authors addressed the promise of painless childbirth for every healthy woman who obeyed nature's laws. (One wonders whether this promise hindered the success of the movement.) Naturopaths advised about prenatal care and its influences on infants' health, vomiting during pregnancy, postnatal care, fetal growth measurements, and the need for home, versus hospital, birth, because of the morbidity and mortality prevalent in hospitals. Pregnant women were told to abstain from sexual intercourse during pregnancy, and men needed to cooperate. Difficult and painful births, miscarriages, and all kinds of disorders during pregnancy were blamed on copulation during pregnancy as well as on "the mother's morbid physiological encumbrances and inappropriate work, especially during the first half of the pregnancy." The value of women's breast milk was indisputable. In 1917 Otto Carque rejected germ theory as the cause of infantile paralysis, arguing that it was caused by improper feeding. He advocated giving infants breast milk only and said that they should never be given pasteurized milk. Women who followed all the right-living methods would be blessed with "Naturopathic babies," pictures of whom appeared in the various journals. As naturopaths combatted attacks by allopaths in the mid-1930s, they decided that they had to become absolutely proficient in obstetrical care. They were convinced of the importance of natural motherhood— untarnished by medicine and science—for both mother and child. Additional proof that natural motherhood was right for humans came in 1959, when an Italian court ruled that artificial insemination was adultery.[26]

Naturopaths produced a strong body of literature on women's proper hygiene and the treatment of female diseases. They viewed women's cyclic physiology as natural, in contrast to the allopathic view that it was a series of medical crises, or a pathology, requiring intervention. Dr. Alice Chase said simply that menopause was not a disease, regardless of common superstitions. Menstruation, pregnancy, birth, and suckling infants were all natural processes. These processes and the uterus itself could be harmed by the wrong foods or behaviors. However, these effects could be remedied with friction baths, an unstimulating diet, and the naturopathic manner of living.[27]

Naturopaths suggested a range of causes and treatments for gynecological disorders, depending upon their training and philosophies. One naturopath-osteopath-chiropractor believed that inflamed cervical glands were caused by poor ventilation, dirty lodgings, and personal uncleanliness; adjustment in the lower cervical region could heal the glands. Unnecessary treatments could also bring on illness. An example was vaginal douches, which were as common as taking a bath. Vaginal flora and secretions protected rather than harmed. Without pathological evidence, use of a douche was uncalled for. It interfered with natural conditions. It injured the mucous membrane, and the removal of protective secretions made infection from outside sources more likely.[28]

Kuhne's influential "Diseases of Women" reemphasized the class-based distinction between the health of hardy women who ignored harmful mainstream trends and the fashionable town-bred lady; the latter's health was by far the worst. Among the most common ailments that beset women was disturbance in menstruation, caused by an "encumbrance," or buildup of morbid matter. It was cured by improved digestion, regular evacuation of the bowels, hydropathic therapies to improve organ functions, and reduction of the abnormally high temperature in the abdomen through cooling baths. Other conditions were fallen womb, injuries resulting from improper use of the pessary, and sterility resulting from morbid matter gathered in the ovaries, fallopian tubes, or uterus. Removing these obstructions in the sexual organs prevented illness. Naturopaths believed that removal of morbid matter was at the core of much disease and that it alone was worthwhile for women, because operations, injections, and other medical procedures not only outraged female modesty but could never guarantee permanent success.[29]

By midcentury, naturopaths discussed cancer of the breast and womb several times a year. Like so many conditions, the diseases were attributed to improper living, processed foods, faulty elimination, and stresses of daily life. There were testimonies from female patients who had undergone breast removal by MDs and still suffered pain. One woman who had taken massive doses of aspirin to no avail was cured in three months' time by natural methods. Dr. Alice Chase wrote a series of articles on the symptoms, diagnoses, therapeutics, and outcomes for breast-cancer patients. One prominent cause was the retained, uneliminated excretions of the cells, which should be treated with periodic fasts, fruit juice diets, and proper hygiene. She rejected what she called breast amputation by MDs and instead called for a diet low in protein and sulphur, as these stimulated the growth of cancer cells.[30]

Other treatments addressed cancer of the womb that had been unsuccessfully treated with radium and x-ray (radiation) treatments, but disappeared after sacroiliac adjustment and other natural treatments. The prominent naturopath E. W. Cordingley treated fibroid tumors of the uterus with warm sitz baths, a diathermy current passed through the uterine area, and abstention from meat. He had found that cancer among vegetarians was rare. He did allow that particularly large or painful tumors might have to be removed surgically but only as a last resort.[31]

SEXUAL DESIRE AND CONTROL WITHIN MARRIAGE

Men and women were only deemed complete when they married, but because of the diversity of approaches within naturopathy, attitudes about marriage ranged from traditional to progressive. Extolling fatherhood as the "new profession" in 1915, one naturopath said that fathers should play a more active role in parenting. Instead of being just wage-earning machines, fathers should participate as actively in family matters as mothers had always done. Not surprisingly, Louisa Lust did not counsel self-deprecation to achieve marital harmony. In 1902 she wrote that keeping a husband required women to be assertive. The first year of marriage set the tone for harmony or lack thereof. While a wife could strive for happiness in the coming years, "no woman ought to surrender her individuality even to make peace in the family." She suggested learning to influence, rather than govern, one's husband while becoming a comfort, a power, and a blessing to him. Here she was consistent with mainstream ideology, in which women were socialized to avoid conflict, a contradictory nature, and a strong personal will. At times she personally rejected these traditional principles, but she saw the value of advocating them—within limits.[32]

Louisa Lust argued that harmony could only be achieved through strictly delineated views of male and female sexuality and amative self-control. Men who exhibited excessive desire victimized women. While this no doubt occurred, this focus perpetuated the culturally imbedded notion that men were more sexual than women and that female asexuality rendered women vulnerable. It also carried the expectation that women police sexual behavior within relationships, monitoring and refusing intercourse. The difficulty of actualizing this, however, went unexplored.

Naturopaths agreed that male sexual desire was a cause of marital discord and could ruin women's health; it spoiled intimate and psychological relations. The control of both male and female desire was imperative for marital success. Naturopaths, like so many middle-class Americans, advocated sex for reproduction only. They believed women had the right to refuse sex for nonprocreative purposes. "Only through marital continence [complete abstention and control over the sexual appetite] can man's welfare, happiness and success in life" be guaranteed.[33]

Early naturopaths, like a number of allopaths, denounced the licentious behavior of masturbation, as shown by Benedict Lust's lecture to New York's Nature Cure Society. Reprinted in the *Naturopath and Herald of Health*, it recommended proactive healthful living as a cure. Specifically, Lust advocated water applications to the genitals, air baths to strengthen prostrated nerves, daily home gymnastics, going barefoot, striking a balance between physical and brain work, and the restoration of a peaceful soul. This advice was not far from what regular physicians advocated, and a variety of devices had been developed to prevent the practice. Naturopaths took a more holistic approach; in Lust's words, "It is the physician's duty to re-establish the harmony of mind, to rouse hope and joy of life."[34]

"Voluntary motherhood," women's right to say no to husbands' sexual advances to prevent pregnancy, had been a radical position of some women since the 1870s. It is important to note that there was no such notion as marital rape at that time; husbands had the legal right to force themselves upon resisting wives. J. Waterloo Dinsdale, MD, echoed voluntary-motherhood activists when he said in 1911 that "until sex slavery is abolished—until a woman is given control of her own body, and permitted to decide when and by whom she shall be a mother," homes and society in general would not have a secure foundation. "Unwelcome motherhood," he argued, "demoralizes the woman and results in degenerate children." Dinsdale said that hysterics in women were the result of the cumulative effects of aggravation, ridicule, and fault-finding that characterized wedlock. Men's "animal passions" and "perverted sexual instincts" imposed misery on women. Intercourse during pregnancy could cause female hysteria or nervous diseases. Women were not weaker than men, nor were they ruled by their physiology. Husbands' unfeeling or ignorant excessive amativeness was imposed on wives and caused their woes. One rather unhelpful naturopath remarked that marital sexual control came with age because of men's normal sexual impotency and menopause, which quelled women's sexual desire.[35]

Typical of naturopathy's opposition to mainstream ideas, some believed that men's excessive sex drive was a sign of weakness, not strength. It was a logic long advocated by female antiprostitution reformers. Whether men's excessive drive stemmed from "disease, environment, or heredity" was irrelevant; it was sexually abnormal and needed to be checked. Men's sexual excess not only fouled the marital bond but fueled what lurked in the background, "the leprous body of female prostitution." Fears of moral laxity during World War I also fueled problems, exemplified by the army's campaign to control soldiers' sexual exploits. At least one naturopath noted that soldiers produced war babies, forcing modern women to protect themselves "from the ravenous appetites of these sexual gluttons who are disgracing the uniforms they wear."[36]

By late in the second decade of the twentieth century these behaviors were called "the abuse of the marriage relation," and their consequences were dire. Books advertised in the *Herald of Health and Naturopath*, some published by Lust, articulated the effects of this "outrage"—a euphemism for rape. A series linking sex and the Bible attributed crime, insanity, and disease directly to excessive venery. One author argued that reliance on Christianity would uphold the sacrament of marriage and preclude these evils.[37]

These arguments changed little in the Roaring Twenties despite, or perhaps because of, burgeoning public displays of female sexuality. New acknowledgments of female sexuality prompted one naturopath to hypothesize that men who demanded passionate wives and married buxom beauties were often disappointed. Women's physical health, not curvaciousness or licentiousness, contributed to a successful marriage. In 1923 one theory was that overzealous sexuality negatively affected health. In women it caused insanity, neurasthenia, anemia, and menstrual disturbance. Furthermore, most marital misery and tragedy came from conflating love with sexual passion. These mistakes could be avoided if the husband was mature.[38]

Naturopaths, like allopaths, addressed men's sexual control and potency and healthy ways to exert self-control and respect one's wife. Advertisements touted cures for impotency, such as the "Erectruss," a belt into which a man inserted his penis in a rubber sheath, then tightened the mechanism to elicit an erection. Naturopaths pointed out the relationship between nervousness and sexual vigor and the use of testosterone to offset the effects of aging and maintain successful work lives. Men, it was argued, must maintain their hormonal ability to compete economically and stave off emotional and mental losses. Another concern, not unlike a concern of Americans in the early twenty-first century, was the decline of men's sex drive with age and a recognition that men had a climacteric similar to menopause in women. This warranted anabolics, treatment to improve abdominal respiration, and after about the fifth or sixth visit, stimulating treatments over the genitalia and the pubic area. Prostatic treatment could be applied as well. The intentional stimulation of men at this life stage made them more fit mentally and physically healthier. No analogy was made for women.[39]

The ultimate price for male licentiousness was venereal disease. Monogamy and abstinence assured its prevention, yet men's damaging sexual needs brought the disease to their wives. In 1908 Dr. Carl Strueh, director of the Naturopathic Institution of the Orchard Beach Sanitarium, near Chicago, attributed male impotency and involuntary spermatorrhea to a licentious life, most horribly demonstrated by venereal diseases. Women suffered from gonorrhea, contracted through either their husbands' or their own immorality. It affected the vagina and the cervix, and in the most dire cases the ovaries and the nympha become ulcerated. Syphilis, the most dangerous venereal disease, contaminated the whole body. Strueh had no compunction about detailing for readers the gory details of pus, bleeding, sores, and the possibility of death if it was left untreated. The cure was regenerative, employing the same principles as those for curing a fever or other excretions resulting from physical crisis. The goal was to expel foreign matter from the body. But Strueh counseled that when essential organs like the eyes were affected, drugs were necessary; even surgery might be called for. Louis Kuhne attributed the spread of syphilis to an uncontrolled sexual impulse and the false cure by allopaths who used injections, dangerous medicaments, arsenic, mercury-based drugs, and iodine. However, both Kuhne and Strueh failed to address basic ideas about contagion and virus. In the following decades naturopaths lambasted allopaths for false diagnoses and ineffective and iatrogenic treatments. They derided the Wasserman test, used to detect syphilis in couples about to be married, which could yield a false positive in 10 percent of cases. This mandatory test was bad for women, because a false positive meant that a woman had to be treated, which would "scare the daylights out of her, break up her home, have her and [the] husband despair[, and] drive them to insanity." When the *New York State Journal of Medicine* in 1939 admitted that the intensive campaign against syphilis had resulted in overtreatment, naturopaths felt vindicated.[40]

BIRTH CONTROL AND ABORTION

Naturopaths shunned notions of birth control and abortion. Birth control, they argued, interfered with nature's order because it prevented conception. The eugenicists' view that the best examples of humankind should procreate drove the belief that birth control disordered society by interfering with the "desirable" race's continuance. Self-control, not artificial means, should be used to prevent contraception. Naturopaths blamed birth control education for the drop in births, believing that it promoted sex for pleasure. It was contrary to the Creator's teachings. It was deemed "revolting and debasing. . . . [Moral integrity] has been flung into the mud and substituted by the unholy vision of a sensuous delirium, with its manhood-sapping slavery." There was no need for sex, said Lillian Carque (as had so many before her), except for procreation.[41]

By the mid-twentieth century, however, naturopaths were advocating for birth control, but for society's "less desirables": immigrants, people of color, and impoverished whites. They quoted Margaret Sanger's eugenic views as proof that the feeble-minded came from the families on relief. The author Elena Slade asserted that a progressive civilization could best take care of both its able and its unable when the more richly endowed were given the opportunity and encouraged to reproduce their kind in greater numbers than the less well endowed. Slade and Sanger both advocated a modified eugenic philosophy that was not strictly class- and race-based biological determinism. Perhaps, Slade said, feeble-mindedness could be improved by including nursery school and better nutrition as part of the public health provisions for the lower classes. She supported the ideas put forward in Sanger's 1920 *Women and the New Race*, which did not advocate forced sterilization but dictated who should reproduce more.[42]

Naturopaths considered abortion the ultimate form of degeneracy. One naturopath wrote in 1908 that any mother who resorted to abortion was a monster. Called the ruthless destruction of the seed of life, abortion was likened to the murder of animals and of society as a whole. "These mothers who abort cry kill! Kill! Kill!!" The public debate about allowing AMA physicians to perform abortions prompted one naturopath to argue that if a regular physician was allowed to end the lives of babies, he could later assume the same gruesome privilege in the case of so-called degenerate adults. Such power would just increase physicians' stranglehold control over American health and society. Naturopaths accused the New York legal system, with its public prosecutors, abortionists, and vice police, of allowing abortion to exist, while covering up the prosecution and persecution of naturopaths.[43]

THE HEALTHY, ATTRACTIVE FEMALE

Active exercise and outdoor activities were provided at Yungborn. Naturopaths advocated them for girls and women to combat ill health. The benefits of outdoor life were linked with ideal motherhood. It was important for urban girls to go to the countryside to restore their health and become educated as future homemakers. In the Chicago area, "a number of girls, young business women, teachers, cashiers and bookkeepers" went camping for two weeks or more each year. They returned to the city regretfully, but "with renewed health and strength from a good time spent in the fresh air and sunshine of the country."[44]

This type of camping was available to more affluent women, but a lucky few working-class and ethnic females had opportunities through the YWCA and the YWHA (Hebrew Association) and through select settlement house programs and other women's benevolent associations. These experiences rejuvenated and strengthened manufacturing and service workers. One naturopath waxed effusive about an outdoor girl in the West whom he had recently met on a trip. She was energized, physically fit, and enjoyed the invigoration of swimming, dancing, walking long distances, and camping. She was not prone to headache, taking medicines, or excessive emotion—all charges levied against the female urban dwellers in the East. These outdoor girls, he said enthusiastically, were the ideal children of nature. The idealistic notion of natural health was that the four main components of childrearing were breathing, feeding, bathing, and exercising. These foundations of naturopathic ideology allowed motherhood to be in harmony with nature. Naturopaths and nature-cure healers advocated being outdoors throughout the twentieth century. As the urban population increased, advice shifted to encourage gardening, walking, and swimming when possible.[45]

The epitome of early-twentieth-century attractiveness for naturopathic women was the natural beauty that came from outdoor activity, a sound diet, and moral behavior. The fount of beauty was food and exercise, not consumer beauty products. In 1904 a naturopath warned that all artificial cosmetics were dangerous to the skin, the hair, and the whole organism. Their "evil effects will appear only too soon." Instead of using cosmetics to enhance beauty, women should turn to bathing, massage, simple foods, swimming, horseback riding, and a simple, temperate outdoor life. Some naturopaths did not critique the desire for female beautification, but the *means* recommended were far more empowering, albeit unachievable for all but wealthy urban dwellers.[46]

From the late 1920s through the 1950s, naturopaths' ideas of beauty standards and physical appearance were more mainstream. However, they continued to criticize mass-produced and synthetic cosmetics because of their expense and unnatural components. Some fashion choices, such as high heels, created orthopedic problems, and smoking and drinking alcohol bred sick children, depleted the complexion, induced unappealing facial features, and compromised one's moral standing.[47]

The naturopathic ideology of the mid-1940s would surely have made Louisa Lust cringe. Healthful eating was likened to glamour; images of middle-class women wearing bathing suits in coquettish poses accompanied articles on weight loss; hormonal facial creams promised youthful skin; and women's beauty was treated as a precious commodity. Naturopaths rejected the harsh chemicals in beauty products, yet natural beauty products were promoted for complexion, hair color, and weight loss. In 1948 a column aimed at debutantes asserted that a woman's beauty was her greatest asset. A large photograph accompanying the column showed a well-to-do woman in a polka-dot dress, arm-length white gloves, hair permed, and face made up, holding a parasol. Her eyes looked skyward. Jewelry, cosmetics, dress, ornaments, hair, and electrical hair removal could produce the type of beauty one desired. All that was missing was a beauty diet. There was no critique of the need to attain beauty. The point was that middle-aged women could retain their youthful beauty by living in harmony with nature's materials.[48]

This new line of advice appeared in the post–World War II era, when American culture embraced poster pin-up calendar girls and the buxom and seductive Marilyn Monroe and Jayne Mansfield. On the other hand, the promise of home, children, and family stability through strict gender roles was a comfort to returned servicemen and a population traumatized by the Great Depression and war. Wives' desire for products that would make them as attractive as the pin-ups also signaled the unprecedented affluence of the middle class. From the naturopathic viewpoint, women were still providers of familial health and food and vital reproducers of the next generation, but the measure of their worth now included their attractiveness to men.

Fat women did not fit into the beauty culture beginning in the 1920s, when female standards of beauty favored a slim, boyish figure. Naturopaths discussed the deleterious health effects and aesthetic displeasures of fat. They argued that fat was unattractive, abhorrent to men, and rendered a woman a disorderly person. Advertisements in which fat was described as disfiguring promoted a youth-giving Mariba Obesity Tea and a belt that reduced the waistline. There was a curious claim to be able to "eat candy and get slim." These beliefs were not new nor unique to naturopaths; mainstream Americans also held these sentiments. In 1926 Purinton wrote that "anybody, of average height, weighing over 200 pounds, should be arrested as a disorderly person." To him, "very fat people are nearly always very earthy, gross and sensual." Naturopaths assumed that most women wanted to lose weight. Once freed "from the millstone of obesity, their personality has a chance!"[49]

Fat loathing escalated in the late 1940s and reflected the new standards for wives. *Nature's Path* ran a demeaning story in which a man visited a friend he had not seen for years. He asked for "that cute little girl you married years ago," and Clara came downstairs. She weighed 275 pounds, and the visitor commented, "What a pig in a poke you got." The author recommended diet and exercise. It was the beginning of the era in which extreme and unrealistic standards of beauty became conflated with notions of health.[50]

LEADERSHIP AND POLITICAL POWER

One interesting and significant phenomenon that shaped naturopathy was the differing viewpoints of the movement's leaders. When it came to women, they argued about the traditional, essentialist, sexually conservative, and reproduction- and beauty-based gender norms. Simultaneously, empowering ideals flourished about female leadership and women as a political force within naturopathy and the nation. There were sage female physicians and authors; wives and mothers determined familial and national health; and there was outright praise for, and encouragement of, physical strength.

In 1915 naturopaths heralded the world's strongest woman. Miss Tarabog, a Hindu woman, was orphaned and raised as a boy for several years. At age thirty she performed feats of strength that were impressive, if not somewhat bizarre. She could bear the weight of a half-ton of stone on her chest while it was struck with a sledgehammer; a carriage carrying several men rode over her chest and arms. Because she had mastered her breathing as well as mental and physical faculties, these feats did not harm her. The message from naturopaths, contrary to other messages of the time, was that gender-crossing and strength were all possible—and desirable. Miss Tarabog embodied the naturopathic belief that self-discipline enabled one to withstand discomfort and overcome pain.[51]

Naturopaths also valued and utilized women's intellectual abilities. In the 1920s women held leadership positions in statewide naturopathic associations; seven of nineteen delegates to a Tampa convention in 1923 were women. Three of eleven officers of the 1935 New Mexico naturopathic association were women, as were three of the thirteen graduates of the American School of Naturopathy in 1927.[52]

When women's leadership and authority flourished, it led to public support for women's rights, for the abolishment of prostitution, and for progressive ideas on expanded women's roles. In 1907 an article, likely by Louisa Lust, stated, "Our time is the woman's age. From slavery and submission she shall free herself by her own personal efforts; but she wants the man to demonstrate her rights before the world at large." These arguments resonated with values Louisa brought into her marriage. Even Purinton, usually conservative, asserted that women should have suffrage.[53]

After women got the vote in 1920, one naturopath wrote that "women are just now entering politics, God bless you in your noble endeavor to better human conditions." Women were warned against political naïveté and being used as pawns by medical science. In support of women as individuals, another message was that men alone were responsible for their unwanted sexual advances; women did not invite them, nor should they blame themselves. Naturopaths recognized the influence of women's organizations and the movement's desire for their support. The social activism of the Massachusetts State Federation of Women's Clubs was important because it urged cooperative efforts among its members. In 1925 the influential naturopath Otis G. Carroll wrote to a woman in Spokane, Washington, to dissuade local women's organizations from aligning with any standardization of medical practice detrimental to naturopaths.[54]

Naturopaths' gender ideology also reflected women's attainment of the formerly male privilege to vote. By far the most radical assertion about sex roles appeared in late 1926 in Purinton's "The Third Sex." He rejected the gender binary, writing that "the girl we call a 'tomboy,' and the boy we call a 'sissy,' should be not the victim of our disparagement and reproach, but the object of our study and respect." He speculated that "each may be the unconscious, perhaps unwilling, forerunner of a new third sex to appear on this planet in ages to come: a man-woman, or woman-man, revealing the greatest and best qualities of each sex, in the guise of another." He said that rigid adherence to old notions of masculinity and femininity meant that manhood makes a brute, and unwavering womanliness, a butterfly. To get to a point of acceptance, he said, much thought about the sexes needed to change. Similarly, Minna Beyer Madsen, ND, argued that the man who wanted his wife to be a household drudge and physical convenience was fast disappearing. Naturopaths continued a decades-old call for dress reform, arguing for lighter, roomier, less restrictive clothing for women and men. For decades, dress reform paralleled the call for less restrictive roles for women.[55]

Women in the naturopathic movement, despite the traditional gender norms within which they worked, cocreated a radical naturopathic world-view that promised human integrity and self-determination. At the national level, Louisa Lust personified early female leadership. Later, Theresa M. Schippell and Louise Nedvidek replaced her and even replaced Benedict himself. Schippell, in Washington, DC, became president of the American Naturopathic Association. Nedvidek served as the association's financial secretary in 1935 and as manager and chair of its annual convention in 1937, at which five of the twelve officers were women. Lora Mae Murray, ND, served as vice president of the American Association of Naturopathic Physicians before it and two other national associations consolidated their membership and resources in Colorado under her leadership to form the National Association of Naturopathic Physicians (NANP) in 1956. In 1957 Murray, a graduate of Connecticut's Blumer College of Natureopathy, became president.[56]

Few women held leadership positions in the International Society for Naturopathic Physicians in the 1950s and 1960s, when gender ideology was more polarized than in earlier decades. A notable exception was Dr. Ellen Schramm, a frequent columnist on obstetrics, who served as secretary of the group and vice president of the Obstetrics Division. Many women helped expand the cause through their leadership in complementary organizations. One was Mrs. Lora W. Little, who edited the *Avalanche*, the official publication of the American Medical Liberty League.[57]

For women who were not practitioners or administrators but were integrally invested in naturopathy, women's auxiliaries arose in the late 1930s aimed at naturopaths' wives. In 1957 the NANP Ladies Auxiliary invited women to the St. Louis convention. The auxiliaries offered leadership opportunities important to the movement: Betty Hughes, elected president of the Ladies' Auxiliary of the American Association of Naturopathic Physicians in 1954, noted that they had 100 percent membership—full participation of wives—and had had a successful fund-raising campaign. They worked to further opportunities for young, earnest naturopathic students.[58]

In the mid-1950s, when competing national organizations merged to form the NANP, their Ladies Auxiliary determined to "take a more active and direct interest in all things concerning the profession of our husbands and its future." Every wife, mother, and daughter was exhorted to join and be active in the auxiliary. However, women's leadership would not reach the level of the 1920s through the 1940s again until the 1970s, as would be the case in US society in general.[59]

CHAPTER 6

Culture Wars

Ideology, Social Trends, and Competition for Clients

As the various medical sects vied for clientele in mid-nineteenth-century America, the competition was as healthy as some of the recommended treatments. The alternative theories and practices appealed to Americans in addition to those of the regular, or allopathic, sect. They fit with the democratic, self-made-man movement of the time, or they at least seemed to adhere to the theory of do no harm rather than using invasive allopathic methods. There were, of course, self-proclaimed doctors who were quacks. Quackery existed among all the health practitioner modes; it fueled the epithets sect leaders hurled at one another in the health culture wars from the nineteenth to the twenty-first century. One thing the regulars had in common with the three most popular irregular sects was that they all combined their scientific understandings with empirical evidence. Homeopathy, eclecticism, and hydropathy all stemmed from documented experiential evidence that the body could be helped to heal itself. At the height of their popularity, these three sects together claimed about 20 percent of the American populace as their patients. Their very existence demonstrated that allopaths had no monopoly on health practice.[1]

Sociopolitical trends favored alternative medicine and healers in their earliest years. By the 1830s and 1840s, each white man was the director of his own destiny, according to the Jacksonian democratic ideal. Anyone who claimed authority as a medical expert—claiming an elite status over and above others in the community—was as likely to be discredited as not during this period of the common man. Average Americans were empowered to challenge those with expertise. In addition, allopathic medicine's reputation had been damaged by poorly trained or untrained men calling themselves doctors. When the American

Medical Association was formed in 1847, the founders, trained at elite institutions, aimed to set themselves apart as the country's leading medical professionals. They created the AMA "for the protection of their interests, for the maintenance of their honour and respectability, for the advancement of their knowledge, and the extension of their usefulness." They drafted their first standards for medical education and created a code of medical ethics. By about 1870 bacteriology was widely accepted as the cause of a variety of illnesses, and regular physicians' treatment based on germ theory was erroneously given sole credit for the reduction in disease. In reality, sanitary reform and sectarian methods deserved equal credit. At the same time that allopathic practitioners set out to make themselves the official standard in medicine, they were under attack by other sects and by significant numbers of Americans. Allopaths had fostered mistrust by delving into forbidden practices, such as body snatching for dissection purposes and vivisection (performed most often without anesthesia) on animals. The public feared the less than scientific use and policies of vaccination, as well as the side effects of toxic treatments. Economic conditions for practicing allopaths were so poor that in the late nineteenth century 75 percent of physicians left medicine within five years to secure a better living. The public also distrusted allopaths because despite their claims of medical authority, they could not curtail the diseases that ravaged the American population.[2]

Yet, the chief hostilities were between regulars and irregulars, not between practitioners and patients. Beginning in the mid-nineteenth century, several forces coalesced to challenge the legitimacy of nonallopathic practitioners, many of whom were women. Prime among these forces was the AMA, with its exclusive criteria for membership. Specifically rejected were hydropaths, homeopaths, and eclectics. Allopaths' establishment of boards, institutes, societies, agencies, and foundations proliferated in the second half of the nineteenth century, asserting the superiority of allopathy. State medical societies teamed up to form the American Public Health Association in 1872. Representatives of twenty-two medical schools met and formed the Provisional Association of American Medical Colleges in 1876. By 1919 all states had health departments and health boards.[3]

Cultural trends also favored allopaths in the late nineteenth century. The middle and upper classes saw professionalization as orderly, efficient, and scientific. America was professionalizing across a wide spectrum of society, in law, journalism, public administration, childrearing, social work, teaching, management, and charity. Not only allopaths but many healing sects agreed by 1870 that licensing and regulated education were necessary.[4]

The desire for orderliness, pragmatism, and security made professionalization attractive. In the last decades of the nineteenth century, the United States was reeling from massive changes. In the 1870s and 1880s, in one of the largest migrations in history, 8 million newcomers entered the United States. There was also a demographic shift away from farms, so that for the first time there were more individuals living in cities and performing industrial wage labor than there were rural agricultural laborers. There was increased mobility through transportation, improved and faster communication, greater mechanization and technology, and fortunes made and lost overnight. Under these ever-changing conditions, the claims of scientific medicine "both reflected and promulgated the belief that it, along with social research conducted by reform-oriented clubs and associations, could bring order and efficiency to the institutions that seemed to be faltering under the burden of coping with the complexities of modern times."[5]

Scientific medical practitioners claimed to be dispassionate bastions of certainty and truth. Scientific innovations, particularly in modern surgical techniques and enforced immunization, helped consolidate professional authority between 1850 and 1920. Equally important, an individual's ability to hire a trained professional came to reflect his or her class status. Simply put, as middle- and upper-middle class Americans sought ways to distinguish themselves from the influx of impoverished immigrants—both in their way of life and in their ideologies—they turned to scientific medicine as one such marker.[6]

Fueling the legitimacy of science were the university-based reforms in medical education. Until the late nineteenth century, students entering medical schools had been ill prepared; admission had been readily granted to young men lacking high-school degrees, the brief two-year programs had no mandated sequence of study, and American medical schools lacked laboratories to conduct original research. Classroom lectures and apprenticeships were the primary forms of instruction. The medical demands of the Civil War and the postwar years highlighted the need for stricter standards in medical education. Not surprisingly, experientially learned healing was discredited in the face of university-derived scientific medicine bolstered by curricula, faculty, facilities, and training regimens. In addition, the elite middle and upper classes, who gained economically from industrialization, embraced a medical analogy that fit their lifestyle. The body could be controlled by scientific doctors, just as scientific management could control labor processes and government, making them more efficient.[7]

By the 1890s, some of the major healing sects, such as osteopathy and chiropractic, were improving their own professional education, paralleling the allopaths' goals of improved medical education and licensing laws. But many naturopaths raged against such restrictive structures and the excesses of capitalism. They confronted the advocates of social control, who obstructed creative health solutions. As allopathy became more focused on germ theory, competing sects moved away from the ideology of one cause, one cure, and embraced an endless array of therapeutic and mechanical modalities. They retained old practices but also experimented with new methods, expanding their range of healing techniques to meet individualized patient needs. At this same time, average Americans, especially women, continued to rely on home remedies, while a steady supply of newly patented "cures" could be bought over the counter or from traveling salesmen.[8]

All of this ran counter to the allopathic consolidation and standardization of practices. That standardization was embodied in the newly created allopathic medical education curriculum between 1900 and 1910. This AMA effort, backed by powerful, well-financed corporate and philanthropic entities, culminated in the 1910 Flexner Report, which ranked medical education institutions and forever changed health care in the United States.[9]

NO MORE "GRANNY MIXES"! FROM PATENT MEDICINES TO PHARMACEUTICALS

As part of their professionalization agenda, allopaths wanted to regulate the production and prescription of remedies. It was logical: anyone at the time could formulate a remedy, and even if it contained little more than alcohol and flavoring, the product could be fraudulently patented as a cure for anything from the common cold to cancer. But in their efforts, the AMA did much more than merely curb the sale of quack cures. A new relationship developed between fledgling pharmaceutical companies and allopaths. The model we take for granted today began at this time; only a doctor could prescribe a medicine that was produced by a company approved by the AMA. Doctors forged alliances with pharmaceutical companies and shaped that industry to bolster the allopathic profession. The regulation of medicinal remedies created profound changes in three cultural arenas. First, it led to a gradual cultural shift away from self-doctoring to the more passive patient-as-recipient model. Second, the AMA successfully portrayed nostrums (patent medicine and home remedies), device makers, and irregular physicians as enemies of their profession—and of their patients. Lastly, public opinion naturally shifted toward acceptance of MD medical expertise and professionalism when patients received medications (purported to cure ailments) only from them.[10]

Prior to the early twentieth century, there was a profound respect for, and reliance on, domestic medical guides. Between 1740 and 1860 both published and private materials challenged professional expertise and valued women's knowledge of medicinal recipes passed from one generation to another. Treatments and remedies were prescribed by women through oral and written traditions experientially learned. They included the complicated chemical processes for extracting the medicinal properties of plants, from the best time and place to harvest to the preparation and applications of tinctures, poultices, salves, and more. This knowledge of domestic healing informed food preparation as well.[11]

Between 1865 and 1900 patent medicine's popularity was at its height. Competition spawned new sales techniques and the medicine show. Traveling entertainers promised relief through elixirs of life, pills, tonics, plasters, and gadgets that promised panacea cures for gout, female weakness, rheumatism, cancer, indigestion, irritability, and countless other woes. They triggered a flurry of muckraking investigations into the veracity of their claims at the turn of the century. These investigations, very much a feature of Progressive Era reform politics, sought to expose corruption, fraud, and profiteering.[12]

In 1905 the journalist and muckraker Samuel Hopkins Adams's serial exposé appeared in *Collier's* and the *National Weekly*. He dubbed nostrum medicine the great American fraud. Adams's articles alerted people to the dangers of self-diagnosis and patent medicines. The AMA in 1905–6 distributed in excess of 150,000 copies. As one medical historian aptly noted, Adams's series was to the proprietary drug makers and advertising doctors what the Flexner Report five years later would be to proprietary medical schools: a withering investigation of deceit by commercial interests, the shutdown of a number of operations, and a consolidation of professional authority. Adams's exposé, among other factors, led to the 1906 Pure Food and Drug Act.[13]

Even more effective was the AMA's new Council on Pharmacy and Chemistry, established in 1905. It set standards for drugs, evaluated their efficacy, and worked against nostrums. The resultant 1912 publication, *New and Nonofficial Remedies*, was widely followed by medical journals to set advertising policies. It rapidly became the sole credible guidebook for prescribing physicians.[14]

While the public benefited from the reduction in fraud and dangerous claims, there was an important gendered aspect to this campaign against patent medicines: lay expertise, particularly women's expertise of home medicinals, was eclipsed in favor of male-centered, institutionally controlled credentialed expertise. Patented, mass-produced products replaced home remedies because so many sales pitches were aimed at women's peculiar diseases and at wives and mothers as caretakers of familial health. Opponents of patent medicines high-lighted the dangers encountered by women who used nostrums. In addition, many of the medicine-show entertainers—burlesque dancers, fortunetellers, and crowd enticers—had been women, adding to the perception that women and products made by women could not be taken seriously.[15]

Many nostrum advertisements reinforced the image of the leisured woman as inherently an invalid. Women's belief in their innate sickliness had created a cult of infirmity that was a marker of social class status. For decades, leisured women—and their physicians—believed they had innate weaknesses that differentiated them from the laboring masses, particularly ethnic women and women of color. These weaknesses were attributable in part to debilitating clothing. As one historian argued, these ailments did not reflect innate female biological weaknesses. Rather, the increased use of nostrums, increased consumption, and poor eating among affluent women were a result of the stressful double standard of morality, idleness, and the dwelling on the aches and pains that accompanied leisured women's boredom.[16]

The customers of Lydia Pinkham, the premiere female manufacturer of remedies for women, were not hapless victims of quacks, however. They consciously chose patent medicines over orthodox treatments they believed were unsafe. Pinkham's advertising reflected women's discontent with allopaths, particularly gynecologists. Self-dosing was not as fearsome; it was inexpensive and logical. Patent medicines, despite their alcohol content, were preferable to the large doses of narcotics that women consumed with alarming frequency. Opiates, in the form of paregoric, laudanum, opium, and morphine, as well as stimulating cocaine derivatives, were sold over the counter. Pipes for inhalation and tablets and powders filled several-page spreads in the mail-order *Sears Roebuck and Company's Catalogue* in the 1890s. These advertisements invariably pictured—and appealed to—women. Women became addicted to pure narcotics, with devastating effects on their health and on their ability to function domestically and socially.[17]

The AMA's control of advertising—a successful application of muckrakers' exposé work—and legislation prohibiting self-advertising by doctors were steps toward consolidation of power and cultural authority. In 1912 the Food and Drug Act was changed to include fraudulent claims of cure. In the 1920s it included newspaper advertising and medicinal labels. Manufacturers of patent medicines increasingly withdrew. The AMA's regulatory system effectively "with[held] . . . information from consumers and rechanneled drug purchasing through physicians."[18]

Patent medicines, categorically dismissed as harmful, had constituted an $80 million annual industry. Now that business was funneled to the AMA and the nascent pharmaceutical industry. The antinostrum exposés continued in what naturopaths saw as a damning AMA power play. Dr. Morris Fishbein's popular health journal *Hygeia*, begun in 1923, routinely attacked patent medicine, along with the non-AMA healing sects and antivivisectionists. In 1927 Fishbein, who was editor of the *Journal of the American Medical Association* (*JAMA*), published a book that decried so-called cults and quacks. In 1933, a book by Arthur Kallet and F. J. Schlink identified dangers found in everyday foods, drugs, and cosmetics. It was reprinted thirty-two times in its first four years and rallied public support to secure passage of the 1938 Food, Drug and Cosmetic Act. By the 1950s two more texts questioned Americans' capacity to self-medicate *at all*. Thus, a near-complete reversal in mainstream cultural attitudes had occurred. By the 1950s the AMA continued to argue—and many lay people shared the view—that people were virtually incapable of making decisions about their own care and that they were content and safe only in the hands of trained experts.[19]

ANTI-AMA SENTIMENT AND SOCIAL CLASS "WARFARE"

Naturopaths, believing in the right of individuals to learn health practices, immediately and consistently identified allopaths as their archenemies and a prime example of monopolistic greed. The conflict went far beyond differing views of disease causality and treatment. Relations between naturopaths and the self-proclaimed experts exploded into a war of words, bringing naturopaths condemnation of their methods and decades of legal exclusions, prohibition to practice even the most benign botanical therapeutics or massage, arrests, and prosecutions. During the first half of the twentieth century, naturopaths referred to allopaths by a variety of derogatory terms, including *medicos*, which they used interchangeably with *medical scientists*. They referred to allopaths as the "Medical Monopoly," the "Medical Trust" (both always capitalized to emphasize their disdain), purveyors of poisons, the real quacks, despots, autocrats, bigots, and tyrannical egoists, among other disparaging descriptions. Naturopaths accused allopaths of monopolizing the federal bureaus and state and local law enforcement; conducting illegal and unjustified raids on drugless practitioners; and monopolizing the healing arts. When the idea of a national health department was reported in the July 1914 issue of the *Naturopath and Herald of Health* (NHH), it was attacked as aimed at securing for allopaths the necessary "power and funds to terrorize people into submission." One author asked, "Are the medical doctors our masters or our servants . . . ? Do we want medical slavery or Medical Freedom?" The journal said that even Jesus Christ healing the sick would be arrested for practicing medicine without a license in New York.[20]

Naturopaths called the AMA's measures brutish examples of pomposity, reeking of social class privilege, profiteering, attempts at medical monopoly and social control. They saw the AMA as an example of capitalism run amuck and anti-Americanism. They saw themselves as advocates of medical freedom and the people's right to choose their practitioners and mode of healing, compared with an allopathic dictatorship based on self-interest.

Naturopaths were not alone in their loathing of allopaths. Other anti-AMA activists included antivivisectionists, antivaccinationists, medical freedom leagues, labor unions, advocates for immigrants and the poor, antigovernment groups, antitrust advocates, political libertarians, and anticapitalists. These groups frequently joined forces in attempts to squelch or minimize the authority and impact of the AMA. After all, as the AMA was on the rise, so was the antimonopoly and union activism made famous by President Theodore Roosevelt, the journalist Ida Tarbell, and the labor activists Eugene Debs and Samuel Gompers. Naturopaths added their voice to those others.[21]

Scorn for regular medicine was in evidence early in the history of naturopathy. In 1908 regulars who had converted to sectarian healing ridiculed their university training, saying things like "It is much better to operate at once and lose than never to operate at all," or "An ounce of pretension is worth a pound of cure," or "A rosy-pink appendix makes a fine mantle piece ornament."[22]

The loathing extended to AMA-sanctioned, philanthropically supported research institutions such as the Rockefeller Institute of Medical Research. One leader condemned its limited patient access and its laboratory animal experimentation, saying that the Oil King's money was being spent in a useless manner. The most profound example of AMA influence was the privately funded Flexner Report, also known as the Carnegie Foundation's *Bulletin Number Four*, published in 1910. Abraham Flexner was hired by the Carnegie Foundation for the Advancement of Teaching. He discussed plans for the study with the editor of *JAMA* and the secretary of the Council on Medical Education. Flexner used his alma mater, Johns Hopkins, an orthodox medical school, as the standard by which to measure all others. He visited and evaluated all 155 American and Canadian medical schools. Because they thought he was on a philanthropic scouting mission for the Carnegie Foundation, looking for potential worthy donation recipients, they gave him access to information and observations that otherwise would have been off-limits. In his report Flexner chronicled their "facilities, laboratory equipment, numbers of faculty and their qualifications, numbers of students and their preparation, the curriculum, patients available as teaching material, income from student fees, and endowments."[23]

Flexner's scathing indictments dealt a death blow to schools, mostly nonorthodox and sectarian, that failed to measure up to the standards of the well-funded Johns Hopkins. His recommendations were direct and simple: strengthen first-class schools to replicate the Johns Hopkins standards; improve some middling schools to reach those high standards; and close the rest. One highly ranked school was the New York Homeopathic Medical School, from which Benedict Lust had graduated. While the report found strictly sectarian schools wanting, medical schools like New York Homeopathic, which conferred MD degrees and incorporated some sectarian methods, measured up.[24]

Flexner's wholesale denunciation of smaller sectarian schools was so severe that contemporary newspapers and medical journals described the report as hostile toward those schools. Small colleges had already been struggling before his *Bulletin Number Four*. Changes in licensing laws and insufficient income had strained their ability to keep up with reforms. In response, some tried to comply, and others turned to fraud, claiming that they met the standards when in fact they did not.[25]

In 1912 Flexner joined the staff of the Rockefeller Foundation's General Education Board. By 1919 he had persuaded Rockefeller to invest $50 million to implement the recommendations in his report. Thus, between 1919 and 1928 American medical education was transformed. In 1906 there had been 162 medical schools in the United States and Canada, many with annual budgets under $10,000. According to Flexner, they could not meet "even in a perfunctory manner . . . statutory, not to say scientific requirements and show a profit." As these folded, Flexner directed Rockefeller and other philanthropic funders (whose contributions were in excess of $150 million by 1936) toward a select group of medical schools. By 1915 the schools in operation had decreased to 95, a much higher number than Flexner's recommendation of 31. By 1928 just 76 schools remained. Of the 46 schools that received a "Class C" (low) rating in the Flexner Report, most were rural and proprietary nonorthodox sectarian schools. By 1918 only one eclectic school, in Cincinnati and one homeopathic school, in Philadelphia, remained. All the others had been eliminated. Those orthodox medical schools that had included homeopathic or eclectic therapeutics phased out that curriculum to comply with the trends in scientific medicine. The fate of schools that focused on educating women was similarly dismal.[26]

In direct response to the licensing laws and the Flexner Report, the naturopathic journal established the column Department of Medical Freedom, which was devoted to defending the right of US citizens, especially members of the military, to choose their healing methods and their practitioners. Medical freedom meant living "without interference from medical dictation [and avoiding] the tyranny of the medical trust that has already succeeded in trampling upon [these] rights." The column was also a response to enforced vaccination of military personnel. The column editor was Dr. Gilbert Bowman, director of the United Schools of Physical Culture in Chicago, who prescribed remedial exercises and corrective diet and offered other general suggestions for improved health.[27]

The flow of money into AMA schools raised Lust's hackles. In 1920 he asked who would give the first million to promote naturopathy. The Carnegie Foundation announced a gift of $5 million to the National Academy of Sciences and the National Research Council, founded in 1916. Average citizens, declared Lust, should resent the vicious propaganda of the medical trust and their research institutions, which brainwashed people to pollute their bodies with bacterial products in the name of curing. Furious, he said that the surplus fortunes of great millionaires were being poured into "the parasitic institutions to promote the most gigantic hoax of modern times—the germ theory." He called an army physician-general administering vaccinations nauseating and compared allopaths to "hypocrite theologians and law-mongers" who endangered the rights of the masses.[28]

Political enmity escalated. In June 1920 the Constitutional Liberty League was created to offer its own candidate for president of the United States, Frederick W. Collins, of Newark, New Jersey, representing the American drugless platform. Collins owned the experimental laboratory and clinic Mecca College of Chiropractic. The league platform asserted that the garden-variety voter must save the country from the autocracy and paternalism of politicians and their henchmen. Americans, and drugless physicians in particular, were urged to adopt the American drugless platform to "stop the exploitation, end political slavery, fight the oppressors, and move into battle." As the Progressive Era's expansion of government bureaus came to an end, naturopaths made a clarion call for less governmental control, fewer laws, personal freedoms, and an end to people's labor being squandered by the wealthy. But naturopaths were selective in their demands to end governmental control. While they wanted an end to controls on health practitioners, they were solidly behind antitrust enforcement, particularly railroad regulation, so that working men would be paid fairly. While they were quick to point out the new party's uncompromising opposition to socialism, the platform criticized all legislation that benefited only the wealthy. This was an

all-encompassing, radical, working-class-based call for political change. Naturopaths aligned themselves with anticapitalists committed to neutralizing the
power of the rich, which the AMA seemed to draw upon at any time. The naturopaths' ire was also aimed at the public health bills before Congress that would
exclude drugless practitioners. In support of Collins's campaign, one naturopath
said that someone needed to stop the AMA as the parent of the Red Cross, the
American Hospital Association, the American Nurses' Association, and the Children's Hospital Association, which were the direct beneficiaries from "the polite
begging schemes of flag day, flower-day, tag-day, and hospital-day collections."[29]

Medical doctors were merely merchants who sold health care to the sick in
the form of medications to deaden pain, claimed naturopaths. In 1935 Benedict
Lust said that the vice-speaker of the AMA House of Delegates would rather let
the poor die than let free medical service exist. Naturopaths accused the AMA
of signing a death knell for twenty-five thousand tobacco workers in Tampa,
Florida, whose medical cooperative, staffed by regular physicians, had been
providing health care at a reduced cost. But a recent decision by the AMA and
the American College of Surgeons had prompted the Tampa municipal hospital
board to deny access to the physicians in that cooperative from practicing in the
hospital. In addition, public discussions about new federally coordinated health
and welfare care had led naturopaths to believe that the laboring classes could
not afford the tax burden that would accompany it. They encouraged working
people to reject "medicos" and their price fixing, which drove up the cost of their
services. They pointed to what they called "Medical Class Legislation" and price
fixing for certain procedures, such as tonsillectomies; maternity cases; and office
x-rays. Their allegiance to working people and the laboring classes led naturopaths to actively link themselves with labor unions. In Illinois in 1940, Charles A.
Toll, PhT, wrote Lust announcing a resolution that he was going to present to the
Illinois State Federation of Labor. It accused the AMA of being "the best organized of all trade unions" and aimed to make the AMA accountable.[30]

Lust, always looking for allies, frequently made conciliatory gestures toward
allopaths who cared about the plight of average patients and were *not* part of the
medical trust. He encouraged them to subscribe to his publications. His softened
position elicited anger from his drugless readers, who wanted no contact with
their enemies. In 1940 Lust said of his olive-branch tactics, "We will educate the
doctors. We are their friends. *Naturopath and Herald of Health* finds its way into
the house of many regular school doctors who send us wonderful letters of appreciation, endorsement and cooperation." Calling regular doctors naturopaths'
friends baffled his readers, who were used to accusations and abusive names be-

ing hurled at the medicos. Challenging Lust on his mixed messages, a reader wrote that he stood ready to fight the self-appointed allopathic despots with whom Lust supposedly had no problems. Others ridiculed the allopaths' recent adoption of long-held naturopathic therapeutics such as nonsurgical relief for sinus trouble, increasing physical capacity through exercise and diet, and avoiding extreme exposure to the sun in California.[31]

Naturopaths recognized early on that their own lack of political savvy was costing them dearly. Allopaths' cooptation of successful naturopathic theories and practices was final proof. All Lust and others could do in their publications was derisively report the details. A flurry of articles in 1915 addressed naturopaths' need for political plans. Dr. B. H. Jones, author of two books on the spine and medical common sense, complained about drugless physicians' political ignorance, which meant that regulars had all the political clout. Edward Earle Purinton said that the only thing wrong with the nature business was "our inefficiency and our deficient organization." Its healers would not be worthy of respect, faith, or money until they could articulate in writing, in a professional manner, their modalities and ensuing patient fees. That fall, a journal subscriber offered to help drugless practitioners professionalize their practices—to double the number of their patients and increase their income and their value to the world. He had successfully aided osteopaths for more than a decade and realized that drugless physicians needed this guidance as well.[32]

BATTLES IN THE PRESS

As early as 1911 naturopaths realized they were losing the publicity battle with allopaths. The *Columbus (OH) Medical Journal* had attempted to bury the *Naturopath* by publishing antinaturopathic articles. Three years later, the San Francisco naturopath A. A. Erz commented on the poisoning of public wells by official medicine in the press. Well-trained writers, he said, pushed stories that glorified the discoveries of scientific medicine, and expert medical press agents left the public unaware of the deceit involved. To rectify this situation, naturopaths tried to get the word out in their own publications. The need was exacerbated by World War I and discussions of the federal takeover of health care for military personnel and the country at large. Faced with compulsory health care that only medical scientists could legally deliver, naturopaths feared that their status would be jeopardized forever.[33]

When the feature film *The Black Stork* was released in 1917, naturopaths hoped it would deal a severe blow to allopaths. It provided the perfect opportunity for naturopaths and others to publicly question the judgment, treatment, and ethics of allopaths. The film told the story of eugenic infanticide carried out by Dr. Harry Haiseldon, a Chicago surgeon, who played himself in the film. Haiseldon persuaded the black parents of an infant with congenital syphilis to let the child die without performing surgery. The child, the film claimed, carried a genetic trait that would leave his progeny unfit to marry. In the film, the mother, persuaded by her doctor, allows the baby to be euthanized. The final scene showed Jesus, in robes, signaling his approval of the act. Naturopaths had their own sympathetic views toward eugenics but believed that its goals should be attained through healthful living and selective breeding, not euthanasia. In 1918 Henry Lindlahr critiqued the film, claiming that it was withdrawn from theaters because of the AMA's concern about its portrayal in the film. He referred to the children as *so-called* defectives. Ironically, this film did much to further the cause of eugenics in America. Because of the *Black Stork* incident, naturopaths devoted a great deal of time and energy to finding articles to reprint that indicted medical science in any way. It kept their battle mentality alive.[34]

Countercharges were inevitable, and one was quite personal in 1918. Benedict Lust, who was of German ancestry, denounced *JAMA* for raising the specter of German propaganda against honest individuals like himself who dared question the misappropriation of Red Cross funds for vivisection work. He also criticized an article by MDs in the *New York Sunday World* that claimed that science saved lives and made men immune to disease. The article defended vivisection, vaccination, and serum therapy. Lust wrote that these claims were false, complaining that the reporting was so one-sided that no naturopathic response would be found in the popular press. Another naturopath objected to every Sunday magazine that contained articles by "some doctor heralding some great scientific discovery." The public should not be duped, he said; the medical trust had bought and paid for them. Drugless practitioners, he said, tried to place their own articles in these magazines, but were invariably refused.[35]

Alternative healers were routinely slammed in the popular press. In the mid-1930s, naprapaths, a breakaway group from naturopaths, had been widely criticized in the Chicago newspapers as medical quacks. Charges had been made against them without official investigation. They had faced discrimination through the Medical Practice Act (which in most cases allowed only AMA-affiliated physicians to practice medicine legally) and had been unable to secure relief through legislation. Naturopaths criticized the *Chicago Tribune* for becoming, in effect, investigator, prosecutor, judge, and jury.[36]

Naturopaths throughout the 1920s attacked allopaths at every opportunity. They speculated about the role of allopaths in the death of President Harding

and railed against allopathic monopoly in every federal, state, county, and municipal hospital, asylum, prison, jail, and education system. Allopaths were criminals,
the "only licensed vendors of Dope and Liquor in the U.S.A.; illegal fee-splitters
with specialists, performers of illegal operations, pseudo scientists, and philan-
derers in 'other rackets.'" In 1924 naturopaths borrowed from a popular political
cartoon of the day and showed the AMA as a medical octopus with a stranglehold on the world. It had to be eliminated along with its toxins. Not only that, but
"we must tear down the tenement building, and make the land free to those who
would use it; we must abolish the sweat shop, we must eliminate child slavery, we
must make the factory and the workshop amenable to life. We must abolish poverty and ignorance and their attendant evils."[37]

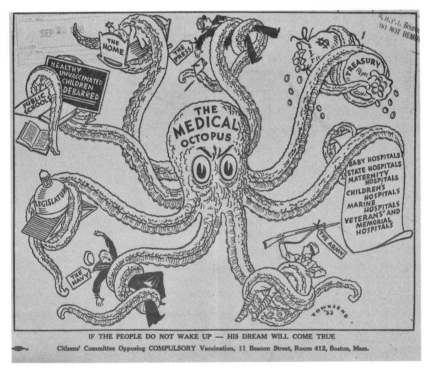

"The Dream of the Medical Octopus." The AMA is shown as a sinister creature
dreaming of control over the home, the press, the treasury, all public and military
hospitals, all branches of the military, the legislature, and public schools (which could
disbar, or expel, unvaccinated children). The Citizens Committee Opposing
Compulsory Vaccination exhorts Americans to wake up, or the octopus's dream will
come true. Box 50, Folder 01, American Medical Association Historical Health Fraud and Alternative
Medicine Collection. Courtesy of the American Medical Association Archives, Chicago.

If coverage of naturopaths in the *New York Times* from 1904 to 1948 is any in-dication, naturopathic fears of being ridiculed and discredited in the press were well founded. A 1904 article portrayed naturopaths as generating faddish litera-ture about their work. The writer, ostensibly reporting on the opening of the Naturopathic Library in New York City, took the opportunity to quip that some popular novelist would probably write a self-aggrandizing volume with a naturo-path as the hero and his student the heroine, using their frivolous cures.[38]

In 1909 the *New York Times* made no attempt at neutrality in a story titled "Held as Bogus Doctor," about Carl F. Starken, a naturopath who had been ar-rested for the third time for practicing medicine without a license. Despite the testimony of credible witnesses, the New York County Medical Society witness had won the day. Undeterred, Starken had vowed to fight the County Medical Society to his dying day.[39]

Other articles lumped all drugless healers together as frauds and dangerous extremists. As the Great War raged in Europe, one Engelbert Bronkhurst, living and working at Lust's Yungborn in New Jersey, where he was building a road to the sanitarium, made the *Times*. He was arrested with six others for conspiring to blow up munitions ships leaving American ports. When he was arraigned before a grand jury, he refused legal representation, saying that he despised lawyers. Bail was set at twenty-five thousand dollars, which he could not muster, and he was jailed. Another road builder for Lust was also arraigned. The second defen-dant purportedly testified that since they had access to dynamite for road con-struction, they had purchased more for nefarious purposes. The *Times* ran no follow-up to the story, leaving the accusations as undisputed truths. The message to readers was that naturopaths, led by the German-born Lust, and those who associated with them were anti-American.[40]

Another front-page story fed the stereotype of charlatanism. It reported a large-scale diploma fraud emanating from Kansas City in which fake high-school diplo-mas and medical diplomas were granted to eclectics. This became an opportunity to conduct a policelike sweep of all chiropractors, naturopaths, and osteopaths. No mention was made of fraudulent naturopaths, but all drugless physicians were classified as fakes. In fact, there *were* fraudulent diploma mills, but a sweep for frauds at hospitals, offices, and institutions was performed without compelling evidence in many cases, and articles damned all drugless physicians as unskilled, mercenary, and dangerous.[41]

The bad news concerning natural healers kept on coming in the *Times*. The nonstop harassment prompted the New York Chiropractic Association to file a petition to stop the persecution. The petition asked for a permanent injunction, principally against the police and the Health commissioners but including all officials in the state, to stop interference in the businesses of chiropractors, naturopaths, and all other drugless healers. Mixer drugless physicians banded together in the fight, seeing new medical practice legislation as a menace and a nuisance. Sectarian practitioners were arrested following visits of undercover policewomen, acts which seemed political and aimed solely at prejudicing the American public against chiropractors and naturopaths, since these cases always made good copy. Four months later the New York County Medical Society refused to approve a Health Department license that had been issued to the Metropolitan School of Physiotherapy. The County Medical Society claimed that "strong political pressure" had led to the issuance, an interesting claim from a group known to regularly apply political pressure. Twenty-three of the forty-two students in the first class of the physiotherapy school were naturopaths. The medical society also claimed that the Metropolitan School and schools like it were publishing misleading, unethical, and fraudulent advertisements and said that all should be investigated. Not surprisingly, the County Medical Society proposed that it issue rules governing these schools, publish a list of the schools that complied, and sever all physiotherapy connections within the Presbyterian, Harlem, Bellevue, and Beekman Street hospitals. The politics behind this story was that chiropractors had put pressure on the County Medical Society, resulting in the health commissioner's more lenient position toward drugless physicians, which in turn had elicited the wrath of the County Medical Society.[42]

Much was at stake in the publicity and power plays. The New York Times published an uncritical review of Louis S. Reed's condemning text The Healing Cults (1932). The authors asserted that the salaries of 36,000 practitioners, who made up "several sectarian groups that do not depend for their healing methods upon the usual medical theories and training," amounted to one-eighth of those of 142,000 trained and licensed physicians. They charged that these drugless healers—osteopaths, chiropractors, naturopaths and allied cults, Christian Science and New Thought healers—were usurping the AMA physicians' rightful business.[43]

Coverage of the ways the allopaths shut down naturopaths could be devastating to naturopathy, but it could be more damaging to the individual. William M. Schreier, a naturopath for twenty-eight years, was convicted of practicing medicine without a license and calling himself a doctor. He was ensnared by an investigator from the State Board of Education who had visited his office. Schreier testified that he had treated 1,043 patients in a four-year period. He received four months in the workhouse and a six-month suspended sentence for using the title "doctor." The court denied a request for a harsher penalty because the defendant was aged and had no previous convictions.[44]

Article after article fueled public doubt about non-AMA practitioners and destroyed the lives of naturopaths who were credentialed and even practiced legally. Virgil MacMickle was a licensed naturopath in Oregon, one of several states that succeeded in securing legal recognition. He was to testify in a California court in a case involving an alleged Communist who faced deportation. MacMickle was to testify against the credibility of a written affidavit. MacMickle and his practice came under attack in the trial. His education was mocked by the prosecution, who said he was only qualified to teach swimming and incapable of offering a medical opinion. After lengthy debate, MacMickle was allowed to testify, but only as a layman, not as an expert. Then his politics were assailed. He was a member of the Friends of the Soviet Union and an executive of the League Against Fascism. He was asked if he was a member of the Communist Party, which he wholeheartedly denied. Then the court of public opinion weighed in. A piece in the San Francisco press accused him of links with the German Bund. He was originally to testify about the mental condition of the accused, but when his own credentials and credibility were, in effect, placed on trial, not only his own reputation but also the image of naturopathy was damaged. Ironically, at that time the naturopathic associations were raising their educational standards, but that was not a topic that sold newspapers.[45]

In a rare nod to the efficacy of naturopathy, the New York Times in 1923 reported an incident in Philadelphia, a stronghold of the system. A naturopath was accused of practicing medicine without a license after he was ensnared by an investigator representing politically motivated medical doctors. Just before the judge pronounced his sentence, seventeen people whom the accused had treated successfully entered the courtroom and testified—impromptu—to his valuable methods. The judge, bound by law, pronounced a two-year sentence, which he then suspended. The Times reporter lamented that the man would be arrested again. In another positive but very brief mention in 1935, the Times reported on Benedict Lust's reelection and referred to him as Doctor Lust. By then his stature in the profession elicited this momentary respect; he had been reelected to his thirty-third year as head of the organization. While the attack by New York regulars and the Times abated momentarily that year, in other parts of the country the battle continued. The St. Louis Star-Times dubbed naturopathy a new medical cult, despite its forty-year history and the fact that the paper openly opposed vivisection by medical scientists—a position it shared with naturopaths.[46]

NATUROPATHS, GERMANY, AND EUGENICS: PROBLEM ALLIANCES

American naturopaths' image was damaged by World War I anti-German sentiment, a gradual postwar awareness of Nazi "social programs," links between American naturopathy's Germanic leadership and support of the German Youth Movement, and calls for human perfection. Naturopaths had a public-image problem regarding both eugenics and sporadic anti-Semitic references. A poetic piece in a 1904 *NHH* article portrayed a Jew as conniving and money-grubbing. Another article discussed the number of times the word *healer* was used in the scriptures, claiming a Jewish monopoly of the profession.[47]

Many leading naturopaths came from Germanic backgrounds. Medical- and health-related advances emanating from the motherland and contributions of German Americans were highlighted in the *NHH* with pride and called for emulation in the United States. At first this linkage benefited American naturopaths, but with Germany an enemy in both world wars, the association was obviously problematic. The residual animosity after the Great War prompted some in the public sector to defend German Americans. Martin H. Glynn, former governor of New York and editor of the *Albany Times-Union*, noted a class of people who might be called "baiters of Americans of German descent" (in boldface in the original). To enhance the Germans' image, he expounded on the work, ideas, and progressive acts of German Americans. Glynn's speech, delivered in Albany, was reprinted in full in the *Naturopath*.[48]

Nazi use of eugenics as a rationale for the Holocaust made things worse for naturopaths, even though American medical professionals of all stripes had embraced eugenics as far back as the first decade of the twentieth century. This ignominy has been overlooked by previous historians of naturopathy, no doubt because it casts American practitioners in a controversial light. There is no evidence that naturopaths were Nazi sympathizers, yet there were indeed naturopaths among the varied medical professionals who advocated eugenics. Eugenics, simply defined, is the science of controlled breeding to increase the occurrence of desirable heritable characteristics. Developed largely by Francis Galton as a method of improving the human race, it fell into disfavor with nondominant groups in the United States and then globally in response to the Nazis' genocidal doctrine and applications. The so-called desirable traits were linked to race, religion, and social class, and affluent Aryan people were deemed worthy to reproduce.[49]

Americans employed eugenics ideology and practices in a variety of settings. There were government agencies, laws, tests, societies, and professionals. In 1907 the first law favoring eugenics sterilization was passed in Indiana, and ultimately thirty-one states passed such laws. The Eugenic Records Office, which existed from 1911 to 1939, closed when the Carnegie Foundation withdrew funds. John Harvey Kellogg founded the Race Betterment Foundation in 1911 and proposed a eugenics registry. The United States participated in International Congresses of Eugenics and hosted some as well. Standardized IQ testing was introduced at Ellis Island in 1914, and the National Congress of Race Betterment was founded the same year. The army administered IQ tests to identify those capable of serving and to sort them by ability during World War I, and the House Committee on Immigration and Naturalization appointed an expert eugenics agent, who served from 1920 to 1924. This was followed by the Immigration Restrictions Act (1924); the founding of the American Eugenics Society (1925); *Buck v. Bell* (1927), in which the Supreme Court upheld forced sterilization of the feeble-minded; and the establishment of the Human Betterment Foundation, which focused on forced sterilization, in 1929. Not until 1967 did the Warren Supreme Court strike down Virginia's 1924 Racial Integrity Act, along with anti-miscegenation laws in fifteen other states.[50]

Within this cultural context, naturopaths' earliest direct reference to eugenics appeared in the *NHH* in 1915. An advertisement for the book *Practical Eugenics*, published by the International Purity Association of Chicago, offered education for intelligent fatherhood and motherhood as a way to "Prove your interest in *Race Betterment*" (emphasis in original). That same year, another article encouraged a more desirable strain of people and called for an organization. The Texan William Lienhard argued that the organization should be led by great men and women, who would collect information on traits that were injurious to humanity so that they could be weeded out, as in the selective breeding of animals.[51]

However, naturopaths did not support sterilization or destruction of undesirables. In fact, even basic birth control was abhorrent to them. Theirs was a modified advocacy in which the best specimens were encouraged to reproduce, while others were encouraged to achieve physical perfection through exercise, bodily self-discipline, and idealized strength—human perfectionism. Physical perfection could best be achieved through outdoor living and strict devotion to self-improvement.[52]

Lust and other naturopaths were interested in social experiments that improved health and morality rather than in those with fascist connections, but the average American might not recognize the difference in German organizations. Beginning in 1927, the German Youth Movement was followed by naturopaths and praised. It was first loosely constituted in 1896 as an educational and cultural movement of youth associations that focused on outdoor activities. The Jugendbewegung began as a hiking club that soon attracted the world's attention and promised to introduce a new cultural renaissance. Naturopaths praised it in 1931 for empowering youth and effecting desirable social change and habits. Germany's ministry of health supported the movement as a great weapon against vice. By 1938, 8 million children were involved, but the Youth Movement ran afoul of the Nazi state, which only allowed the Hitler Youth.[53]

The frightening politics of individual practitioners also created problems for naturopaths. In 1936 a naprapathic practitioner, Arne L. Suominen, DN, was considered a distinguished colleague. It was not unusual for naturopathic publications to report on the practices and methods of the naprapathic school, which was based largely on curing human ailments by manipulation. Suominen traveled to Germany to learn about the drugless professions' positive status. But his timing was poor: he arrived during the Berlin Olympics, which are now notorious for Germany's display of military might, for Adolph Hitler's unabashed assertions of Aryan physical superiority, and for the public shunning of the African American Olympic athlete Jesse Owens. Suominen was not a naturopath per se, but given the American press's tendency to lump all alternative healers together, the similarity between the spelling of the two systems, and his high standing among American naturopaths, his actions were seen as representative of them. He had enjoyed naturopaths' public support. He met members of the Third Reich with the aid of a German government interpreter and made the rounds of the clinics, colleges, and sanitariums. He chronicled the long and impressive list of natural healers of German decent. He did not comment on the state of German politics and policies, but he spoke glowingly of Germans' openness to his healing methods. To be fair, this was before global awareness of the extent of Nazi campaigns—their routine medical torture and experimentation and genocide. It also preceded US entry into World War II. Yet the racist, anti-Semitic Nuremberg Laws and the Nazi treatment of Jews in Germany was known by 1935, prompting President Roosevelt to slightly loosen requirements for Jews seeking immigration to the United States that year. Suominen's praise was soon seen as offensive and suspect, once again placing naturopaths in a negative light.[54]

In 1937 the *NHH* reprinted a *New York Times* article about a meeting that year of the Congress of the German National Health Movement, which supposedly had 6 million members, that advocated a return to natural living and healing practices. However, the brief report quoted only one Congress speaker, who called Hitler "history's greatest physician" and said that he had "rescued a whole nation" from a despairing mind-set. The speaker said that "the health of our nation during the past has shown how far the Jew already has corrupted and poisoned us." This reprint was even more damaging than past reprints published in the *NHH*, because by the time of its publication the journal was calling itself the *official* journal of the American Naturopathic Association and the American School of Naturopathy.[55]

Given naturopaths' progressive ideas about freedom and self-improvement, Lust's decision to publish these reprints and contributors is puzzling. Lillian Carque, an occasional author in the journal, wrote in 1939 that "we are multiplying morons." She supported eugenics and likened disease-free humans and those without criminal tendencies to cactuses without thorns and persimmons without pucker. In plants, she said, inbreeding has reproduced many undesirable traits; now these traits can be weeded out. The same was possible with humans.[56]

The naturopaths' German roots and sympathies led to a connection between antinaturopathic sentiment, Germanic people, and Nazi sterilization. In 1942 a scathing *New York Times* article on Nazified medicine accused German doctors of "falling back on Hitler's eugenic ideas." The German professor Hans Reuter commented that "racist measures must be carried out on a wide scale particularly sterilization as a means of preventing the reproduction of the socially unfit." The article blurred the distinction between American and German naturopaths and indicted them for their stances against vaccination, serum, and bacteriology. The author asserted that these stances were based on naturopaths' belief that "'non-Aryan' and 'Aryan' protein [might be mixed] to the great detriment of the German race." American naturopaths were slandered alongside their German counterparts without distinction or evidence; they were presumed to support Aryan supremacy. There was a self-righteous loathing of German eugenics in the press despite ample evidence of American eugenicist activities outside naturopathy. Naturopaths' principled opposition to vaccination was reduced to merely a eugenics-based justification. However, some of the naturopaths' own articles had opened the door for this public condemnation.[57]

GENERATING POSITIVE PRESS

By 1949 Benedict Lust, aware of these public-image problems, emphatically pleaded for naturopathic veterans of both world wars to identify themselves and create a high-profile veterans' affairs committee to publicly demonstrate their patriotism. Show of national support was also necessary because earlier in the century naturopathic articles had advocated pacifism and rejected arms exportation to Europe, and proworker sentiments had been interpreted as socialism by some. Again, these attitudes were not unique to naturopaths—they were shared by other Americans—but taken together, these stances all added fuel to the negative publicity against them.[58]

Lust had to make good press a priority. The national naturopathic movement looked to Lust's journals as the voice and outlet that combatted the medicos' dominance and asserted naturopathy's place in the health professions. They were the platform for sharing therapeutics, for professionalization and sociopolitical efforts. Of course, they did not have the impact and widespread influence of the *New York Times*. But among nature-cure practitioners, patients, and supporters, Benedict and Louisa Lust's journals had the widest circulation, the greatest number of advertisers, contributors from countless states reporting their news, and convention previews and notes. International coverage came from France, South Africa, China, Borneo and Java, Britain, India, Scotland, the British West Indies, Austria, Canada, and Australia. They also contained Dr. Lust Speaking, his ever-popular editorials, and endless coverage of complementary alternative health and political movements.[59]

Discussions about how to get the word out about natural healing and how to enhance naturopaths' image had been ongoing since the early years. In 1915 the "mixer" chiropractor-naturopath J. W. Bush had advised practitioners to place articles in popular magazines and use the newspapers to enlighten the public, but that was easier said than done when the press openly ridiculed nonregulars. The exposé article written that same year by Mrs. Franconia Benzacry (whose name also appears sometimes as Frances Benazcry, sometimes as Franconia Bonzacry), a detective for the New York County Medical Society, showed just how much work had to be done. One naturopath in particular did his part. After O. G. Carroll treated and cured the sick wife of an influential newspaper man, positive articles about drugless healers began to appear in a few mainstream papers and quieted attacks against natural practitioners, who were able to practice in peace for a while.[60]

From 1902 to 1940 naturopaths and their allies tried to publicize themselves in two ways, by promoting their own practices and by actively assailing the propaganda of the AMA. They were coached on how to advertise effectively. Purinton said that advertisements increased the number of patients and the value of naturopathy through "continued repetition of their names and claims, through . . .

newspapers, magazines, bill-boards, street-car appeals, trade packs and packages and pictures. . . . No advertisement, no advancement."[61]

In 1918 naturopaths hoped that their call for people to spread the word about naturopathy was minimizing the impact of the AMA. Satisfied patients, political allies, and practitioners volunteered to promote naturopathy. A so-called missionary could pick any issue—such as compulsory health insurance, which was important to working people, or vivisection—and through word of mouth help counteract AMA accusations. One volunteer, a merchant, who was not a drugless practitioner but had benefited immeasurably from treatments, said her participation was a way of giving back to the system that had helped her.[62]

By 1920 a beleaguered Benedict admitted that the AMA was winning the press war and called for the formation of a publicity committee within the American Naturopathic Association. At the ANA's annual convention he noted the need to widely publicize naturopathy's principles and methods. First, they had to create a catechism for naturopaths so that general readers could understand the principles and fundamentals of naturopathy. It worked. Lust was pleased that the press treated the 1923 annual convention liberally and justly. Then in 1926 the *Los Angeles Times* devoted a seven-page section to naturopathic news and nature-cure subjects written about by an active authority in the field. The Hearst newspapers were also generally favorable towards naturopathy because of William Randolph's opposition to vivisection.[63]

Lust used his journals to critique American norms such as consumerism, processed foods, toxins in everyday life, tobacco use, and sugar consumption. The advertisements alone spoke volumes about the movement: they marketed herbs, books, magazines and pamphlets, sanitariums, mechanical devices, salves, tablets, decaffeinated coffees, bakery products, health drinks, posture belts, colonic irrigators, soybean products, anabolic foods, rectal dilators, massage machines, laxatives, vibrators, diet products, items related to astrology, and countless other products. Some of the accompanying images were subdued, and their texts reasonable; others border on the ridiculous and charlatan. It is clear that neither Lust nor the staff filtered or researched the claims of advertisers. At times this made his naturopathic publications contradictory, with serious articles by practitioners detailing disease causality and treatment appearing on the same page as advertisements for contraptions and claims of panacea effectiveness. One even pictured crystal balls and Svengali-like practitioners coaxing people into the cosmic realm.[64]

NATURE'S PATH

A turning point in the culture wars was Lust's creation in 1925 of *Nature's Path*, a lay publication out of his headquarters in New York City. All previous naturopathic publications had been geared toward practitioners. A staff of assistants, coeditors, and helpers were engaged. The time had come for a magazine aimed at the general public about natural living, nature cure, and the science of naturopathy. The idea had been around for decades, and the decision had been announced at the 1924 convention of the ANA in Los Angeles. *Nature's Path*, Lust believed, would help increase the naturopathic patient base and make converts and friends. The first edition numbered fifty thousand copies—a risk, Lust acknowledged, but one to benefit all naturopaths. He asked each member of the ANA to subscribe to the magazine and place a permanent monthly order for one hundred copies for the cost of $1; the fee was $20 if the copies were mailed directly to patients and friends. Those who complied were promised a free listing in all subsequent issues. The tactic seemed to work. The next year the New York State Association of Naturopaths announced that it was two hundred strong and had scheduled a convention in June for the purpose of spreading lay propaganda.[65]

Nature's Path's masthead in 1927 was a marketing tool. It stated the publication's purpose, intended readership, and its origin. It was a "frank but clean exponent of the attainment of what human beings want most, through better ways of living, healing, thinking, planning, working, saving, hoping, loving, conquering and achieving." The magazine was devoted to the proper care and use of natural living through knowledge, development, and enjoyment of life. It was the official journal of the Lay Department of the American Naturopathic Association, the American School of Naturopathy and Chiropractic, "and several other Societies and Movements devoted to Natural Life, Nature Cure and Medical Freedom." In the years that followed, Lust liberally published correspondence and praise for *Nature's Path* from lay people, mostly women, and practitioners. They found it interesting, sensible, informative, and worthy of circulation, and they followed its advice.[66]

In typical all-inclusive Lust fashion, by 1931 the journal masthead was a five-paragraph itemization of all the attendant sociopolitical and health-society links contained within the journal. In addition to the standard affiliations, it now was the official journal of "several other Societies for Nature Cure, Diet, Chiropractic, Physical Culture, Osteopathy, Anti-Vaccination, Anti-Vivisection, Vegetarianism, Medical Freedom, Youth Movement, Natural Life, etc."[67]

Nature's Path became a primary vehicle to entice a new generation of lay women into the movement, which was critical for the survival and growth of naturopathy. The women's articles and their accompanying images from the 1920s on were designed to appeal to middle-class women and recruit them to become health and food experts for their families.

Over the decades, the content of *Nature's Path* revealed the ways in which naturopaths were ahead of their time in challenging the excesses and side effects of industrialization. By midcentury they saw disease as largely the result of environmental toxins. In the words of one practitioner, "The use of chemicals, fertilizers, devitalized, demineralized foods, chemicals, poisonous sprays, occupations which subject one to fumes of all sorts have changed the whole story of health." The traditional emphasis on therapeutics and advertisements for colonic irrigation and other system-cleaning methods continued, but the recognition of environmental toxins led naturopaths to point out the ways in which people ingested toxic substances, from plant fertilizers and saccharine to radioactivity. Tobacco, with its potent toxicity and addictive nature, was constantly vilified. Meat was now viewed as a toxin in that it contributed to the incidence of fibroid tumors. Alcohol was another toxin, destroying the body, the family, and society.[68]

In the late 1940s the trendsetting power of Hollywood film stars was made use of in *Nature's Path* to broaden appeal. They embodied the postwar sense of hope and economic well-being for the middle class. Articles on health and longevity, some of them written by health and beauty advisers to the stars, portrayed aging as a problem to be dealt with through healthy living. The age that prompted concern fell from fifty in earlier issues to forty. Advertisements in *Nature's Path* promoted a wide array of goods, such as eliminative products, medical vibrators to soothe and relax tight muscles and taut nerves to induce sound sleep, herbs, coffee substitutes, denicotinized cigars, juicers, ear drops, publications, and a legion of other items. Also advertised were health resorts in Santa Fe, New Mexico; the Isle of Pines, Cuba; Milwaukee, Wisconsin; and Rhinebeck, New York.[69]

While the magazine could be an effective tool in the health culture war, it did not always help to solidify naturopathy's core concepts. There was often a disjuncture between articles' content and advertisements. Lengthy pieces warned of the danger of salt use, only to be followed a few pages later by an advertisement for salt pills. Complaints about anti-aging and beautification treatments were juxtaposed to advertisements for natural hair color or hormone-based facial creams. The reader was left to sort out these conflicting messages from advertisers, whose revenue no doubt kept the publication afloat. *Nature's Path* also advertised various health-related schools and correspondence courses, although the ANA had already abandoned correspondence courses. The validity of schools' claims went unchecked and likely contributed to the disrepute of the movement.[70]

Nature's Path was so influential for decades that Lust reconsidered the format of the *Naturopath and Herald of Health* in 1943. At the ANA's annual convention in New York City, Lust brought this question before the attendees. Some suggested that it be combined with *Nature's Path* and the price lowered. In January 1944 Lust announced that the two would not merge but that the *NHH* would contain a less technical laymen's section to recruit patients and increase revenues. To generate popular appeal, the May 1944 issue illustrated an article on sensible living with a photograph of a smiling baseball player in uniform. In a bold move to distance naturopathy from charlatans, it would only be sold to "bona fide, actual practicing Naturopaths-Chiropractors, Osteopaths, Physiotherapists, and other recognized branches of the natural healing art." This was impossible to enforce, of course, but Lust understood the importance of publications for a professional image. The *NHH* was expanded to include articles for parents, more disease-specific remedies, features on women and children, and columns aimed at female readers. Interestingly, both Lust and readers referred to the revised *NHH* as *The Naturopath*.[71]

Naturopaths held fast to their mid-nineteenth-century roots in individual responsibility and their belief that the body could heal itself and that healers should do no harm even as they struggled to compete for patients in the twentieth-century health culture war. The vastly different practices within the movement helped them garner new followers, while they alienated other segments of Americans. When Lust announced in 1931 that antivivisection and antivaccination were issues of key concern, he was highlighting one of the many battles in the allopath-naturopath war. There had been explosive debates about these issues for decades. These two practices, according to naturopaths, were unscientific, disease-producing, and contrary to nature's laws. The debates about them, like all the arguments between the two sects, pointed to the disparity in philosophy and praxis that drove the rivalry.

CHAPTER 7

Medical Monsters

Vivisection and Vaccination

Naturopaths' abhorrence of vivisection and of vaccination were philosophically linked. Both violated what naturopaths defined as natural laws; they disordered bodily and cosmic synchronicity. Both practices were seen as prime examples of destructive allopathic intervention. Naturopaths believed that vivisection—performing experimental surgery or other research on live animals—exploited animals and doomed them to suffering; vaccination parlayed the data from that research into deadly toxins introduced into the human body. In the words of one practitioner in 1914, "WE HOLD vaccination, vivisection, poisonous drugs, gluttony, intoxication and the like to be abhorrent to nature and destructive to the well-being, health and happiness of mankind."[1] Naturopaths believed that neither practice furthered healing or science but that instead they secured two additional venues in which allopaths could claim expertise and impose their methods.

Naturopathy served as the perfect umbrella for the oppositional campaigns—as it had for other movements. Opponents of both methods gained shelter, support, and legitimacy within the mass movement for alternative medicine. At the same time, because of their opposition to vivisection and vaccination, naturopaths were ultimately discredited as American society came to accept the practices. Medical science ultimately won broader support through organized vaccination efforts.

Antivivisection and antivaccination activism were examples of how nineteenth-century health reform continued into the twentieth century. They were antiallopathic and rejected the animal research model of scientific medicine.

"A PLEA FOR THE DUMB ANIMALS": ANTIVIVISECTION

From naturopathy's beginnings, a core tenet was that animals should not be used for medical experimentation. This was in keeping with a commitment to harmony within nature and to clean bodily systems that also led them to embrace vegetarianism. They saw no need for experimentation on animals for vaccines or surgeries, advocating instead the autotoxemia theory—that the human body would heal itself through its own ability to throw off disease through "crisis." Any foreign matter disturbed that process.

The protest against vivisection pointed to the rights of individuals among both humans and animals. This was one of the reasons why antivivisection also became a realm for female leadership. Supportive correspondence was exchanged among antivivisectionists in varied organizations. By challenging unfettered human dominance over the natural world, naturopaths likely also expanded the number of allies who sought naturopathic care. In embracing antivivisection together with vegetarianism, naturopathy—for a while—led the way in resistance, with positive effects.

When naturopaths began their protests, they were contributing to a long history of arguments about animal experimentation. They joined a vibrant, articulate, and effective movement at its height in popularity in the 1890s. As early as the seventeenth century the eminent French philosopher René Descartes had asserted that animals were soulless and unable to feel pain in the same way as humans, so that the notion of animal pain was morally irrelevant. This assertion was used as justification for vivisection and the superiority of man. Naturopaths, however, defined medical experimentation on animals as unchristian, cruel, and abusive, despite its justification for creating human vaccines. Naturopathic antivivisectionists used the morality of early-nineteenth-century concepts of Christian perfectionism and Christian nationhood in their arguments. The author Maud S. Weeks argued that the church should oppose vivisection. It was, she said, a "perversion of the Christ Spirit-Self Sacrifice [because] the helpless suffering animals in the medical laboratories are not Willing Sacrifices for the tortures they endure." Opponents lamented the morally corrosive influence of vivisection on those who practiced and witnessed it—medical students and schoolchildren. Weeks wrote, "Children should be taught the sacredness of life and a merciful spirit, and will the vivisection knife do it?" A corollary concern was that acceptance of animal cruelty—even applauding it—to advance medical science would lead to medical experimentation on humans.[2]

Naturopaths built upon a long-held and articulate antivivisection position from England. Records of medically based animal cruelty on conscious animals dated from the mid-1600s. Between 1870 and 1900 the British practice of vivisection went from being an offshoot of anatomy to becoming an internationally known experimental school in its own right. Supporters argued that the beneficial results outweighed the *necessary* suffering of the animals.[3]

In 1876 the Victoria Street Society was founded. It later became the Society for the Protection of Animals Liable to Vivisection and is now named the National Anti-Vivisection Society. A year earlier, Parliament had passed the Bill for Regulating the Practice of Vivisection, which prohibited the use of live cats, dogs, and horses in experiments and required the use of anesthesia in all other species. Yet in practice, exemption certificates were easy to come by, and the animals were not protected; in its one-hundred-year history there were no prosecutions under the act. Even Queen Victoria was disappointed and spoke out against vivisection in 1881. By 1910 studies showed that ninety-five thousand live animals had been caused pain through experimentation.[4]

Meanwhile, two striking events fueled antivivisectionists' arguments in the United States. In 1903 Mark Twain published a story in *Harper's Monthly* that was told from a dog's viewpoint and protested vivisection and animal cruelty. In it, the dog saves the family's baby from a fire in the nursery but later sees her own puppy blinded and killed by her scientist-owner's medical experimentation. Provivisectionists said it was overly sentimental, but antivivisectionists applauded the publicity for their cause.[5]

The experiments of Ivan Pavlov in 1904 also fueled antivivisection when he won the Nobel Prize in Physiology or Medicine in 1904. Pavlov had severed a dog's esophagus and sewn the loose ends to its throat, leaving two side-by-side holes connected by separate passages to its mouth and stomach. The dog was left hungry, harnessed to a wooden stand, and given raw meat to eat. It could not get full no matter how much it consumed, as the meat leaked through the esophageal opening and back into the bowl; the dog lapped it up. Meanwhile, a glass tube attached to the dog's stomach allowed gastric secretions to collect in a bottle. When filtered and analyzed, the secretions were sold to the public as a remedy for dyspepsia. Ironically, Pavlov admired dogs' intellect, found them touching, and approved a statue in St. Petersburg, Russia, to honor his dog subjects. One bronze plaque depicts dogs on laboratory tables, tied to wooden frames, with their fistulas open.[6]

The earliest naturopathic journals condemned laboratory vivisection as bad science. In 1902 Benedict Lust's front-page article in the *Naturopath and Herald*

of Health (NHH) declared, "Vivisection or Scientific torturing of animals con-
sists in torturing live animals for the benefit of science. They may be either
burned, stewed, roasted, cut-up, or a hole may be bored into them, or they may
be made to die of hunger or thirst." He said that his plea was "for our second best
friends, the dumb animals." Lust had passed out and actively shrunk from vivisection in medical school and been mocked for it by his classmates. Naturopaths
opposed vivisection for three main, scientifically based reasons. First, severe pain
or anesthesia so disordered bodily systems that any conclusions drawn under these
circumstances were useless. Second, conclusions drawn from another species did
not necessarily apply to humans. The third argument involved a standard of medical ethics: man had no right to inflict suffering on animals; it was dishonorable,
unethical, and immoral. In 1908 Lust insightfully and prophetically linked vivisection with the hunting and ultimate erasure of species. He decried the killing of
whales, seals, penguins, and elephants for man's use; he predicted that these acts
would eventually lead to their complete annihilation.[7]

Vivisection was just bad science, according to naturopaths. To prove its extreme cruelty, naturopaths discussed the acts levied upon animals in their
opposition to the vaccination campaigns that justified them. Authors hoped that
by listing atrocities they could keep Americans from being duped by medical or
state authorities who argued the so-called benefits for humanity without providing details of the procedures. According to naturopaths, the torture showed no
tangible results. It was time to stop. By listing repulsive tortures, naturopaths
backed up their moral argument and their point about the impact of trauma to
bodily systems on research outcomes. Scientists used "every conceivable kind of
animal torture," such as "pouring boiling water into intestines, distending stomach by air pressure, sticking needles into hearts and brains, putting mustard oil
into eyes, enforcing ceaseless activity in cages until death come [*sic*]." The
means were horrific, the creatures abused did not benefit from the procedures,
and the information gleaned was useless. Almost none of the animals used in
these experiments were anesthetized, and when they were, the anesthesia used
was unlike the surgical anesthesia used on humans.[8]

Some naturopath antivivisectionists approved surgeries to prolong human life
but adamantly opposed experiments conducted upon "men, women, and children, not for the benefit of the organism operated upon, but for . . . the good of
medical science." This opposition was extended to experiments on animals.
They slammed the medical scientists hailed as pioneers in the second decade of
the twentieth century. Naturopaths' position was that if everyone lived more in
harmony with nature, surgeries would be less necessary. Citing slavery and other

human atrocities over the centuries and the role of Jesus as a moral guide, George Allen White asserted that the public will never justify cruelty.[9]

Authors also chronicled forms of animal cruelty that they believed contributed to humans' insensitivity to vivisection. Meat eating, for instance, distanced humans from the lives and suffering of animals. If people were aware of the cruelty involved in killing animals to eat, surely the demand for meat would decrease. Man was created vegetarian; it was not until he indulged his earthly sensuality and devilish ways that he craved animal meat. In 1920 a poem titled "Carried to Slaughter" painted a mournful scene of cattle going to slaughter when so much other bounty existed. It began "They passed by my train / I looked into their sad features / They were carried to be slain / To feed human creatures." Dr. Alice Chase wrote about the tragedy of full-page advertisements in magazines showing infants eating cans of pulverized meat, since smugness and indifference to animal pain were not far removed from indifference to human pain. The naturopathic leader Dr. Jesse Mercer Gehman associated consumers' meat dependency with the profits of packers, butchers, and advertising companies, who told Americans that meat was vital to the human diet.[10]

Naturopaths linked animal cruelty to human vanity, consumerism, and fashion. As a 1908 protest statement against the exploitation of animals put it, "We destroy the birds and the beasts by the thousand to envelop ourselves in furs and adorn our hats with feathers. What care we that their blood is spilt; what is their suffering to us! If they are trapped, and snared, and hunted, and harpooned and shot at, and clubbed, scared, terrified and tormented, and done to death; what is that to us? . . . Man is the lord of the beast and the bird!!"[11]

The point was simple: it was the duty of humanity, and health providers in particular, to preserve life, not destroy it. Naturopaths found in the New York Anti-Vivisection Society a ready source for collaboration. The society had formed in direct response to the new Rockefeller Institute for Medical Research, which was to be a center for vivisection. With naturopathic support, the society put two bills before the state legislature in 1908 to curb animal cruelty, but the institute had such strong backing from powerful entities that the bills failed.[12]

GENDER AND ANTIVIVISECTION

Prior to naturopaths' entry into the debate, mid-nineteenth-century American women led and joined antivivisectionist efforts, as Englishwomen had. Because of women's perceived moral authority, the movement provided significant opportunity for female leadership. In the 1860s both American and British female leaders had argued that experiments were cruel, unnecessary, and destroyed the

morals of the brutal vivisectors. It was not difficult for some women to identify
with the animals, since they shared three experiences with them, in the realms
of gynecology, pornography, and literature. The vivisected animal conjured
women's fears of sexual surgery, producing visions of women strapped down and
gynecologists standing over them with knives. Invalidism and neurasthenia were
rampant among middle- and upper-class European American women, and "corrective" gynecological surgeries routinely involved oophorectomies and clitero-
dectomies aimed at restoring gender-normative behaviors and desires. Episodes
in pornographic novels between 1870 and 1910 portrayed women tied down,
subdued, and thus "mounted" more easily. Literature in this period depicted
animals in similar dire circumstances, brought to ruin by male overuse. Most
notable among these was *Black Beauty* (1877), a standard read for adolescent girls
that discussed "breaking in the horse" and forcing a bit between its teeth. (The
bit was standard fair for restrained and brutalized women in Victorian pornography as well.) The issues of indecent exposure, restraint, dominance, pain, and
bodily destruction were embodied in animal experimentation—and well understood by Victorian women. It was, as one historian put it, "the abuse of authority,
the delight in the spectacle of pain, and the sexual subjection of the weak by the
strong" that created an empathetic bond.[13]

Initially, the American leadership was made up of both women and men, but
by the 1890s antivivisection was a woman's movement and cause. Opposition to
vivisection led to the founding of the American Society for the Prevention of Cruelty to Animals (ASPCA) in the 1860s. Antivivisectionists used the term *zoophily*
to encapsulate the moral base of their argument. It was more than love for animals: vivisection contradicted *mercy*, the great cause of civilization. Antivivisectionists believed that medical science, through animal experimentation (which
they called torture), was based on cruelty, the antithesis of mercy. It "demoralized practitioners and retarded the advance of civilization." Experimenters, in
response, denied that cruelty was involved and focused on their research's beneficial results for mankind. They pointed out antivivisectionists' inaccuracies and
ridiculed them.[14]

Sympathetic opponents reacted speedily to the first animal experimentation
laboratories. The American Anti-Vivisection Society was founded in 1883 at a
meeting called by the Women's Branch of the Pennsylvania Society for the
Prevention of Cruelty to Animals (PSPCA) by Mary F. Lovell and Caroline E.
White, who later edited the society's *Journal of Zoophily*, founded in 1892.
Female leaders drew members from among the urban, eastern, middle-class
women, many of whom were interested in temperance reform through the

Women's Christian Temperance Union (WCTU), while many others not affiliated with any other political or social reform joined as well. It was a female demographic that overlapped with naturopathy's profile.[15]

Thousands were mobilized in response to the accounts of the sheer brutality
of the experiments: violent blows levied on dogs to dislocate their limbs while
they howled in agony; rubbing dogs with turpentine and setting them on fire;
breaking of spines; severing of nerves; and more. The use and abuse of animal
mothers most disturbed Victorian women. Women were charged with protecting
and fostering the sanctity of motherhood and home. Experimentation on a mother
dog whose spine was severed and her pups put before her—pinched and hurt—to
determine whether she would still try to protect them was a scenario that horri-
fied and repulsed them. Not only were these actions cruel and immoral but
women feared that under certain horrific circumstances they might be used on
human mothers. Fifty years later, this fear would broaden among Americans of
both sexes.[16]

Not all condemnation of vivisection came from the written word. A clever yet
disturbing cartoon pictured a variety of dogs wielding knives, forceps, and sabers
over the tied-down body of a terrified and awake man. Captioned "The Dream of
the Vivisectionist," it spoke to the terrors man would know if only the atrocities
were committed against him.[17]

Just as naturopathy was developing, activist women linked antivivisection,
motherhood, and children's moral upbringing and made the 1890s the strongest
era in antivivisection legislation. They did this in concert with naturopathy's ally
the American Medical Liberty League, whose platform included antivivisection
as a major component. In some cities, women seeking more leadership and
control of the debate formed female-only organizations such as the Women's
Humane Societies. They proposed legislation, ran educational campaigns, and
brought public pressure to bear. Carriage horses received particular attention. In
1920 naturopaths likened the ruining of a horse through overuse and negligence
to laboratory experiments in which scientists destroyed dogs by cutting eyes, destroying hearing, and cutting nerves.[18]

In the midst of these activities at the turn of the century, Dr. S. Weir Mitchell,
called by some the father of American gynecology (but since reconsidered by
some as a notorious exploiter of enslaved women for medical experimentation),
asked Carolyn White, of the Women's Philadelphia Society for the Prevention of
Cruelty to Animals (WPSPCA), for the unwanted dogs in their Philadelphia
shelter for use in experimentation in his research hospital. Alarmed and furious,
White called a meeting of her organization's executive committee that resulted

in a public protest against vivisection. In 1909 the WPSPCA, supported by other antivivisectionists, passed legislation that forbade the selling or buying of disabled work horses. The WPSPCA also opposed all animal blood sports, such as rooster and dog fights and animal baiting—tying up an animal and letting others attack it.[19]

During the movement's heyday, attempts to discredit antivivisectionists were rife and routinely aimed at naturopaths. Particularly damaging was the 1909 creation of the disease category zoophil-psychosis by the American neurologist Charles Loomis Dana. Dana used animals' nervous systems to research neurasthenia, a nerve-based disease diagnosed initially among soldiers, then largely among women. In women, a diagnosis of neurasthenia was often based on the belief of their mental and biological frailties and nervousness. Dana claimed that an elevated concern for animals was a form of mental illness called zoophil-psychosis. Who could take antivivisectionists seriously, when they suffered from insanity? Since the constituencies of the organizations against vivisection were primarily female, this resulted in the charge that women were particularly susceptible to the disease. Accusations that they were more concerned with animal welfare than with the welfare of their own children further discredited them.[20]

ANIMALS' LURE

Another reason why naturopaths loathed vivisection was that destruction of animals also destroyed their knowledge and the public's ability to benefit from it. They explored animals' physicality, their intuitive abilities, their bravery, their self-sacrifice, their capacity for affection, and their ways of living within nature's laws. Antivivisection found particular public favor after World War I as the nation expressed gratitude and affection for heroic war dogs that had assisted soldiers. One estimate is that 16 million animals "served" in World War I, including carrier pigeons, dogs, the Camel Corps, and others. The popular press described their exploits and heroism, and they were chronicled in movie-house newsreels and immortalized in literature. Popular fascination with—and genuine admiration for—animal film stars and real-life animal heroes helped the naturopaths' arguments. German shepherds were particularly popular in the silent film era, and none more so than Rin Tin Tin, who was a Warner Brother's film star for eight years. He starred in twenty-three movies, beginning in the 1920s, and earned his owner-trainer Lee Duncan more than $5 million from packed movie houses. He was internationally feted by heads of state and local politicians, and he received thousands of fan letters each week. His character was portrayed as noble, compassionate, loyal, and vulnerable. These traits called into question the practice of vivisection in the public collective consciousness. In 1925 another dog, Balto, achieved fame by leading a sled dog team and human mushers to Nome, Alaska, to deliver serum to combat a diphtheria outbreak that threatened that isolated community. This heroism is reenacted yearly in the Iditarod Sled Dog Race, also known as "The Last Great Race."[21]

While naturopaths did not directly mention either of these hero-dogs (after all, Balto was delivering a vaccine), their popularity in the press influenced naturopath's arguments nevertheless. In 1925 the *Naturopath* printed the touching tale of a dog that died of a broken heart, a loyal spaniel who guarded his drowned master for one week before a friend arrived. Rescued, the dog refused both food and water and died within a week. Above all else, the fact that a dog could die of a broken heart put dogs on par with humans.[22]

Naturopaths emphasized admirable traits that came with animals' pure living. Animals followed their instincts and needed no pills, powders, tonics, vaccines, hospitals, sanitariums, or free clinics. This made them health specialists of sorts. This romantic view of animals, immune from physical distress and disease, while oversimplified and incorrect, tugged at people's hearts. To reinforce the point (and perhaps contradict it) in 1902, the *NHH* introduced a Naturopathic Veterinarian column, which offered veterinary advice promoting natural methods. The larger point was that animals, because they lived more in harmony with nature, were less prone to disease and less subject to damaging cures. Horses' eating habits were invoked to demonstrate that their health would be destroyed if they ate white flour as humans did. They would not be able to haul wagons or win races. They would be too unattractive to sell and too sickly to work. The naturopath Jesse Mercer Gehman quipped, "If food is such an important factor in raising good healthy horses, chicks, dogs, cows, or even pigs, certainly it is equally important if not more so in the upbuilding or the deterioration of the human animal."[23]

In the 1930s, naturopaths received a considerable boost when data about the number of animals experimented upon without anesthesia emerged from England. Anecdotal evidence had existed for years, but the British Home Office reports for 1933 revealed 18,185 vivisections with anesthetics, 575,055 without. "That is, for every animal vivisected in England, with even imperfect anesthesia, 19 animals are experimented on without anaesthetics!" Naturopaths suspected that the numbers in the United States, a larger country, were far worse. They believed that pressure was placed on MDs to support vivisection—that if they did not, they would be blacklisted. To make their case, naturopaths had only to reprint medicos' scholarly articles about medical experimentation, and reports of those on monkeys were most disturbing. The *American Journal of Physiology* in 1938 chronicled recent experiments at Yale University in which interference with the nerve supply to the primates' faces had resulted in paralysis of certain muscles. Researchers had observed contractions brought on by frightening the monkeys. They had provoked them with sticks, black rubber hoses, monkey-catching nets, and electricity. Terrified, the monkeys had tried to escape as researchers noted their spasms and contractions. They had also been injected with acetylcholine, which induced drooling, teeth grinding, heart-rate escalation, and diminished activity. Some had lain prostrate, eyelids open, nostrils and upper lips twisted to the side with paralysis. The *NHH* report concluded with the note that of the twenty-seven articles in the *Journal of Physiology* for March, twenty-four dealt with the vivisection of animals. No further commentary was offered—or needed.[24]

By the 1930s, to bolster their case, naturopaths had learned to exploit the growing cultural fondness for, and increased numbers of, dogs as household pets. It was well known that dogs were stolen or bought and sold illegally for medical experimentation. In stories taken from mainstream city newspapers, dogs were being rounded up and shipped to medical laboratories for profit. A snippet in the *Boston Globe* in 1939 exposed a dog-stealing gang of thieves in Massachusetts. The *Globe* also ran an advertisement for carload lots of mongrel dogs, placed by the same person who advertised for fifty mongrel dogs for breeding purposes. A newspaper investigator had evidence that the sale of animals to medical schools had been going on for years. Also in Boston, an advertisement reading "WANTED AT ONCE—Twelve mongrel puppies approximately six weeks old" was traced to an intern in a prominent Boston hospital. In Chicago, 6,299 dogs were sent from city pounds to medical laboratories. Since few laws existed restricting the practice of vivisection, and those that did exist were difficult to enforce, this trade in animals was not illegal.[25]

In 1940 antivivisectionists celebrated Boston's decision to stop vivisection of small animals by high-school children in biology and zoology classes. Naturopaths supported the Boston decision, calling for increased activism. "Can you guarantee that your little pet will not some day find itself in the clutches of a mad vivisector?" They pleaded for people to start community-based groups to help end "the ghastly orgies."[26]

In 1944 naturopaths reported on a bill that the newspaper magnate William Randolph Hearst was writing. Hearst was an outspoken adversary of vivisection and used his papers to protest it. But the onslaught of World War II stalled those efforts. What *should not* be put off was getting control of unclaimed dogs in public pounds, even those under jurisdiction of humane societies. The reason for urgency was that New York's Mayor LaGuardia wanted to abolish the ASPCA. In this New York would be following the example of Chicago, where strays and lost pets were rounded up and turned over directly to medical laboratories. Naturopaths drew on people's personal affection for pets by publishing stories like one compelling tale accompanied by a photo of a young boy in shorts and shirt petting his dog's head. The caption below the image warned that "in every community there are degenerate human beings who encourage the stealing of dogs. And these stolen dogs end their days on the torture racks of the scientific laboratories."[27]

Naturopaths' concern was not limited to domesticated animals. Antivivisectionists feared—rightly so—that scientists would gain access to the Bronx Zoo's cages and experiment on the zoo animals. One article pictured the sultry face and piercing eyes of a leopard peering out at the reader, framed by the query, "Will the Bronx Zoo become a hell?" The author condemned the damage caused by penning up wild animals, as evidenced by decreased health and vigor, from daily torments of boys throwing stones at them, lighted cigarettes fed to the ostriches, and the showers of cheap candies thrown by unthinking children into the monkey cages. She also offered proof that "the dregs of the scientific slime has [sic] determined that the largest zoo in this country shall be turned into truly an 'animal hell on earth.'" She called for refusing vivisectionists access to the zoo and emphasized the need to work to dismantle zoos. Images continued to be used effectively. One pictured a white horse, head and neck turned aside, sporting bridle and a jauntily placed straw hat. The text was familiar, chronicling the horrors of vivisection, but notable was the sentimental anthropomorphism employed—this image was the first to show an animal in human clothing. Other stories addressed the care of animals as well, proudly contrasting humane care with the inhumane use of vivisection.[28]

POWERFUL FOES AND SHIFTING TIDES

Lust and his colleagues were aware that scientists were not their only foes. The powerful US Departments of Health and Agriculture and the US Army and Navy, all influenced by the AMA, supported vivisection.[29] Then an addition to the Social Security Act was introduced in Congress in 1943. The Wagner-Murray-Dingell Bill proposed a national health act that would include health insurance in Social Security. It met with opposition from various groups and did not receive support from President Roosevelt. Naturopaths adamantly opposed the bill on the familiar grounds that it promoted authoritarian medical expertise, violated individual choice, and codified government intervention. They warned that it would place physicians on the government payroll; worse yet, one section of the bill stipulated that a portion of the proceeds of this compulsory insurance tax would be used to subsidize medical research. It was also opposed by the AMA. The bill did not pass, but it served as a crucial rallying point.[30]

Undeterred by the massive political forces they faced, naturopathic antivivisectionists pressed on. But they were handicapped by an American public consumed with war-related worries and losses. One naturopath commented in 1944 that the American public had lost track of the realities of vivisection, erroneously thinking that it had been virtually abandoned. Unaware of the realities of the practice, the public believed the rhetoric, he said, that one child's life was worth the lives of tens of thousands of dogs, and for "dog-owners, this sainted axiom of medical voodooism falls on deaf ears." In their zeal, naturopaths sometimes lashed out against their medical opponents in ways that harmed their case. In the middle of the war, a naturopath chafed at the irony of a surgeon practicing devilish tortures on monkeys to create the polio vaccine, while shedding copious tears at the Japanese tortures of American prisoners. It was not an argument that would gain much sympathy from the American public.[31]

Historians have proposed three historical eras in antivivisectionist thought: from 1860 to World War I, from World War I to 1970, and from 1970 to 1990. From 1860 until World War I, oppositional sentiment was at its height. After World War I the US government became a major sponsor of scientific research based on the animal experimentation model. The budget of the National Institutes of Health (NIH), the main agency funding research, grew exponentially and provided more monies for research in the public sector. Concurrently the discoveries of penicillin and streptomycin tremendously expanded pharmaceutical research and the size of the prescription drug industry. Increased governmental spending and drug discoveries increased the demand for laboratory animals. From World War I until 1950 there was a distinct shift among Americans to *support* animal-based experiments. It was during these years that animal research developed as a way to discover new biological data en route to potential cures. Opposition was consistent during this period, albeit less public, and ultimately had an impact on policymakers in the next decade.[32]

Fueled by the increased powers of the AMA, military medicine, public health, and numerous legal victories, biomedicine enjoyed great support in the 1950s. Naturopaths' credibility diminished because they opposed vivisection and because they challenged allopathy. They appeared to be antiresearch. When cultural attitudes shifted to support animal use, the word *vivisection* was replaced by the more euphemistic *animal research*. This shift in attitudes occurred despite continued protest within the natural healing community and from the powerful William Randolph Hearst. In 1949 a National Society for Medical Research poll found that 85 percent of Americans approved of using animals in research, while only 8 percent disapproved. An ASPCA poll might have yielded different statistics. Despite increased governmental spending, naturopaths' unshakable opposition opened the way for other groups, such as the Animal Welfare Institute (1951) and the Humane Society of the United States (1954), to carry on their work.[33]

What did the shifting attitudes mean, in real numbers, for laboratory animals? In 1957 the International Committee for Laboratory Animal Science estimated that 17 million laboratory animals had been used that year. A decade later this number had increased to 40–50 million annually. During these years NIH funding increased sixfold.[34]

But this cultural acceptance of laboratory animal use plummeted after mid-century, and once again cultural attitudes were more in line with naturopathic beliefs. As they had been decades earlier, dog theft and the poor care of lab animals were public issues. In a watershed case in 1965, a Pennsylvania family's Dalmation, Pepper, was stolen and sold to a research hospital in the Bronx, where its chest was cut open in an unsuccessful test of a new cardiac pacemaker. Congressman Joseph Resnick introduced a dognapping bill that required governmental licensing for laboratories and dealers that traded in dogs and cats; animal theft became a federal offense. Passed as H.R. 9743, "Pepper's law" broke a stalemate between animal activists and the biomedical industry. Release of the 1961 Disney film 101 *Dalmatians*, which featured dognappers attempting to steal and kill the spotted puppies for a coveted fur coat, preceded the law. The film placed the breed squarely into the hearts of Americans and highlighted the dangers faced by innocent pets. Immediately after the bill, in 1966, *Life* magazine's exposé entitled "Concentration Death Camp for Dogs" forced Congress to act. By summer of that year the Laboratory Animal Welfare Act was passed and made law. What had begun as a measure to prevent pet theft became "the most comprehensive animal-welfare legislation" in American history.[35]

Between 1970 and 1990 animal use decreased, perhaps by as much as half, despite ever-rising funding. This was because the cost per animal was nine times higher than before. In addition, improved medical techniques reduced or replaced animal use. The claims that animal research decreased by half came largely from US Department of Agriculture (USDA) annual reports, but many critics have said that these reports were unsubstantiated and exaggerated.[36]

Most important from a naturopathic viewpoint, concern for animal welfare bloomed in the 1970s as naturopathy and alternative healing regained popularity and credibility. The 1970s and 1980s witnessed dramatic changes regarding *which* animals the public approved for use in research. There was much less support for the use of dogs (55%) and primates than for the use of mice and rats (88%). There was a direct correlation between tolerance of the research and its use. Opinion was in favor of animal use for cancer or diabetes research, but 27 percent opposed animal testing for allergies. Increasingly, citizens cared deeply about the pain, suffering, and distress of laboratory animals.[37]

In 1970 a new version of the Laboratory Animal Welfare Act, the Animal Welfare Act (AWA), included research institutions. It mandated that animals receive adequate veterinary care. In 1985 the Public Health Service created Institutional Animal Care and Use Committees for institutions receiving funds for animal research. Against this backdrop, the work of several animal rights organizations, such as the Animal Liberation Front, Greenpeace, and People for the Ethical Treatment of Animals, altered American opinion about the use and abuse of laboratory as well as free-ranging animals. In the 1990s the USDA cracked down on dog and cat acquisition through dealers selling them to laboratories, once again correlating to an increase in naturopath practice and status.[38]

In the early twenty-first century the legacy of the movement from the previous two centuries persists. In 2007 the Pet Safety and Protection Act sought to amend the AWA to require that all dogs and cats used for research be obtained legally. The bill did not succeed, and its defeat is attributable in no small part to medical lobbying. Experts believe that in future decades the critical goals will be improved veterinary care and animal health, reduction of the number of animals used, monitoring of pain and distress, the use of primates (animals rights groups argue that none should be used), and genetically modified animals.[39]

Naturopaths co-led the powerful antivivisection work for a century. The primary rationale for vivisection was the advancement of scientific medical knowledge, but naturopaths vehemently opposed allopathic claims to authority and the right to inflict harm. Naturopaths also rejected the intellectual integrity of the animal model as it was applied to humans, believing that the results were neither relatable nor justified. Antivivisection was also another example of naturopathic liaisons forged with political health-conscious dissenters.

At its apex in the period 1870–1920, legislation by the antivivisection movement was blocked by a powerful medical lobby. Other forces weakened the movement during these years and beyond; the desire for social efficiency in the Progressive Era had glorified laboratory medicine, eventually entrenching it and making it nearly untouchable. But the movement was never eradicated. The

American Anti-Vivisection Society remains an active and effective organization to the present day.[40] Antivivisection activism expanded the ranks of naturopathic allies and potential followers of naturopathic healing.

VACCINATION: "A LITTLE SHORT OF LUNACY"

The story of naturopaths' opposition to vaccination in the twentieth century parallels the vivisection debate. By the 1890s naturopaths had joined and then led a chorus of alternative healers and health organizations, including the American Anti-Vaccination League, the American Medical Liberty League, the International Medical Freedom Association, and state leagues against compulsory vaccination. The antis' position became marginalized, but they were vociferous and united in opposition as scientific expertise won public confidence by midcentury. Antivaccination activism bonded disparate branches of the naturopathic system. It also solidified the support of naturopathic followers, practitioners, patients, and advocates.

The idea of vaccination struck naturopaths as absurd. They argued that it did not stimulate bodily protection but deeply compromised it. It could, and did, lead to death. Simply defined, a vaccine is "a substance that introduces a whole or partial version of a pathogenic microorganism into the body in order to train the immune system to defend itself when the organism threatens to cause an infection through natural means." But naturopaths did not see it as a boost to immunity; in their eyes, it was a bodily invasion. Their distrust of vaccines continued a long history of both British and American resistance to it.[41]

The introduction of vaccines in the late eighteenth century generated both support and fear. Many histories begin the story of vaccines with Edward Jenner, a country doctor from Gloucestershire, England, who introduced a smallpox vaccine derived from cows between 1770 and 1798. By 1801 nearly 100,000 Europeans had been vaccinated.[42] In the American colonies, however, 1721 was a watershed year for variolation, a method in which the smallpox virus, usually taken directly from the pustule of a sick patient, was scratched into a person's skin with a knife or other sharp object. Variolation was controversial because it was not always successful, but the practice continued. Later, American revolutionary leaders, including George Washington, Benjamin Franklin, and John and Abigail Adams, spread the practice more widely. By the time Jenner created his vaccine, inoculation through variolation had become well known in America. In 1809 Massachusetts became the first state to grant city boards of health the authority to require smallpox vaccination.[43]

Not all regularly educated physicians believed that vaccination was a cure. Benjamin Rush, a leading American physician, used intense bleeding and purging with calomel in response to the yellow fever epidemic in Philadelphia in 1793. In both Britain and the New England Puritan-based colonies, opposition to vaccines came from those who faulted the logic for several reasons: the belief that vaccines were unchristian because they came from an animal; distrust of the medical profession; disagreement about how disease was spread (some thought smallpox was caused by miasma, pollution, or decaying matter in the atmosphere); and the belief that government-required vaccination violated personal liberties.[44]

In 1800 vaccination spread in the United States. That year an outbreak of smallpox in Marblehead, Massachusetts, caused the townspeople to use the vaccine. The results were disastrous: the vaccine was either contaminated or inactive. Smallpox sickened a thousand people and killed sixty-eight of them. In the first half of the nineteenth century, physicians and some government officials generally favored vaccination, but not so the American public. Nevertheless, in the early nineteenth century Boston led the way in compulsory vaccination, fueled by fears that disease would spread through public education. By 1855 vaccination was compulsory in Massachusetts.[45]

The vaccine medical experiments on diseased cows completely appalled nature doctors, who valued the sanctity of animal life and rejected animal medical research. The vaccine was produced by infecting cows with virus harvested from other cows. Nature doctors also did not believe that animal and human bodily systems responded in the same way. Opponents' efforts led to the establishment of the Anti-Vaccination Society of America in 1879, the New England Anti-Compulsory Vaccination League in 1882, and the Anti-Vaccination League of New York City in 1885.[46]

What distinguished naturopathic involvement was its emphasis on the relationship between living conditions and disease causality and on right living to prevent disease. They were not the first to oppose vaccines, but their comprehensive written and verbal campaigns were unrelenting over the decades. Naturopaths invoked and recruited like-minded leaders to strengthen resistance to medical authority and what they saw as misguided science. In naturopathic publications the vast majority of titles of articles about vaccinations appeared in boldface and capital letters to emphasize their importance. The common naturopathic theory in the early twentieth century was that fever and disease should be allowed to progress once they took hold but that they could be prevented. The point that won favor with so many was that disease was exacerbated by unsanitary living conditions—along with poor nutrition and lack of clean air and sunshine.[47]

Naturopaths' philosophy at this time was consistent with their advice during the 1918 influenza epidemic: the patient should be put to bed in a warm, well-ventilated room, cleansed internally and externally, and offered a liquid diet, little milk, and much rest. In essence, "all acute diseases . . . scarlet fever, pneumonia, small-pox, diphtheria . . . are but the efforts of Nature to purify and rejuvenate the system if not interfered with by drugs, serums, etc. [They] will usually run their course, leaving the patient clean, but weakened."[48]

The prolific naturopathic author A. A. Erz, of San Francisco, wrote a lengthy article laying out the reasons why vaccination was antithetical to naturopathic principles. Perfect health yielded perfect immunity to all disease. And while vaccination might not immediately cause death, it left dangerous taints that would emerge years later. Naturopaths believed that vaccination could, in fact, *induce* disease, as evidenced by the eruptions brought on by the use of the calf vaccine against syphilis. One author called vaccination a gigantic, useless delusion. Unfortunately, naturopaths' vicious battle with allopaths over their right to be health practitioners obstructed their ability to see any benefit of vaccination at all.[49]

The AMA and ensuing health boards were also their own worst enemies when it came to selling vaccination as a worthwhile practice for communities. The heavy-handed, even draconian methods used to force an imperfect, invasive procedure on people went against the grain of democratic thought at the turn of the century. In the wake of the democratic Populist Movement, and in the midst of social justice reforms, the AMA expanded its control of health with methods that smacked of oppressive dictatorship. The AMA's public health doctors were convinced of a need for social control in their reform efforts at a time when the power of monopolies was under scrutiny.[50]

By the late nineteenth century, immigrants, people of color, and the poor were crowded into notoriously unsanitary tenements. These living situations were further "proof" to some white middle- and upper-class Americans that the poor and ethnic minorities were inherently diseased and lesser beings. There was also a moral aspect assigned to ill health: immorality caused disease and poverty; those with health and wealth had adhered to moral laws.[51]

Eastern cities struggled with the problems resulting from overpopulation in the poorest neighborhoods, home to record-breaking numbers of immigrants, along with migrants from rural areas, who had come in search of industrial work. In 1900 more than 2.3 million out of 3.4 million people living in New York City, or 69 percent, lived in crowded tenement housing. Tiny rooms often had no light or air circulation, much less sanitation. Health conditions were so bad in these buildings that, in the words of one report, "from the tenements there comes a stream of sick, helpless people to our hospitals and dispensaries, few of whom are able to afford the luxury of a private physician, and some houses are in such bad sanitary condition that few people can be seriously ill in them and get well."[52]

In the second decade of the twentieth century, naturopaths continued to bring home the point that *living conditions*, more than any other single factor, caused epidemic disease and that vaccination did not provide a cure. A physician from Texas said in 1914: "You grafters may swat the fly, bat the rat, kill the mosquitos and inject poison antitoxin and vaccine virus into the hides of innocent people, but if you fail to clean up your dirty, filthy nests and inaugurate a better sanitary and hygienic system, . . . no improvements will be known 'til doomsday." The belief in environment as the primary cause did not lose sway over the decades. In 1937 a naturopath wrote that man was born with natural resistance to disease, attributed in large part to his ancestors' physical vigor, but resistance could be radically changed by environment.[53]

The environmental impact upon public health was one of the very few points on which naturopaths agreed with allopaths. The public health specialty developed in the late nineteenth century. In 1847 the newly formed AMA conducted an investigation that found living conditions in American slums to be even worse than those in Europe. Hygiene and living standards needed improvement. This, the leadership thought, necessitated the collection of statistics, which led to the establishment of the Bureau of Vital Statistics at the turn of the century.[54]

Public health efforts were hastened by the tragedies of the Crimean War and the American Civil War. The needs of war brought about the rise of nursing as a profession, and battlefield nurses offered improved sanitary conditions that helped stop the spread of disease. The Civil War was a watershed in terms of concern about epidemic disease. The US Sanitary Commission—largely an effort designed and executed by women during the war—was founded to ameliorate soldiers' mortality rates. The commission confronted the grim fact that for every one soldier killed in battle, two died from dysentery, diarrhea, typhoid, or malaria. The American Public Health Association was founded in 1872, during the postwar period. By the next year the number of local boards of public health had increased from 4 to 123. They were institutionalized under the auspices of the AMA.[55]

Public health officials gained power and influence during the US smallpox epidemic in 1898–1904, but the early emphasis on sanitation gave way to policies of enforced vaccination during the epidemic. The egregious procedures and actions of officials prompted an outcry demanding medical civil liberties among Americans, prime among them the American Medical Liberty League naturopaths. There was certainly a tremendous need to curtail the spread of the disease, but one problem was that the vaccines were not uniform in quality, efficacy, and safety.[56]

Even AMA leaders made no secret of the violent coercion used to vaccinate people of color and immigrants. At the annual AMA meeting in 1899, a St. Louis bacteriologist bragged about the number of inoculations he had overseen. After one African American man had come to a city clinic with smallpox, the Health Department had "vaccinated the whole male negro population of the city, and as many women as could be captured." Vaccinators had stormed "barrel houses . . . low, filthy saloons [where black laborers] sleep on the floor" and forcibly vaccinated them. Border officials at Ellis Island and on the Mexican border vaccinated anyone who did not have a vaccination scar or show evidence of smallpox. In Philadelphia, hundreds of Polish, Italian, and African American laborers at work were yanked aside and vaccinated. There were similar scenes in other American cities—in schools, factories, and on railroads. In 1901 a smallpox vaccination raid was staged in New York City in the middle of the night. Two hundred fifty men descended on a tenement house in the Italian area and forcibly vaccinated everyone with whom they came into contact. Police held down men in nightclothes while they were vaccinated. Inspectors searched each room looking for children with smallpox and took them from their mothers' arms to place them in the city pesthouse for smallpox victims. In Middlesboro, Kentucky, police and vaccinators rounded up African American men and women, handcuffed them, and vaccinated them at gunpoint.[57]

As part of their antivaccine activism, naturopaths published accounts of tragic tales, personal disasters that resulted from vaccines. Real or fiction, precise or exaggerated, these victims' stories kept the battle alive. One such story was that of Lucilee Sturdevant, a healthy girl who died thirteen days after being vaccinated. Her father sued for $25,000 in damages—unsuccessfully. Broken and discouraged, he took to drink, and her mother was placed in an asylum. Countless letters to Lust as the *NHH* editor echoed the cry to halt vaccination.[58]

Graphic detailed accounts of deaths caused by vaccinations heightened rhetoric, fear, and resistance. In the winter of 1914 the Philadelphia-based Anti-Vaccination League sent the *NHH* details about twenty-five cases of vaccination disaster. The cases were drawn from Connecticut, California, Iowa, Pennsylvania, New York, Vermont, and Massachusetts. For each case, the person's name, location, and the medical condition that caused acute suffering or death after vaccination were provided. The conditions included smallpox, spinal meningitis, typhoid, acute diabetes mellitus, heart and kidney failure, abscess, tetanus, infant paralysis, and poisoning. Judging from the letters sent in response, these gruesome details had a strong effect on readers. As the decades went on, the tragic children's stories persisted, ending with "the wasted body of the child victim . . . on a slab in the hospital's morgue."[59]

For naturopaths, the egregious practice of forced vaccination was made more inexcusable by the fact that the government did not check the effectiveness or safety of the serums; pharmaceutical manufacturers were ungoverned. In 1902 R. Swinburne Clymer, MD, summarized naturopathic sentiment when he complained about the lack of proof of the vaccine's efficacy. He cited statistics from Cleveland, Ohio, where in 1901 seven people died of lockjaw from the vaccine. Clymer noted that as a result of these cases, the health officer had immediately stopped the crusade and instituted in its stead a house-to-house disinfection by sanitary officers. The officers had burned garbage heaps and sticks of formaldehyde and disinfected shops, factories, and schools, and they had placed those already infected in quarantine, but no one was inoculated. Because no cases had been reported in Cleveland in the last five months, he attributed the health improvements to cleanliness and disinfection.[60]

In 1905 a US Supreme Court case exacerbated the hostility between antivaccinationists and public health officials. Henning Jacobson had refused to be vaccinated, citing the Fourteenth Amendment. He was fined five dollars (the equivalent of roughly $139 in 2014). He said he had the freedom to judge what was right for his own body and that in the past he and his son had reacted poorly to vaccination. The fine was upheld on appeal. In *Jacobson v. Massachusetts* (1905) the US Supreme Court held that the state had authority to require vaccination during an epidemic that endangered the community. While it was not in the state's power to vaccinate by force, the Court said that "it was the legislature's prerogative to determine *how* to control the epidemic, as long as it did not act in an unreasonable, arbitrary or oppressive manner." The Court recognized that there were limits; children who had a certificate signed by a registered physician stating they were unfit subjects for vaccination could be exempted. The opinion held for a hundred years and was the constitutional foundation for governmental actions that limited personal choice in the name of public health. The most famous story of a forced public health action was the quarantine of the Irish immigrant Mary Mallon (1869–1938), better known as Typhoid Mary. Portrayed as a monster by the press, she was asymptomatic and may have infected as many as fifty-three people, three of whom died, through her profession as a cook. When she would not stop cooking for a living, she was forced to live in isolation, and in the end she spent twenty-six of her remaining years in isolation on North Brother Island, New York, where she died. She had been given no recourse to a free and normal life.[61]

The case of Typhoid Mary and other public health cases fed into the racism and xenophobia of the 1920s. Despite the ruling that vaccination could not be forced, public health officers used their authority to legitimize raids in the poorest neighborhoods and vaccinate when communities were under the threat of epidemic. During the twenties and beyond, the legal theory invoked in *Jacobson v. Massachusetts* was cited for vaccination and even forced sterilization procedures. The eugenics movement was growing amidst antiimmigrant legislation and the rise of the Ku Klux Klan. Against this backdrop, the 1927 Supreme Court case of *Buck v. Bell* allowed the sterilization of the "unfit," drawing upon evidence from *Jacobson*, which had upheld the use of compulsory vaccination. As the historian Arthur Allen put it, "Both public health and eugenics, in this instance, set their sights on 'final solutions' for eradicating infections and hereditary diseases." Between the antiimmigrant backlash and court decisions, naturopaths were up against powerful foes.[62]

These compulsory-vaccination measures led early naturopaths to cite cities' records in the journal when harm resulted from vaccination, particularly to children. They reprinted the original newspaper articles in which the information had appeared. Outraged, one author argued that when public schools required children to be forcibly vaccinated, schools should be abolished.[63]

In 1908, naturopaths made a big push against vaccines, publicizing case after case of injustices and the deaths that justified their position. When six-year-old Charles Holloway died after a tetanus shot, Dr. Randall, of Malden, Massachusetts, was named—a method of public shaming, implying that he and the policy that protected him were responsible for the child's death. Louise Jenkins was a public-school student whose father, not believing in vaccination, refused to comply. She was expelled from school, prompting a lawsuit to compel the school to readmit her.[64]

An editorial in 1908, likely by Benedict Lust, railed against the "State Medical Imposition" mandating vaccinations in the public schools. Students went to school for education, he asserted, not for medical treatment. "The doctrine of vaccination . . . is false in theory, and its contentions are refuted by experience. It is cruel in practice and deadly in its effects. It is a little short of lunacy to advise such a ruinous practice." As authorities in each state contemplated laws requiring vaccinations similar to those upheld in court, naturopathic authors followed their work and argued in each case against their imposition. In 1911, at the height of public health legal activism, several articles in the *NHH* decried the uselessness and dangers of forced vaccinations.[65]

A WAR OF WORDS—AND RESULTS

Throughout the second and third decades of the twentieth century naturopaths' arguments, not surprisingly, increasingly linked the AMA's vaccination campaign with its monopoly on medicine and eradication of sectarian practitioners. In response, many regular medical doctors mocked the naturopaths' position. In 1917 a lecturer at a meeting of the Philadelphia branch of the American Pharmaceutical Association held at Temple University, charged opponents of vaccination with being "ignorant . . . using underhanded methods of fighting . . . and championing the cause of pestilence."[66]

Naturopaths responded by calling allopaths lawmongers and undertakers' best earthly friends. In 1920 Professor Gilbert Patten Brown, a life member of the International Medical Freedom Association and holder of five medical degrees, railed against MDs' power to undermine sectarian healers. More laws meant less liberty for the masses. While condemning the AMA, naturopaths happily documented the practitioners who rejected vaccination—or who supported states that eased harsh enforcement. J. W. Hodge, a former public vaccinator, wrote that he had originally believed that vaccination prevented smallpox or lessened the severity of the disease. He had also advocated revaccination to ensure complete immunity from smallpox. But his work during an epidemic of smallpox in Lockport, New York, "drove all those stupid beliefs out of my cranium." He said that he had seen those he had vaccinated contract the disease and die from it, while some unvaccinated people had thrived.[67]

In 1914, to bolster naturopaths' hope in the face of health department successes, the New York attorney Harry Weinberger highlighted cases in which citizens' rights had been violated and public health powers had been curtailed, if only temporarily. Weinberger worked on behalf of an array of political causes involving individual rights, having counseled Emma Goldman and other anarchists. In the case of *Smith v. Health Commissioner Emery* (1896), William Smith's quarantine for possible exposure to smallpox was found unlawful on appeal. Smith also won a separate civil suit against Emery for damages in lost wages as a result of his quarantine. Smith, who operated a delivery service in the city's worst-infected districts, had refused to be vaccinated. The case revealed the disconnect between the reasoning of health department doctors, the law, and personal freedom. The public health lawyers argued that by law the health

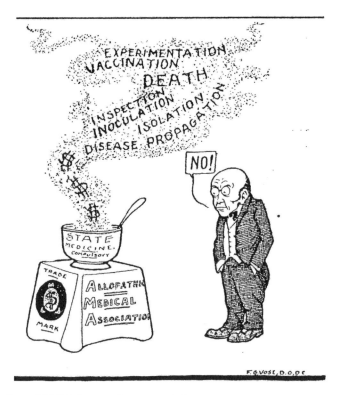

"Vaccination—NO!" A frustrated naturopath fumes over state compulsory vaccination campaigns originating from the AMA concoction, which spews money for experimentation, inspection, and inoculation and causes isolation, disease, and death. *Naturopath* 28, no. 9 (Sept. 1923): 505.

commissioner, with the approval of the mayor and the president of the local medical society, could "require the isolation of all persons and things infected with, or exposed to, such disease." They said that the rapid spread of smallpox and its dangers had justified Smith's quarantine. The court disagreed. Since Smith had not been diagnosed with smallpox and there was no proof that he had been exposed to it, the quarantine was indefensible.[68]

Another significant, if singular, victory for naturopaths was won in 1911, when the California Anti–Compulsory Vaccination League succeeded in ending the mandatory inoculation of schoolchildren. This was confirmed in 1920, when the Christian Scientists and the San Francisco Anti-Vivisection Society successfully defeated a compulsory-vaccination bill. But there were many exceptions: in the event of an epidemic one had to submit or go into isolation, and all students entering state universities had to be vaccinated.[69]

In 1943, in the middle of the World War II, H. B. Anderson wrote in the *NHH* about his 360-page *Facts against Compulsory Vaccination*, published in 1929, which had given naturopaths a voice outside their profession to back their beliefs. Written especially for average citizens, its aim was to end the campaign. Anderson noted with satisfaction a recent article by an MD discussing the dangers of vaccination in the *Journal of the American Medical Association*. By this time, however, vaccination was mandatory in New Hampshire, Massachusetts, Rhode Island, New York, Connecticut, New Jersey, Pennsylvania, Delaware, Maryland, Kentucky, South Carolina, and New Mexico and optional in Oregon. That the majority of states in which it was mandatory were clustered on the Eastern Seaboard and that naturopaths' headquarters were similarly located fueled naturopaths' animosity toward allopaths.[70]

Anderson highlighted the profitability of vaccines for medical doctors and pharmaceutical companies. His points were not new. As early as 1911 protesters had pointed out that vaccinations were utterly useless but provided fees for doctors and enriched serum manufacturers and undertakers. Benedict Lust added his weight in a diatribe: "So gross and greedy has modern medical practice grown that whole communities are now urged [to be vaccinated and have unnecessary surgeries]." Dr. Jesse Mercer Gehman estimated that physicians pocketed tens of millions of dollars annually from enforced vaccination.[71]

Even when naturopaths were unsuccessful, they managed to put organizations on the defensive. Members of the American Boy Scouts were required to be vaccinated before they headed to the World Jamboree in Holland in 1937. Naturopaths were joined in their outrage by the National Chiropractic Association and the National Health Foundation, as well as *Plain Talk* magazine. The order apparently came from the US surgeon general and the health officer of the District of Columbia. The naturopath E. W. Cordingley wrote an open letter to the Boy Scouts, telling them that most Chicagoans rightly believed that drugless healing was superior to medicine and that forced vaccination was foolish. The next month, James E. West, the chief scout executive, was compelled by the considerable outcry against the vaccination decision to write to *Plain Talk* and defend the decision, stating that it had been made by the executive board.[72]

Throughout the 1940s the column Anti-Vaccination News or Anti-Vaccination Reports was a regular feature in the *NHH*. In 1943 the journal denounced Iowa's compulsory-vaccination bill. To highlight the need for continued activism, the

NHH published a chart showing statistics for fifteen eastern states between 1938 and 1942 that showed a correlation between their mandatory vaccinations and death rates. Naturopaths concluded, based on the data, that it was folly to attribute elimination of disease to compulsory laws. Naturopaths also cautioned about the deluge of bills requiring vaccination against smallpox and diphtheria based on an emotional pretext that these diseases *might* reach epidemic proportions because of wartime conditions.[73]

THE RABIES SCARE: VACCINATION AND PROFITS

Humans were not the only sufferers of mandatory vaccination: naturopaths associated vaccination with the erroneous diagnosis of canine rabies and the profiteering from its treatment. Rabies was also called hydrophobia because the afflicted were believed to be afraid of water and unable to quench their thirst. It was a disease naturopaths felt was grossly misunderstood by medical scientists. Early naturopaths believed that the disease was transmitted not only from animal to animal but possibly by vaccination as well. They saw a link as early as 1911 between the fear of rabies and profits for drug manufacturers, medical scientists, and politicians.[74]

The cultural context in which the rabies scare occurred is revealing about American beliefs about dogs and the human-animal bond. It also reveals fears of potential cultural mayhem and urban dangers. In the early twentieth century, dogs became increasingly popular as household pets and as cultural icons. At the same time, a host of American urban ills were blamed on immigrants, from a multitude of diseases such as smallpox and tuberculosis to the Red Scare of 1918–19. Fear of immigrants fueled the eugenics movement and ultimately a set of anti-immigration laws by the mid-1920s. Loose dogs on city streets contributed to a sense of uncontrollable urban changes. *Rabid* dogs on the loose were alarming if it meant that a person's most loyal companion could be fatally infected and turn on him or her—an image used by antivivisectionists to gain support.[75]

Scholars have made some connections between the rabies scare and cultural trends. One argues that in the second half of the nineteenth century humans bitten by rabid dogs and cats were reported to take on animalistic behaviors that revealed a belief in spiritual connections between humans and animals. Only the destruction of a rabid dog or cutting off its tail was thought to protect the person from hydrophobia. "Canine hysteria" was a disease category used by veterinarians in the Western world during the 1920s and 1930s. Animals suffering from this disease, which was also called "fright disease," reacted as if they had seen a ghost and rushed away terrified. Michael Worboys speculates that this "disease" may have been a result of human hysteria caused by anxiety and stress, or a possible reaction to long-term domestication and new pet-keeping practices of the early twentieth century. The *Lancet* advised treatment remarkably similar to the prescription for human neurasthenia: rest and quiet, nerve sedatives, and— uniquely—a cereal-free diet. Some owners of dogs with hysteria feared that vaccines against distemper (among other vaccines) induced the behavior. Few agreed on its cause and cure. Given the remarkable resemblance between the rest-diet treatments described and those prescribed for middle-class women diagnosed with neurasthenia at the turn of the century, canine hysteria seemed to signify an anthropomorphic extension of female stress and despair or hysteria. In women, the condition so disrupted familial life and social order that many medical doctors prescribed the rest cure, a bland diet, and in many cases gynecological surgeries.[76]

With all these beliefs surrounding dogs and their relationships with humans, naturopaths and drugless healers absolutely opposed rabies vaccination and the killing of hydrophobic, or rabid, dogs. They blamed vaccinations as a primary cause of the disease, which is not surprising given that naturopaths were in the throes of an all-out war with allopaths over their right to practice, theories of disease, and barbaric vivisection practices.

Naturopaths argued that the widespread fear of mad dogs had been brought about by corrupt doctors and politicians who wanted to "diddle in the public purse." One antivaccination MD noted that of 56,000 stray dogs picked up in Philadelphia in 1911, none had rabies, but he failed to note that since pound dogs were sought for medical research, the threat might have been exaggerated by medical doctors so they could round up dogs.[77]

WARTIME IMPOSITION OF VACCINES

Antivaccination protests prompted Utah and five other states to abandon mandatory practices, what one naturopath called "this abomination of modern surgery," by 1917. But these gains were overshadowed by military policies instituted during World War I. All soldiers were required to be vaccinated against typhoid, smallpox, tetanus, diphtheria, and yellow fever. Military personnel who refused to be vaccinated were court-martialed. The efforts to protect soldiers from diseases during both world wars increased US medical authority at home and abroad. This combination of military and medical authority struck at the core naturopathic beliefs in individualism and antimonopoly. Naturopaths' high-profile opposition to military vaccination was supported by some Americans and mocked by many.[78]

Naturopaths saw forced military vaccination as medically dangerous, politically motivated, and yet another loss of medical freedom. Dr. Gilbert Bowman, director of the United School of Physical Culture in Chicago, was among them. A teacher of exercise and corrective diet, his column Department of Medical Freedom in the *Herald of Health and Naturopath* argued for the rights of citizens, including those in the army and navy. In his view, people should not be forced to submit to any type of medical treatment against their will. Others feared that the practice would lead to forced vaccination for a variety of ills. Other naturopaths called vaccination blood poisoning and blamed it for causing infectious diseases.[79]

In 1937 naturopaths explored worldwide reports of deaths from vaccines. They found that countries without mandatory laws had fewer deaths than those that enforced vaccination. Such disparate international policies also caused problems for well-to-do individuals who regularly traveled, garnering some publicity for the cause.[80]

As American involvement in World War II loomed, naturopaths and other antivaccinationists remained uncompromising. They were distressed and suspicious when in 1940 the Rockefeller Foundation board of directors sent a representative to Spain, France, Portugal, and England to meet with health officials and researchers from the American Red Cross and the American Friends Service Committee. Anticipating American involvement in the war, the Rockefeller Foundation in 1941 produced fifty thousand doses of vaccine per week. For opponents, this foreshadowed more unsavory and profit-driven bonds between big money and big government. The Rockefeller board recommended vaccines for influenza, typhus, and yellow fever, along with adequate food supplies and nutritional supplements.[81]

When the United States entered the war in 1941, the military required unprecedented numbers of vaccines for its personnel, reigniting naturopathic concerns about issues of medical freedom and therapeutic efficacy. By March 1942, 42,000 cases of vaccine-related hepatitis were reported among the troops in California, caused by nine lots of tainted vaccine; 84 cases were fatal. In 1942 the accidental infection of 300,000 military men with hepatitis B and the deaths of more than 100 from contaminated yellow fever vaccine exacerbated doubts about the efficacy of vaccines. One dissenting MD said that tainted vaccines had failed because they were too old or impotent or had been administered incorrectly. Naturopaths published a picture of two smiling servicemen, a sailor and a soldier, with the caption, "Men like these are being poisoned by dangerous vaccines and deadly inoculations."[82]

An irony of this vaccination policy was that President Roosevelt was advised by the White House physician *not* to be inoculated prior to his journey to Casablanca. How odd, noted one naturopath sarcastically, given the importance of vaccines for all service personnel. The same naturopath asserted that soldiers and sailors discharged from military service because of inoculation-related impaired health should receive compensation and pointed to tragic British military deaths induced by vaccinations.[83]

The author H. B. Anderson inspired the Citizens Medical Reference Bureau to write a letter to President Roosevelt and members of the Council of National Defense in 1943. They argued, backed by naturopaths, that those in the military should be free to be treated by chiropractors or osteopaths or the practitioner of their choice when they were at home. Secretary of War Henry Stimson was quoted as saying in 1942 that "once in uniform, the individual must be treated by medical corps doctors and surgeons who are qualified for membership in the American Medical Association." Several months later, in July, Albert Whiting, ND, bemoaned the inability of naturopaths, who were experts in nutrition, to determine whether certain service men needed increased rations.[84]

Throughout the war, naturopaths and individual medical doctors insulted and counterattacked each other. The psychiatrist and investigative reporter Albert Deutsch was a critic of the care that servicemen and veterans received. He denounced Dr. Morris Fishbein, editor of the *Journal of the American Medical Association*, and other AMA leaders. Fishbein had argued that only medical doctors could provide effective care. Fishbein then fired off an insulting and vitriolic response: Deutsch, he said, was completely without medical training and ignorant of both medicine and statistical methods. He said it was a pity that readers depended on his kind of medical information. Fishbein had dismissed Deutsch's considerable self-education and his seminal scholarly books on the psychiatric treatment of the mentally ill and the welfare of veterans. Fishbein's personal attack was typical of the kinds of attacks that occurred when the sects clashed publicly.[85]

During and after the war these salvos against military practices were mediated somewhat by naturopaths' patriotic gestures. *NHH* articles counseled ways to save rubber to aid the wartime effort, and numerous snippets urged readers to buy and wear a Buddy Poppy, funds from the sale of which benefited disabled veterans, as a way to honor the dead and help the living.[86]

THE POLIO VACCINE

Today there are two interpretations of the polio vaccination campaign. Some see it as a case of disease obliteration brought about by science; others see it as receiving far too much credit for the reduction and cessation of this health crisis. In the 1940s the nation's heartfelt efforts to eradicate polio were related to the image of, in the words of Arthur Allen, "the polio-crippled man who led the nation through the Great Depression and the Second World War, who had created the polio foundation while searching for help for his own tragic predicament." Inspired by President Roosevelt, the National Foundation for Infantile Paralysis, relying on the contributions of 100 million Americans, managed the attack on polio and funded most of Jonas Salk's research to create a vaccine. The attention was disproportionate, as deaths from both tuberculosis and pneumonia outnumbered deaths from polio at the time.[87]

Ironically, sanitation improvements had likely facilitated the disease's virulence; in earlier times maternal antibodies had offered protection, and infected infants had been spared from polio's worst effects by their own immune response. Polio epidemics in urban America overlapped with the availability of clean filtered water, precluding babies' routine exposure to the virus. In the polio epidemic's worst years, 1952 and 1953, 57,000 cases of polio were reported, with 3,145 deaths and roughly 20,000 cases of mild to severe paralysis.[88]

In the mid-1940s naturopaths took a decidedly unpopular stance against the polio vaccine and the American icon at the helm. Franklin D. Roosevelt inspired a presidential fundraiser in collaboration with the March of Dimes in 1944 that aimed to secure funds for the monkey-based polio vaccine research. Naturopaths responded with outrage, saying that research would involve the torture of thousands of helpless monkeys. They criticized the polio immunization program, saying that the live, active polio vaccine could cause polio—a correct claim that was evidenced in the cases caused by the vaccine. They also accused medicos and pharmaceutical companies of profiteering and described the use of polio vaccine as a "Wall Street–controlled medical tyranny and monopoly which raked in millions of dollars in profits at the cost of the lives of our children." Critiques of the vaccine were in keeping with naturopathic beliefs and an oppositional stance to nearly all AMA-derived methods. Yet naturopaths' inability or unwillingness to acknowledge statistics showing that vaccines clearly curtailed the disease cost them credibility and may have compromised their patients' health.[89]

Naturopaths' distrust of the polio vaccine culminated in condemnation of the March of Dimes campaign. In 1944 Benedict Lust claimed that the American people were "being taken for a buggy-ride." He questioned whether the National Foundation for Infantile Paralysis was a real charity or a money-grabbing scheme. When donations were made, Lust argued, those who had given—including him—had anticipated that good would come from their dimes; but after years of collecting, he wondered whether the foundation was truly philanthropic or a private clinic for a bunch of fair-haired medical boys who had not produced discernible results. He chronicled the successes of the nurse Elizabeth Kenney, of Australia, who had developed a highly successful method for treating the disease without medicine but had been denied permission to use her method in America. Lust's criticism of the foundation was hyperbole laced with some truth. He was correct in saying that some grant recipients spent lavishly. It was also the case, as Lust claimed, that animals, particularly rhesus monkeys, were captured, experimented on, paralyzed, and died or were killed for research purposes. Lust condemned what the foundation spent its money on: research trials, treatments, and testing the Salk vaccine in trials with children.[90]

The foundation was the mover and shaker behind the polio vaccine. It was pressured by Americans who had contributed money and time and wanted results. Then, to make the vaccine available, corners were cut when vaccine production was turned over to the federal government, with private pharmaceutical companies producing tainted vaccine when they failed to follow the foundation's laboratory criteria. Lust, a formidable orator and author, exploited these crises to reinforce his cause. He fully discredited the foundation despite its positive results.

The abhorrence of polio studies and of the presumed trickery of the March of Dimes was compounded because of the nefarious ways that research animals were procured. By the 1950s naturopaths did not equivocate; their journals railed against both the vaccine itself and the vivisection research that produced it. Rhesus monkeys were captured and used to grow strains of the vaccine; deaths and paralysis were documented among people who worked with them, and strains of cancer-causing virus from the monkey hosts caused new illnesses. By the mid-fifties more than one hundred thousand rhesus monkeys caught in the Philippines and India were being caged and sent to virology labs each year. They arrived diseased, screaming, dying, biting, and throwing their feces at their medical experimenters.[91]

In 1955 inactivated polio vaccine (IPV) was licensed, followed by the monovalent oral polio vaccine (OPV) in 1961 and the trivalent oral polio vaccine in 1963. Problems with the vaccine preceded this approval. Nationwide reports of induced infant paralysis *caused* by tainted vaccines emerged, as did reports of deaths. These results were known to the federal government and the National Institutes of Health, but the program was not stopped. Other nations, testing their lots, did stop the campaign. A Lilly scientist by 1960 had counted forty monkey viruses encountered in the labs. The fortieth strain was SV-40, "which demonstrably caused cancer in laboratory animals" and did not belong in a vaccine for children. Some researchers think that SV-40 may have caused mesotheliomas, an aggressive chest-wall cancer linked with asbestos poisoning.[92]

Problems with the vaccine have been largely written out of polio history, but they prove that naturopaths' consistent albeit often blind opposition to vaccines was not without merit. The sixtieth anniversary of the injectable vaccine was in April 2015, and discussions bear out the vaccine's history. Polio is still an international issue, and debates revolve around whether to use the cheaper oral vaccine, made from a weakened live virus and causing infection in about 1 in 750,000 cases, or the injectable version, made from a killed virus and still used in the United States. After years of using the oral version in low- and middle-income countries, injectable vaccinations are now on the increase in the hopes of eradicating the disease globally.[93]

1970–2015: THE DEBATE CONTINUES

Naturopathic views gained ground from some unlikely sources in late-twentieth-century America. In 1971 the president of the Infectious Disease Society of America, writing in the leading journal of that field, said that the decline in tuberculosis, diphtheria, pneumonia, puerperal sepsis, and other diseases was not attributable primarily to medical care and vaccinations but correlated with better socioeconomic circumstances and less crowded population conditions.[94]

In 1972 the US government ended mandatory smallpox vaccination, largely because smallpox had been nearly eradicated. During that decade, most diseases that could be prevented with vaccines had been curtailed, and controversy flourished. Oppositional voices claimed that vaccines were unnecessary, dangerous and caused brain damage and increased chronic illness. In 1976 fears of a global swine flu pandemic led to the vaccination of 40 million Americans, and then the pandemic did not occur. The government paid $110 million to citizens claiming to have contracted autoimmune disease as a result of vaccination. In 1986 a National Vaccine Injury Commission Compensation Program made it possible for children injured from vaccines to file for government payments. Undeterred, advocates continued their efforts. In 2002, President George W. Bush staged a media moment in which he received a smallpox vaccine. This act was based on perceived threats of bioterrorism as the United States prepared to invade Iraq. The politicization of vaccination was not new, of course.[95]

In the late twentieth century ambivalent and conflicting views of those educated as scientific naturopaths, on the one hand, and lay naturopaths and natural healers, on the other, flared on the issue of childhood vaccination. Even among scientifically trained naturopaths there was not universal agreement. In 1991 the American Association of Naturopathic Physicians published their *Position Paper on Childhood Vaccinations*. It was necessary to take a professional stand—to distinguish themselves from lay practitioners, to set guidelines that would address their differing viewpoints about vaccination, and to ward off accusations by MDs that naturopathy's philosophy was delusional, not scientific. The position paper asserted naturopathy's strong moral and legal commitment to upholding public health laws and to preventive medicine—to protect patients from the consequences of infectious disease. It stated that naturopathic primary care physicians should be legally able to administer vaccinations wherever they had legal licensure. Further, because of morbidity rates and questions of efficacy, physicians were ethically obligated to give parents current and accurate information about the risks and benefits of childhood vaccinations so that parents could give informed consent. The association's position was that safer, more effective vaccinations should be developed; that physicians should consider conditions that contraindicated vaccine use; and that parents should sign consent forms that fully disclosed the pros and cons. The paper, which was updated in 2011 with the same position and new nomenclature, has triggered much dissent.[96]

Naturopaths trained at accredited schools have good reason for a well-thought-out professional stand, despite dissent in the ranks. Two MDs writing on Quackwatch in 2011 openly ridiculed the official naturopathic position on vaccinations. This Internet site monitors and determines what one MD considers good and bad medicine. In December 2010 the site had 11 million hits on its home page. Its influence, therefore, cannot be underestimated. The site misrepresented the naturopathic position, writing that the position paper had said "that vaccinations are dangerous and unnecessary and that parents should be encouraged to avoid them. This is one reason why NDs are . . . 'held in low regard.'" There is also contradictory information on the site. One author said that NDs educated in accredited schools did not oppose vaccination.[97]

Among a significant minority of Americans, suspicion and rejection of vaccines has grown. According to the historian Michael Willrich, in 2003 the Centers for Disease Control reported that "22 percent of American parents of young children were refusing one or more vaccines for their children. Five years later that percentage had nearly doubled to 40 percent of all Americans." Willrich believes this increase means that the vaccination debate is of prime importance and challenges public health practices.[98]

Some prominent pediatric disease researchers are also less certain of vaccination's efficacy. One author reported that the American Academy of Pediatrics was reconsidering the carte blanche use of polio vaccine; he echoed the position of naturopaths and noted that 130 vaccine recipients had been severely compromised or died from VAAP (vaccine-associated paralytic polio) in a period of seventeen years, while there was not a single case of wild polio in the United States. He reported that several countries had begun abandoning the vaccine for school-age children as early as 1957.[99]

In San Diego County, California, the number of parents opting out of vaccinations for their children beginning with kindergarten in 2010 was four times the number who had done so in 1990. Parents made this choice based on religious beliefs or distrust of the vaccines. The county's exemption rate was 2.64 percent in 2010; that is, 10,280 children were granted exemptions. One mother who had opted out preferred to rely on breast-feeding, proper nutrition, and exercise for her three children, practices suggested by naturopathy 120 years earlier. At least one local pediatric infectious disease specialist has evidence that such practices led to increased outbreaks of measles and whooping cough.[100]

Vaccinations are still required for some US populations. As of 2006, thirty-one vaccinations were required for US military personnel. These range from single doses to annual doses to one dose every ten years. As Arthur Allen quipped, in America "vaccination is the first act the state requires of a person; without it, or a legal exemption, a kid can't even get into nursery school."[101]

In 2015 there continue to be questions about the mandatory use of vaccinations. As of this writing, the antivaccination movement has a wide range of supporters, from high-profile celebrities to the tens of thousands of claimants who have blamed autism and bowel disease on the MMR vaccine or on thimerosal, the vaccine preservative that contains ethyl mercury, even though the research linking that cause and effect has been thoroughly discredited. Reuters reported back in 2010 that *Lancet* had formally retracted faulty 1998 research, and the physician who published it is no longer allowed to practice. California has been the leader in the public debate, which took a sharp turn with the passage of controversial state legislation in June 2015 requiring that all schoolchildren be vaccinated against measles and common preventable illnesses. It eliminates the "personal belief exemptions," which allowed parents to opt out of vaccination.

Bringing the topic into the public eye was a measles breakout at Disneyland and in Santa Monica. Officials had been warning of infectious disease perils since whooping cough in 2010 and 2014 affected eleven thousand people, causing illness and death. As one of only twenty states allowing parents to opt out, and one depending upon tourism, it was inevitable that the Disneyland breakout would cause lawmakers to act.[102]

Recent public distrust of vaccination has caught up with arguments naturopaths made more than a century ago. It continues to be a key concern for practicing naturopaths. What began as wholesale rejection of the mandatory practice and the medical and political liberties it violated has evolved into a more complex, evidence-based position. It is still viewed with caution and is a cause for dissension among naturopaths on the national and international levels, just as it has been in the public at large.

Publicity about opposition to vaccination obscures the obvious tremendous benefits that have resulted from vaccines. They have curbed and in some instances eradicated diseases that killed countless people throughout history. Human losses before them were immense; and while their enforcement by public health officials was heavy-handed, sometimes racist, and often unwelcome, the result was a decline in fatal diseases. The means did not justify the end, but vaccination is one of the most important public health measures in recent times.

Vaccination, like vivisection, is entrenched in current orthodox medicine. Both are recognized as furthering scientific knowledge and promoting human health. But naturopaths have left their mark on the practices, serving as America's watchdogs in solidarity with other activists to ensure humane and transparent processes and debate. As medical researchers today argue the need for mandatory vaccination and continued vivisection experiments, naturopaths will continue to weigh in on environmental factors and research alternatives to be addressed at the same time.

CHAPTER 8

Legal Battles

Democracy or Autocracy?

Naturopaths were constantly faced with arrests and lengthy trials as they ran afoul of ever-shifting local laws. They spurned legal bureaucracy and authority, adopting an independent, grass-roots, self-help approach to healing. They believed health was a personal matter involving personal choices. Because they embraced this identity as outsiders to professional medicine, they were lumped together by the AMA and by governmental regulatory agencies with a plethora of dangerous quacks, charlatans, and snake-oil salesmen. The irony was that naturopaths' democratic beliefs about medical freedom and individual personal responsibility led to their legal persecution during and after the Progressive Era (1890–1917), one of the greatest periods of democratic and antimonopoly reform in US history.

Benedict Lust was personally and painfully familiar with legal persecution. A stalwart protector of other sectarian healers struggling to practice, he saw an urgent need for professional cohesion. He strove for organized unification against the AMA, and he worked for the passage of legal statutes to protect naturopathic practitioners. In his railings against the AMA and its "Medical Trust," he used Progressive Era language and the antimonopoly, anti-big-business ideas of progressive reform. He meant that allopaths, in the manner of Standard Oil and U.S. Steel, were amassing great power and wealth and asserting their dominance to force out competition and centralize power. At the same time, AMA physicians were also key players as progressive reformers. They realized that government could be the key to realigning power in the public sector. Like other reformers who were using government to protect citizens and to gain control of corporate behavior (e.g., in creating the Food and Drug Act), the AMA made use of government agencies, law enforcement agencies, and multiple medical organizations. In the case of the AMA, teaming with government agencies allowed it to centralize its power by targeting and persecuting outsider medical practitioners.[1]

Therefore, both naturopaths and the AMA saw themselves as champions of progressive reform. That was not unusual at the time. Many progressives wanted to increase the influence of the common people, while other reformers wanted to expand experts' authority. Naturopaths championed the cause of unfettered self-determination against the monopoly of the AMA; the AMA saw itself as the single legitimate scientific body of experts that could clean up public health. But naturopaths' democratic logic had blind spots. They argued against restrictive health practice regulations but could not arrive at their own standards because of the number and diversity of their movement's constituents. They risked underprotecting the public from harmful practices and bogus cures. Naturopaths' initial inability to agree upon standardized expertise made their patients vulnerable.[2]

Naturopaths also refused to consider scientific medical efficacy because allopaths' claims to superior knowledge were sometimes unfounded. Their treatments could be more damaging than those of healers in the popular health movement. Allopaths' successful innovations increased *after* their campaigns for professional legitimacy and authority. In the twentieth century, allopathic medicine turned to controlled clinical trials. While the early attempt at controls was sometimes more theory than practice, effective knowledge was emerging from scientific medicine, and it was ignored by naturopaths. The AMA's aim for total dominance blinded early naturopaths to all things allopathic. The naturopaths had argued that living conditions and nutrition had an enormous impact on public health, but they believed that the medical trust had the power and authority to pronounce their own truths and advance their own agenda. Naturopaths rejected much of the scientific innovation, then, because of their longtime distrust of allopaths and their core philosophical differences about health and disease. In short, naturopaths' unyielding refusal to accept successful scientific advances, coupled with incessant wobbling on their own standards, compromised their efforts to gain political and legal legitimacy.[3]

Lust, who was the foremost naturopathic leader, came to believe that in order to advance the naturopathic mission in the early twentieth century, legal legitimacy was a pressing and primary goal. His tireless will and frequent trips to other states to lobby for legal recognition earned Lust credit for securing protective laws in more than twenty-five states, most notably the naturopathic statutes still in place today in Oregon, Washington, and Connecticut.[4]

The successful albeit sporadic and uneven establishment of laws for naturo-paths helped them survive in increasingly hostile sociopolitical circumstances. For many years nearly all naturopaths labored with a difficult and exclusionary legal status. Success depended upon two things: the ability to successfully orga-nize a viable local campaign and the popularity of naturopathy and its practitio-ners in a given place and time.

THE AMA AS A LEGAL LOBBY

Naturopaths railed against what they believed was the controlling political ele-ment of the AMA working in concert with affiliated medical organizations. One article urged cooperation among the vast array of fighters who opposed medical rule. They saw a collusion between the AMA and state boards of health, which recognized only allopathic physicians as legitimate medical practitioners. To make things worse, patients risked public exposure about their condition when legal charges were brought against drugless practitioners they employed.[5]

Naturopaths lambasted the 1920 "Better Babies" campaign, an intensive coun-trywide campaign by, in their view, political doctors to bring the nation under autocratic medical control and provide harmful treatment to the new genera-tion. Naturopaths said that the weigh-ins of babies falsely declared some of them undernourished and that allopaths urged mothers to give their children devital-ized milk and starch to fatten them. While these programs may have been well intentioned, they triggered deeper, darker concerns. Would allopaths keep com-ing up with ways to classify people as normal or abnormal? Left unchecked, the AMA and state boards might appoint themselves to "end the existence of degen-erate babies." The 1917 film *The Black Stork*, promoted by AMA eugenicists, had emphasized the point. Who would decide who was a "degenerate"?[6]

Naturopaths saw ample evidence of a dependent relationship among health organizations that created a self-fueling engine for social control. They saw the AMA as the "parent" of the Red Cross, a fundraiser for many programs, and they saw the AMA and what they called its other progeny—the American Hospital Association, the American Nurses Association, and the Children's Hospital As-sociation—as puppets of the medical trust and players in the nationwide vaccina-tion efforts. "The AMA leadership talk like lawmakers, say what THEY want, and pretend to decide what shall and what shall not be done concerning the public welfare." In short, they were police, judge, jury, and legislature all rolled into one under the guise of public advocates.[7]

Naturopaths also decried injustices at the hands of insurance companies. An Illinois practitioner, complaining about companies rejecting billing from drugless practitioners, suggested boycotting them. They should be reported to state and local naturopathic associations, he said, so organized action could be taken against them. By the early 1920s, naturopaths had clearly identified the AMA, public health boards, and insurance companies as enemies of drugless practitioners.[8]

Lust and his followers targeted the AMA and those they dubbed its legal henchmen. As he wrote in 1925, "There never was a cult, a religion, a political party, a government, a king, a soviet or an institution so narrow, so egotistic, criminal, materialistic and selfish as the medical trust." This tension had been mounting. For years naturopaths had harped on the medical monopoly and its seemingly antidemocratic elitism.[9]

Inherent in these critiques were key goals also articulated in the twenty-first-century debate over health coverage: patients' right of choice, freedom of expression and competition, limiting for-profit insurance companies and medicine, and defining the parameters of scientific expertise. Early naturopaths believed these goals, and the values they represented, were encompassed in "Medical Freedom" (capitalized to denote its importance). Naturopathic schools taught the *art* of healing but not the *business* of healing. The allopaths, said naturopaths, were focused on their own corporate capitalism, to the detriment of the public welfare. They used and abused laws to keep the sick from receiving the best care. This made them frauds who damage patients' health. Other reformers had successfully argued about the greed of big business at the turn of the century. It was used effectively by Lust and others against allopaths who had charged them with quackery. "Medical Freedom" became the clarion call among the various sects. The author Bernarr MacFadden (1868–1955), known as the father of physical culture, argued eloquently in 1920 that people had religious freedom and freedom of speech but not medical freedom.[10]

A meticulous record of state legislation enacted by the so-called medical trust to expand its power was a feature of naturopathic publications. Articles detailed laws such as the 1922 Texas law prohibiting all except regularly licensed medical men from practicing the healing art. Publishing the laws to outrage fellow naturopaths, natural healers, and their patient followers was one way of gaining support for an organized legal plan of their own.[11]

Naturopaths attacked what they saw as medical scientists' strategies to gain control. These included the AMA's links with the Rockefeller Institute of Medical Research; the exorbitant fees MDs charged to appear exclusive and to fuel their lobbying; prosecutions (often personal) of sectarian healers; state medical interference; submission of stories to the press; restricted medical education;

proposed national health insurance coverage that would privilege only the AMA; and the burgeoning bureaus of public health, which mandated vaccination. There were several methods of fighting back: asserting the legal rights of drugless practitioners at their arrests and trials, forming state associations to design protective laws, gaining public recognition of their professional schools, creating the Legal Department column in the *Herald of Health and Naturopath,* and calling for standardization in naturopathic education and ethical practices.[12]

Mindful that drugless healers were lumped together with a whole host of profit-seeking quacks, naturopaths had to walk a fine line when criticizing the persecution methods of allopaths. How were they to fight persecution, maintain their stance as outsider independently trained practitioners, and at the same time distance themselves from dangerous swindlers and protect their patients' health? After all, when charlatans were persecuted, so were the naturopaths, since all were outside the bounds of AMA acceptance.[13]

LEGAL STANDING AND HARASSMENT ARRESTS: THE EARLY YEARS

The legal standing of drugless practitioners was naturopathy's Achilles heel. By 1921, according to Benedict Lust, the legal situation in many states was deteriorating. No naturopathic school could stay open. In addition, the ANA was being infiltrated by defectors from the medical trust and by upstarts who undertook natural medicine after failing at everything else. This complicated the group motivations for legal and political action. Naturopaths used the term *drugless practitioners* interchangeably with *naturopaths* in the early decades of the twentieth century, which in itself was a problem for them. Many, if not most, used botanicals, for which they could be arrested under the charge of practicing without a medical license. What constituted drugless was up for interpretation by practitioners, allopaths, and enforcement agencies alike. But Lust's philosophy was all for one and one for all: any self-proclaimed nature healer was welcome to join forces with naturopaths against the centralizing power of the AMA. The umbrella term *drugless practitioner* was useful when employed by state legislators to grant legitimacy to practitioners; it was a problem when quacks employed it. When local agencies clamped down on one sect or another, referring to them all as quacks, it also spurred infighting among the different sects.[14]

Writers and advocates for drugless practitioners in the second and third decades of the twentieth century fought for legitimacy on a number of fronts. One tactic denounced states' medical practice legislation, which limited legitimacy to allopaths. Only when a modality was unnamed or specifically exempted from an act (e.g., massage or Swedish movement) could natural healers make a diagnosis and treat the patient. This rationale fueled mutual cooperation among drugless practitioners, since the rights accorded to one group could theoretically be extended to another. A particular point of contention in the acts was the insistence by allopaths that all drugless healers, including naturopaths, must apply for new licenses that would be granted only to those holding diplomas from recognized allopathic schools. This would be analogous to asking all "ministers of all denominations to take their examinations under a Baptist Board of Examiners." After much intense lobbying at considerable expense, some bills were amended and the old licenses remained valid.[15]

Another tactic of resistance was voiced by the Constitutional Liberty League and the National League for Medical Freedom, among others, which rejected the need for any licensure. The Constitutional Liberty League rejected *any* regulation of drugless practitioners on the basis of states' rights, a representative democracy, and a total denouncement of the growing powers of government at all levels. This was a compelling theoretical argument, but one detached from the nitty-gritty battles being waged in early-twentieth-century legal arenas.[16]

As the number of laws asserting allopathic dominance increased, the prosecutions against drugless healers, naturopaths in particular, escalated accordingly. The most common charge was practicing medicine without a license. The charges were so pervasive that they shaped the naturopathic movement. The accused were often fined, sometimes jailed. Many successfully argued that their methods did not replicate those of the medicos; some argued that since they did not charge for their services, they were not practicing medicine.

It was not always the treatment of patients that constituted a crime. One New York City man was charged and convicted of practicing medicine without a license in 1908 for manufacturing and selling health foods. After two years, with numerous appeals and at considerable cost, the charges were dismissed on appeal. The man maintained a lively correspondence about his legal battles with the *Naturopath and Herald of Health*. When vindicated, he wrote of his joy in "establishing a precedent in favor of the constitutional rights and liberty of the citizen and against medical tyranny." Benedict Lust saw the importance of the precedent, saying that it had established the right of an American citizen "to use his own intelligence and scientific knowledge to relieve suffering and better the condition of his fellow men."[17]

There were other wins. Charles McFadden, ND, in Pomeroy, Washington, had been convicted of manslaughter in the death of a child. He had treated *and cured* the child of "summer complaints" by naturopathic means. Three months later the child was taken sick again; after treatment by a medical doctor, the child died. The allopaths urged the family to charge McFadden with manslaughter, claiming that the child had died from his naturopathic treatment. Convicted, McFadden, who was not a rich man, with the support of friends and acquaintances took the case to the Supreme Court, where the decision was reversed. This ordeal, commented Lust, cost McFadden thousands of dollars and immeasurable stress. This victory was held high as proof that naturopaths who did not promote themselves as MDs and did not prescribe medicines were practicing legally. Similar victories were won in the state supreme courts of New York and North Carolina.[18]

Some judges also sided with naturopaths when they were deliberately entrapped. In 1908 the New York osteopath and naturopath Carl Conrad was charged with practicing surgery without a license by the county medical society. He was arrested and convicted, and then his sentence was overturned. The charge was so broad that it implicated any and all aspects of restoring health by lay people. The crimes Conrad was charged with included giving massages, practicing water cure, giving dietetic or physical fitness advice, and giving footbaths or hair washes, all of which were called medical practices for purposes of the arrest. Conrad, like many others, had been entrapped by private detectives working for the medical society, who posed as sickly and weak and requested treatment. One judge wrote that convictions thus obtained were not justifiable.[19]

Naturopaths tried to address the political and personal realities of these arrests. The medical practice acts were originally intended to set standards for allopaths whose education was questionable. Only recently had these acts been rewritten to apply to drugless physicians—this time, in naturopaths' view, for a medical monopoly. The personal toll of these battles was high, since natural healers who subscribed to traditional methods were not getting rich from treating patients. One naturopath, married with six children, was serving a year's jail time with hard labor for applying natural methods of treatment. His wife was unable to pay the mortgage or provide food, coal, and necessities for their children. There was a lengthy list of contributors for his $550 fine ($13,400 in 2014) in the NHH.[20] In a 1923 retrospective article, Benedict Lust called the period 1898–1910 the worst time for prosecution. His belief was well founded. In 1906 the AMA created a Propaganda for Reform Department to track and compile reports about critics of allopathy and outsider practitioners. One author has called the department a "surveillance machine, encouraging networks of physicians to infiltrate and report back on the 'forces of evil' emanating from 'nostrums,' 'patent medicines,' and 'quacks.'" On many mornings during that time more than a dozen drugless doctors were lined up in the criminal court in New York City, each found guilty and fined $250 ($6,430 in 2014). Lust was proud that no naturopath had ever pleaded guilty. The year 1915 was a pivotal year. Lust wrote about his own countless legal persecutions, his fines paid and time in jail. That year two drugless practitioners, a neuropath and a mechanotherapist, were arrested; three chiropractic cases from 1914 were pending; and according to one author, the police department had become the enforcers for the medical trust. Another drugless physician was arrested for advising a patient to take a bath once a week, "use Epsom salts or castor oil about as often, and keep away from rum holes and bad women." Lust liked to report on the persecutions faced by drugless healers who did not self-identify as naturopaths. He saw the fate of one inextricably connected with the fate of all—a theory that would ultimately threaten the naturopaths' legal standing and public credibility. In 1915 public opposition rose against these events. The influential and widely read *Life* magazine sympathetically remarked that "it will soon be illegal to come into the world—without the aid of a doctor! It will also be considered prima facie evidence of crime if you die—without the aid of a physician!" The *NHH* reported that a gathering of prominent citizens was considering legal proceedings against the New York courts and the medical trust for depriving them of their right to choose their own healers. There was no followup to the article, but clearly the public had noticed the powerful tactics of the AMA.[21]

In cases brought to trial against naturopaths, practitioners generally represented themselves and claimed legal entrapment by agents of the AMA, who had sought advice for an ailment or purchased an item such as a salve or food. One Florida naturopath reported in 1920 that he would fight to have his fine and conviction of practicing unlawful medicine overturned because he had been framed. The defendants all admitted that medicine was the purview of the AMA and that drugless healing did not employ AMA therapeutics. These were effective arguments in some instances, but most of the accused were discredited, fined, and/or jailed. Drugless healers who outmaneuvered the legal system were often subject to repeated arrests, escalated charges, and larger fines.[22]

Legal persecutions thus ranged from individual sagas like these to efforts to delegitimize specific groups of drugless practitioners. In 1914, in Duluth, Minnesota, drugless practitioners fought an ordinance that was intended to rid the city of quacks. The medical and osteopathic doctors claimed that quackery included drugless methods such as naturopathy, chiropractic, neuropathy, and mechanotherapy. This case highlighted the swing position that osteopathy occupied. In some states it aligned with allopathic practitioners, while in others it too was deemed quackery by MDs.[23]

As a naturopathic leader and politico-legal activist, Benedict Lust became a prime target for multiple arrests. In all, he was arrested sixteen times by New York authorities and three times by federal agents. Once a newspaper headline simply read, "They Have Lust Again." For his part, Lust publicly attacked the AMA's most notorious private detective who entrapped drugless healers: Mrs. Frances, or Franconia, Benzacry. In 1915 the *NHH* carried an exchange between Benedict Lust, writing on behalf of chiropractors and the American Naturopathic Association, and the *Ladies Home Journal*. He was angered by the *Journal's* inclusion of the article "What I Found Out as a Medical Detective," written by Benzacry, named as a detective of the New York Medical Society. Lust detailed her unscrupulous tactics of entrapment, faking symptoms and requesting treatment, and he offered to provide proof of her perjured court testimony. In response, the *Journal* expressed regret that Lust had read her article as discrediting all chiropractors, saying that that had not been the article's intention. In fact, the correspondent wrote, chiropractic had done much good. Benzacry, after years of entrapping drugless healers as an AMA undercover agent, later published her life story in the *New York Times*. She admitted that none of the drugless treatments she had received had done her any harm. Her *Ladies Home Journal* article appeared to be just another self-aggrandizing activity.[24]

Lust had publicly insulted Benzacry in a variety of ways over time, but when he wrote in 1921 that she was "a disgrace to American womanhood and to the free soil of America on which she treads," he was arrested for criminal libel and released only after a $10,000 bond was posted. Lust prevailed in this libel suit, which was important because the AMA had sought to prove that it had the authority to prosecute drugless practitioners. But the AMA had no charter to exercise police power. The case was dropped, and prosecutions against practitioners slowed down, much to the relief and satisfaction of Lust, naturopaths, and the other sects.[25]

Lust learned another important lesson, one that he did not always remember in the coming years: that the term *naturopath* was solidified in the process of the libel suit and was a powerful tool for professionalizing. Naturopaths could not use the words *cure, healing, therapist, physician,* or *doctor* in reference to themselves or their practices. *Naturopath* seemed to be the only term having to do with natural methods that they could safely use, and by "this term we did not hold ourselves out as practitioners. In one year we had more than fifty arrests. The word 'Naturopath' was the magic word that set us free." Unfortunately, legal expediency in the coming years led Lust to embrace other terms in a variety of scenarios.[26]

EXCLUSIONARY LICENSING LAWS

Naturopaths galvanized themselves regionally, albeit in very different ways, against the abundant legislation against them. At the core of legislative attacks were the basic science laws. Before sectarian practitioners could take a licensing test in their field of practice, they had to pass a general examination in anatomy, physiology, pathology, and other areas of science—usually with knowledge obtained from an allopathic medical school education. In actuality, the laws were designed to prevent naturopaths from obtaining licenses. As Lust put it, "Most states make no provision for examining our kind. Their laws say that in order to qualify for taking the examination, we must graduate from a regularly organized Medical College." Naturopaths argued that preventing them from taking the tests was unconstitutional. They supported the Declaration of Medical Liberty Rights, which emerged in a variety of venues and argued for medical freedoms in all areas, including testing.[27]

Naturopaths' attempts to be legitimately licensed were chaotic, since so many modalities fell under the massive umbrella term *naturopathy*. In California, as in other states, infighting among drugless practitioners ensued in the early 1920s, each eager to blame the faults of the others for their legal battles. Articles in the *Naturopath and Herald of Health* cautioned practitioners that such fighting weakened them in the eyes of the laymen and "puts a sword in the enemy's hands. There has been entirely too much mudslinging, ridicule and false accusation." As a result of the infighting, the old California naturopathic association split up. Some of the practitioners even claimed legitimacy to dispense drugs under a narcotic law. This was a black eye for all of naturopathy and those licensed under the 1913 California Medical Practice Act, allowing for a "drugless practice certificate," which sheltered naturopaths. One author urged leaders to support the act as the preferred licensing method and to rebuild the American Naturopathic Association. He was willing to embrace chiropractic and osteopathy as branches of naturopathy if they would confine their practices to their own schools and refrain from disreputable advertisements and rash promises.[28]

Complicating the legal infighting were Lust's occasional overstated claims. In 1927 he wrote that naturopathy had been legalized, directly or indirectly, in twenty-eight states and that those states had good schools with excellent standards. Yet legal recognition existed alongside ongoing persecution and harassment that belied any assertion of legal legitimacy. He also overlooked naturopathic schools' widely differing curricula and quality. At the 1927 national ANA convention Lust misleadingly portrayed reciprocity among natural healers, saying that "co-operation was also given to the newer off-shoots of the old Nature Cure Methods." In reality, at the time there was vicious internal strife and self-aggrandizement among straight chiropractors and osteopaths—at the expense of naturopaths. As a result, a growing number of naturopaths had come to distrust the straights and wanted to separate from them completely. It is likely that Lust portrayed harmony within and surrounding the naturopathic movement to present a united front to its critics. The division of naturopathy after Lust's death in 1945 into at least six different national organizations was proof that Lust himself had buffered the tension.[29]

There was a gaping disconnect between Lust's strategy for standardization and legal legitimacy, on the one hand, and his traditional view of naturopathy as a champion of all self-healing or nature healing modes, on the other. He continued to define naturopathy in 1927 as "a practice without fences or arbitrary boundaries. For it includes any and every method of preventing disease and healing the sick that is in accord with Natural Law." While this was his definition, it was not one his colleagues always supported. Even he vacillated on it increasingly as frictions escalated. As early as 1915, ambivalence about sectarian cooperation was visible in snippets in the *NHH*. Articles praised eclecticism for its fundamental broad-mindedness; a congenial visit between the New Jersey College of Chiropractic's dean and graduates of Yungborn was reported; and a speaker on physical culture was praised and admired. Yet writings before and after these pieces disparaged an osteopathic amendment in New Jersey that had written out the other drugless schools. One piece complained about straight osteopathy's disdain for chiropractic; another objected to an Ohio licensure bill's accidental adoption of the term *neuropathy* instead of *naturopathy*, which the bill originally was supposed to protect. Another article claimed that chiropractic was superior to osteopathy, mechanotherapy, neurology, magnetic healing, any form of massage, or any motion cure or muscular exercise. If these messages were not garbled enough, readers could go on to read a report about osteopathic alignment with regular medicine or yet another on the value of eclectic osteopathy.[30]

THE NEED FOR STANDARDS: "DOUGH WITHOUT YEAST"

By the mid-1920s one consistent message emerged in these legal reports: the practice of naturopathy should be standardized to achieve licensing legitimacy. The problem was to determine who could claim to be a naturopath and how to arrive at standardized measures. In general, the 1910 Flexner Report accelerated changes already under way. Most people in sects who had completed a rigorous course of study agreed with allopaths that uniform standards were needed to curtail quackery. What the naturopaths, homeopaths, hydropaths, eclectics, and others had not counted on was that an improvement in their standards would prompt allopaths to more formally exclude them, leading to the demise of some of these sects.[31]

Standardization of medical education and licensure came through basic science acts passed at the behest of state AMA affiliates. These acts decreed that only standardized basic science learned at AMA-approved schools qualified one to sit for an exam. The basic science acts, according to the naturopaths, originated with standards established by Chicago's Rush Medical College in 1907 requiring two years of high school and two years of standard medical training. Some naturopaths complained that the cost of meeting these standards made medical education unaffordable to men and women of moderate means. One author called the acts "adverse legislation."[32]

The Flexner standards were based on curricula and examination, and following the Flexner Report, only those schools deemed legitimate flourished. Naturopaths were of two minds: those who graduated from reputable schools came to embrace these criteria; those who claimed the title "naturopath," "nature doctor," or "drugless healer" but had no formal education (all of whom were embraced by Lust) had little to gain from these standards—and much to lose. While Lust and others concluded that educational pathways seemed the best defense against accusations of quackery, the debate continued for decades.

As early as 1904 naturopaths worried about the danger posed by incompetents. The profession was on the lookout for them and warned the public about them. In 1915 no fewer than eight articles in the *NHH* addressed this topic. For example, Edward Earle Purinton blamed drugless physicians for their arrests and persecutions, saying that many were unfit and ill prepared. He asserted the need for uniform education, graduation, and registration for all healers, including naturopaths, as practiced by medicos. These standards should be required of every practitioner who prescribed herbs, drugs, foods, or affirmations (positive thoughts). His colleagues applauded Purinton's stance. One wrote that there were too many charlatans and misfits in the healing business and that they should be purged. Standards also meant a code of ethics for naturopaths.[33]

The development of standards was controversial. When, in 1911, the newly formed Pennsylvania Association of Naturopaths set as a goal the elevation of the profession through standards, not everyone agreed that it would gain naturopaths recognition from the general public or the state. One Californian advocated complete divorce from medicine and the state and called for a rejection of all regulation of the healing arts. It was clear that the greatest obstacle to standards was the diverse training, methods, and philosophies in nature cure. In the second decade of the twentieth century, a coalition of these diverse groups was committed to the lifestyle of healthy choices and opposed the medical trust. If standards were adopted, some practitioners would be left out, and the coalition would crumble. Oscar Evertz, editor of the regular *NHH* column Efficiency Department, lamented that it was impossible to bring all the drugless methods together under one large evangelistic tent. He advocated for reform of the profession through legalization and higher curricula standards. To ignore criticism about their lack of standards, he said, was suicidal. But few had the leadership to create cohesion and action. Evertz wrote, "We speak and write with enthusiasm about Organization of Forces; Standardization of Methods, Elimination of Incompetents, etc.—all worthy ideals, but mere dough without yeast." The yeast needed was legislation, but the drugless schools varied too widely to be covered by a single standardized-practice law.[34]

Benedict Lust always championed unification. In 1915, after several authors spoke *against* the likelihood of all drugless healers joining forces, Benedict shot back with upbeat details of his trip throughout the West. He had met leaders of drugless movements in every branch of physiological and mental systems. These were dependable men and women of character ready to defend and protest for drugless healers' rights. That same year, the nationally known naturopath and chiropractor H. Riley Spitler, president of the Ohio Association of Neuropaths (whose members emphasized mechanical therapies but were distinct from chiropractors), weighed in. He described a fortuitous bonding between sects in his state. The word *naturopathy* had been stricken from the state licensure bill, but inadvertently the state medical board had left in the word *neuropathy*. Ohio naturopaths decided that the practice of neuropathy included naturopathic methods. Three days after the law went into effect, neuropaths, naturopaths, and others formed a state neuropathic association and applied for licensure. The medical board admitted that naturopaths had "'one over on them.'" Thus naturopaths took every path to legitimacy where they could find it. But such liaisons and strategies added to the confusion about titles and practices. Spitler worried that members of his neuropathic association might be mistaken as adversaries by naturopaths. He reassured Lust that they were not antagonistic but had found a way to get legal recognition with little resistance.[35]

Naturopaths in the 1920s understood that confusion about titles and methods meant that patients had different understandings of naturopathy. While some saw it as a system of baths, others might see it as similar to chiropractic or to Christian Science. One author quipped that in the eyes of the public, naturopaths were as different as cats, dogs, chickens, and horses. They needed to agree on the details of a distinct science replete with training, core concepts, and a creed that engendered patients' respect. This would allow naturopaths to embrace the label with pride. In the 1930s, naturopaths continued to complain; they knew they had a problem, for naturopathic practitioners used at least five different titles. They could not unite, because they were not the same, and various laws had also forced them to adopt different labels. While there was a growing desire to protect their patients from fraud and incompetence, disagreements stemming from practitioner differences stonewalled their attempts.[36]

There were some gains. In 1929 two regions seemed to achieve a united naturopathic voice. After a year of efforts, Minnesota naturopaths had failed to get legislation passed to protect their practice. However, in the process they had developed a distinct identity and called for educational standards based on the most advanced scientific knowledge. And while there was respect for all licensed branches of the healing arts, none was more deserving than true naturopathy. In later years Lust called Minnesota's position ideal. A more influential gain—and beacon of hope— was passage by the US Congress of acts in 1929 and 1931 legitimizing naturopathy within the District of Columbia, placing into the *Congressional Record* a solid definition of naturopathy and the diagnoses and practices it allowed.[37]

By the late 1930s discussions about naturopathic standards persisted as basic science laws proliferated to weed out charlatans. Naturopaths modeled their strategies on allopaths' efforts to distance themselves from quacks. The South Carolina Naturopathic Examining Board set a high educational standard in 1937, and qualified naturopaths were encouraged to set up practice there. The board's goal was to promote and advance naturopathy through rigorous scientific naturopathic education. The subjects for examination included anatomy, physiology, pathology, chemistry, toxicology, dietetics, diagnosis, and the principles and practice of naturopathy. The well-organized South Carolina Naturopathic Association and Board of Naturopathic Examiners articulated licensure and membership requirements, meeting protocols, officer duties, committee formation, and procedures for changing bylaws. The exam demonstrated some naturopaths' desire to be bound by regular medicine's disciplines, albeit as defined and regulated by themselves.[38]

Amidst these calls for a more narrow, distinct professional identity, Lust defined naturopathy that year with *more* diversity. While incorporating basic sciences such as those listed above, he also included no fewer than *twenty-seven*
agents of healing. He said that all these modalities and standards had received
the seal of the ANA, noting that four thousand hours of coursework were required.[39] This was typical of Lust. He was always far more inclusive of complementary therapeutics than his peers, even though it undermined efforts to
legitimize naturopathy. The conditions of naturopathic standards and licensure
had become so muddled that beginning in the late 1930s, one hundred or
more questions and answers about state boards appeared in the *NHH*. They were
comprehensive but focused largely on traditional aspects of scientific medicine.
Naturopaths had been publishing information and dates for state exams over the
years, revealing the sometimes obscure practices (e.g., spondylotherapy, zone short
wave therapy, heliotherapy, chromotherapy, etc.) grouped with naturopathy.[40]

The years 1938–40 were a watershed period in the standards debate. Naturopaths saw an irony in the fact that they were struggling for legitimacy, while the
allopaths' uneven enforcement of their own criteria among themselves failed to
create a consistent cohort of competent professionals. One ND noted that refu-
gee doctors (medicos from other countries) had won their case to practice medicine in New York State without being examined. That confirmed for naturopaths
that the "State Board Exams do not qualify or disqualify, but only uselessly restrain and despotically control." Meanwhile, the ANA adopted a new plan at the
1938 congress in Seattle. Membership criteria reflected Lust's sweeping inclusivity for naturopaths, physiotherapists, osteopaths, chiropractors, and "any and all
those engaged in the Natural Healing Arts." The Membership Committee could
accept or reject applications; those rejected could appeal to the Board of Trustees
and the Executive Committee. Membership promised legal and financial protection in accordance with state law and was a step toward professional identity.[41]

Despite national attempts to clarify an identity, ambivalence toward naturopathic standards persisted. In 1939 a meeting of the newly formed Amalgamated
Chiropractic and Naturopathic Society of Pennsylvania focused on those two
groups, but any drugless practitioners were welcome. The society's goal was to
protect the licensure of all properly qualified drugless doctors in Pennsylvania.
The *Chiropractic Journal* noted that this new society had been formed to pass a
composite bill protecting all, but the professional bonds between the sects were
often hazy and confusing, sometimes even hostile. That same year, Utah's proposed bill did the opposite of what Pennsylvania's did. It stated unequivocally

that anyone wanting to be a naturopathic physician had to graduate from a legally chartered naturopathic college.[42]

As naturopaths continued to debate whether to have a traditional, inclusive strategy or a newer strategy exclusive to naturopaths, Lust reached his choking point. The man who embraced diverse modalities admitted that there were fakers in naturopathic ranks. He added his outrage about pretenders who adopted any and all modalities. The unworthy, those who should not be protected as naturopaths, "have either to reform and qualify and become honest if they want to stay, or they will have to look for another job." No more could the ANA be used by opportunists and exploiters. The criminal element must also be purged. He then borrowed the allopaths' language: "These snake charmers are the men that have been supported by these so-called 'drugless' side-shows and similar movements to bore within the ANA." These combined evils had resulted in many state licensing defeats and had given naturopaths' enemies reason to oppose them. According to Lust and other formally educated naturopathic leaders, a faker or charlatan in the naturopathic movement took different forms. Because chronic disease was the result of suppressing nature's efforts in acute illness, the use of aspirin, alcohol, quinine in fevers, headache powders, and pills was suppressive. Anyone who made a living off of supposed cures using these suppressives, tobacco products, hypodermic injections, or scam devices was a quack. In the 1930s he said that many within the sectarian ranks were parasites, in it for the money, and compromising health and practice by means of quick fixes. These ne'er-do-well doctors "have no conscience when they . . . stoop to illegal work, condemnable surgery and other practices" that have "wrecked the standing and honor of our profession, have lowered its reputation," and violated the naturopathic Hippocratic oath.[43]

In addition, Lust gradually came to believe that all the single-method systems should go their separate ways. Naturopathy had compromised itself by encouraging their growth. He attacked straight chiropractic as a passing fad that had grown and thrived on advertising. He thought chiropractors' philosophy was a hoax because they made no mention of hygiene or natural living. He also denounced osteopaths, who he said over time had adopted so many allopathic practices—as well as their licensure privileges—that they no longer resembled an alternative. When these practitioners adopted naturopathic philosophies and modalities, he would be willing to embrace them as naturopaths, but he believed that anyone with a one-dimensional cure who claimed to be a natural practitioner was a fake.[44]

Because of naturopaths' need for standardized professional training to gain legal legitimacy, by 1940 four years at a naturopathic college with a minimum of four thousand hours of coursework were required. Within three years the ANA Committee on Education named five schools that met the rigorous standards required for highly skilled naturopaths. These were Lust's American School of Naturopathy in New York City; the Nashville College of Drugless Therapeutics in Tennessee; Metropolitan College in Cleveland, Ohio; National College in Chicago, Illinois; and Western States College in Portland, Oregon. In each issue of the *Naturopath and Herald of Health* the ANA committee reported progress, and it suggested that a subcommittee be formed to oversee standards. There were plans for evaluation and accreditation and for cooperation between institutions and the states, along with other signs of cohesiveness. But many of these plans sputtered, came to pass in part, and then were derailed when Lust died in 1945.[45]

LEGISLATIVE EFFORTS AND TURNING POINTS

Over the decades, naturopaths achieved significant legislative successes through coalition building, savvy local leadership, patient support, and perseverance. Legislation was inherently contradictory. Laws inevitably excluded some who considered themselves naturopaths and cut against the democratizing, inclusive rhetoric under Lust's leadership.

Paving the way for successful legislation was the Association of Naturopaths of California (ANC), which created a charter in 1904 under the leadership of Dr. Carl Schultz (1849–1935), whom two historians called the "Benedict Lust of the West" and the "Father of Naturopathy in California." Members of the ANC were accorded the legal right to practice their profession in 1909. Legislation was designed to stop the harassment of nature doctors and to allow people to choose their own healers. A proposed bill in New York in 1914 promised extended protective powers to those who practiced dietetics, hydrotherapy, physical culture, dynamic breathing, massage, Swedish movements, structural adjustment and those who prescribed sun, light, and air baths. It also covered Kneipp and Just curists and those advocating fasting. The proposed law asserted that naturopathy was a system in its own right.[46]

By the second decade of the twentieth century, other state naturopathic societies had written similar legislation. Some sought legal recognition, and others sought to regulate the practice of naturopathy in their states by creating state boards of naturopathic examiners. Bills defined the powers and duties of these boards and provided penalties for violating provisions of a medical practice act.

By constituting a naturopathic board, peers, not a panel of medicos, would decide who were credible naturopaths. Such boards were created in Florida in 1923 and Indiana in 1925. Advocates of licensure argued that state-run, self-regulating boards must control and regulate practice to eliminate abuses. At the 1927 ANA convention, Lust declared that the "butchered and drugged public" needed to force state legislatures to legalize the disciples of nature cure despite the medical trust's control.[47]

Naturopaths' legislative plans were met almost universally with blockades and counterplans by the AMA-backed authorities. While there were some successes for some naturopaths, their moderate attempts to gain legitimacy did not cut deeply enough into the medical monopoly. Some felt that a full frontal attack was needed, as evidenced by the 1925 Declaration of Medical Liberty Rights and ensuing formation of medical rights leagues. A few heretical practitioners of naturopathy called for a US constitutional amendment guaranteeing medical liberty so they would not have to argue for naturopathic laws in each state. The proposed amendment provided for the people's right "to be secure in their persons, health, houses, and effects, against the enforcement of unreasonable medical regulations and restrictions." Specifically, it prohibited forced vaccination and quarantine and provided for the right to choose a physician and to rely on self-help without the aid of a physician. It stated that there would be no prosecution of the insane without due process; that no schools of healing art should have a monopoly, nor should any school be disallowed by Congress; and that all physicians could employ the methods of their choice. As the historian James Whorton points out, this case for medical freedom highlighted a desire to protect two types of constitutional liberties, citizens' right to choose their own treatment and each person's freedom to pursue a career. The call for an amendment was in step with the times. Within the past dozen years, four amendments had been passed about income tax, election of senators, prohibition, and women's suffrage. In addition, there was public support for the charge that, as one eclectic put it, "medical orthodoxy has always been as intolerant and bigoted as religious orthodoxy, and about as ready to torture and destroy."[48]

The proposed amendment reflected naturopaths' concerns for the past quarter century, and in 1926 they threw down the legal gauntlet. The Chiropractic Naturopathic Association in New York City brought a complaint against the city's police commissioner and the health commissioner. Frustrated and angered by nonstop legal harassment, these practitioners sought a permanent legal injunction restraining state officials from interfering with chiropractors, naturopaths, and all other drugless healers. No laws regulated chiropractors or naturopaths. In their complaint, which was ultimately unsuccessful, they called actions taken against them politically motivated, intended to turn the public against them.[49] The naturopaths who sought state laws and the two varieties of radicals—those who rejected all professional standards and those who wanted a federal constitutional amendment—represented distinct approaches to legal power. It is likely that the radicals made the more moderate goals of state advocates seem less threatening in the medical world.

In 1929 and 1931 two breakthrough legislative acts passed in the US House of Representatives to legitimize naturopathy in the District of Columbia. These successes were accomplished through tireless political lobbying, public support for alternative healing, and strong public opinion in favor of choice and against monopoly. The acts made naturopathy a distinct healing profession on par with osteopathy, chiropractic, and allopathy in Washington, DC. The bills were only applicable to that small jurisdiction, but the House debate—and the House record documenting the passed bills—defined naturopathy in a public venue that strengthened naturopaths' legal arguments. The 1929 act placed many diverse therapies, treatments, and foods under legitimate naturopathic purview. But the act required clarification, and the amendment passed in 1931. Because the acts covered many modalities under the umbrella of naturopathy, Lust rejoiced, for they achieved the inclusivity that he had so long advocated. His 1931 "President's Proclamation," delivered to the annual ANA convention in Milwaukee, stressed that naturopaths had reinforced their constitutional rights. After all, Washington, DC, was a world stage.[50]

This was Lust's finest hour. He interwove an eloquent history of the long and illustrious path of natural healing with a call for more organizing on a global level. Hospitals were needed, as were clinics and more trained naturopaths. The list of future possibilities was endless and seemingly achievable. Lust did not take personal credit for the victory; he credited the legislation to Theresa M. Schippell, an ND based in Washington, DC, who was an effective and innovative leader and organizer. Schippell worked closely with Lust as his right hand in his later years and was secretary-treasurer of the ANA. After his death she served as president and resumed publication of the *Naturopath* as a professional trade journal. Schippell, along with Congresswoman Katharine Langley (R-Kentucky) and powerful friends in Congress, had lobbied, secured financing for, and sustained the eight-year fight.

The congressional acts were seen as a turning point by both naturopaths and allopaths. Naturopaths saw opportunity, while allopaths saw a growing threat of illegitimate practitioners pushing their way into medicine. As the New York commissioner of public health put it in 1927, the licensing of various "cults" was "a hydra-headed monster. You destroy one and others rise up against you." After passage of the acts, the legislative campaign against naturopaths ramped up considerably.[51]

The *Congressional Record* entry for the 1931 Amendment of the Act to Regulate the Practice of the Healing Art in the District of Columbia reveals why naturopaths were so elated and what fueled the allopathic backlash: "that 'naturopathy' as a branch of the healing art is placed upon the same basis as medicine, osteopathy, and chiropractic . . . that the term 'naturopathy,' as used therein, is self-definitive to the same extent as are the terms 'medicine,' 'osteopathy,' and 'chiropractic.'" Section 2 defined allowable acts, practices, and usages. Fourteen distinct modalities were included, and practitioners could also prescribe botanics. The rights and definitions were expansive. They legitimized naturopathy but also other modalities. The legislation succeeded because of coalitions, and all of them except, notably, chiropractic and osteopathy were included in the final product.[52]

The passage of the 1931 amendment that defined naturopathy unleashed a legal bombardment against naturopaths because of the precedent it set and the powers it promised naturopaths in high-profile Washington, DC. In 1932 the AMA's Committee on the Costs of Medical Care urged the centralization of regular medical care around hospitals, limiting participation to those already qualified, excluding so-called cultists (2,500 naturopaths among them), on whom the public spent countless thousands. The battle had become a battle about business interests and profits: if patients could choose health care from the naturopathic cadre, allopaths lost revenue.[53]

In 1935 Lust proclaimed that the medical trust was on the warpath again and described a proposed bill in New York that would attack the establishment of colleges of natural therapy. In addition, allopaths' most destructive tactic in the states gained momentum. After 1935, new acts were introduced and the old, haunting charge of practicing medicine without a license was levied continuously. Naturopaths and their supporters organized to defeat these laws in North Dakota, Maine, and Rhode Island. Exasperated, one naturopath said that naturopaths' disorganization and lack of unity, coupled with the countless practitioners who laid claim to the title "naturopath," stunted their legitimacy as they tried to fight back.[54]

Despite the attacks, naturopaths sought passage of laws to protect themselves and secure licensure in Minnesota, Texas, New Mexico, Georgia, Michigan, and Maryland, among other states. In Oregon the Healing Arts Constitutional

Amendment was soundly defeated in 1934. The setback was particularly demoralizing because the famed chiropractor B. J. Palmer had urged the people to vote it down. In California the naturopaths likewise failed to pass a new bill, in part because straight chiropractors had turned on their mixer colleagues, thus weakening the ranks. Another reason for these bills' defeat was that they were poorly written and did not weed out the charlatans. By 1939 naturopaths anticipated being excluded from the Social Security Act, as well as from the new Washington, DC–based United States Medical and Surgical College. This was to be the only college to train army, navy, and public health personnel.[55]

The backlash escalated. Otis G. Carroll was a strong ally of Lust's and a nationally respected practitioner. In the Washington State Supreme Court case of *Kelly v. Carroll* he was prosecuted as a drugless healer (not as a naturopath) for practicing medicine without a license. Then, also in Washington, a graduate of Lust's school adopted the word *sanipractic* to distinguish himself from naturopaths. In 1939 the Tennessee State Medical Association took suppression of natural healing to a new low: the state legislature forbade the publishing, then the advertising, of the herbalist Jethro Kloss's book *Back to Eden* for therapeutic purposes. Legislation defined the practice of naturopathy as a gross misdemeanor, punishable by up to a year in jail.[56]

Naturopaths fought back with their two primary means of organization: they established naturopathic licensing boards, which were successful in Florida, Illinois, Hawaii, and Utah, and they passed bills to regulate the practice of naturopathy, such as those in Kansas, Alaska, Idaho, and Texas.[57] Despite noticeable successes and relentless work, constituents continued to blame naturopaths' problems on their lack of organization and solidarity. One criticized the naturopathic leadership for their failure to secure basic licensure. Hurting the cause were the diverse degrees a nature doctor might hold, individual versus collective efforts, and breakaway practitioners seeking personal glory. How could naturopathy possibly distinguish itself in the morass of degrees that included doctorates in natural therapeutics, sanipractic, physcultopathy, physiological therapeutics, and naturopathy? The result, he said, was the licensure of osteopathy in every state in the union, chiropractic in nearly all states, and naturopathy in only seven. These criticisms irked Lust. He urged the critics to join the cause and put their energy into the work at hand.[58]

Naturopaths built on their successes when they could. In 1933 Arizona passed a bill that gave naturopaths constitutional rights to treat the sick as would physicians in other branches of the healing arts. Then in 1940 the AMA and two affiliated medical societies were found to be in violation of the Sherman Anti-Trust

Act. They had conspired to restrain trade and to impair or destroy the business and activities of a local nonprofit cooperative that was providing medical care and hospitalization for its members. Morris Fishbein, MD, editor of the *Journal of the American Medical Association* and a rancorous critic of naturopathy, was singled out as the primary culprit. For naturopaths the finding was a small public acknowledgment of the methods of the medical trust.[59]

CASE STUDIES

While a full-length study of each region has yet to be conducted, an overview of legislative efforts in California, Illinois, and New York provides insight into regional distinctions and notable successes and failures. Future state histories will surely reveal more lessons about the struggles to professionalize naturopaths.[60]

State-by-state licensing efforts encapsulate all of the contradictions, tensions, aspirations, successes, and blind spots emblematic of the movement itself. Coalitions among alternative healers were both empowering and damaging, and while at times they lent organizational strength, they also impeded a consistent and coherent definition of naturopathy. Last week's allies were this week's traitors. Liaisons also meant vulnerability; naturopathy's fate rested with chiropractors, osteopaths, drugless healers, and others—the list was endless—as they all experienced setbacks. Factions made each group's position weaker.

Lust's interactions with and respect from vegetarian societies, sanitarium owners, massage and water operators, health leagues, the American Medical Freedom League, and antivaccinationists all buoyed his efforts, but they also made his work more difficult. Few joined the ANA, paid dues, or heeded its precepts. They formed new groups that laid claim to naturopathy but did not necessarily support the national agenda. This splintered the self-proclaimed naturopaths, who made accusations against one another.

State legislative efforts shared in common the spotlight naturopaths shone on allopathic abuses; shifting definitions of practicing medicine and naturopathy; and reliance on, and vulnerability to, local leaders' political abilities. They worked within cultural climates of both acceptance and suspicion, and there were internal power struggles over the leadership of state boards. Finally, naturopaths had to decide how many curriculum hours were required for a degree and whether to adopt certain allopathic standards. This work seemed insurmountable at times. But these successes and failures kept alive the nineteenth-century tradition of health reform and shepherded it into the twenty-first century.

These state efforts reflected a growing nationwide base of naturopathic followers and showed that sectors of the American public continued a wariness

of allopathic therapeutics that had existed for more than a century. State successes validated naturopathic philosophes and treatments once deemed fringe.

NEW YORK, CALIFORNIA, AND ILLINOIS

Lust's experience in New York was a microcosm of the movement. Lust's school,
bakery, publications, and one of his practices were based there. Because of the
personal connection, Lust could exploit his painful local political battles on a national stage, but he was also vulnerable to challenges from within the ranks. The
hub of activity in New York also reflected the almost schizophrenic relationships
between the sects, with articles gloating over the misfortunes of chiropractors or
osteopaths on one page and a story of a friendly, beneficial exchange between
them on the next. The *Naturopath and Herald of Health* cautioned practitioners
about the consequences of failed loyalty between groups. In 1914 a bill submitted
to the state legislature jointly by naturopaths and chiropractors to protect their
common interests had been sabotaged at the last moment by straight chiropractors introducing their own bill, with the result that both bills had been defeated.
Since that time, all attempts at separate legislation had been fruitless.[61]

Chiropractors and osteopaths are key players in these stories. The mixer
practitioners—those who incorporated naturopathic modalities—were often
staunch naturopathic allies. Naturopaths also hoped that successes that accrued
to mixer chiropractors and osteopaths would serve as legal precedents for them.
However, straight chiropractors and osteopaths sought legal recognition for
themselves by excluding other natural healers from their bills, thus making
them as much enemies of the cause as allopaths were. They were seen as medical science wannabes—an abhorrent breed. Naturopaths gloated about straight
osteopaths in Illinois who had tried to gain recognition from the AMA, only to
be assailed by them. "The spider coaxed the fly into the web," one author wrote,
"and the devouring process [is] at hand."[62]

In New York, naturopathic goals became increasingly difficult to attain as the
definition of practicing medicine morphed over time. The 1881 New York appellate case of *Smith v. Lane* defined it as applying and using medicines and drugs
with the goal of curing, mitigating, and alleviating bodily diseases. These applications were limited to manual operations, usually done with surgical instruments or appliances. Beginning in 1918, the law was interpreted to apply to
anyone who claimed to "diagnose, treat, operate or prescribe for any human
disease, pain, injury, deformity or physical condition." Practitioners who did not
register were guilty of a misdemeanor. Those practicing under assumed names
or impersonating a physician were guilty of felonies. As a result of the precedent,

"Every Cloud of Autocracy Has a Silver Lining." The nasty political struggle between the AMA regulars is represented by the wolf with the Monopoly man and the ANA healing sects. Alternative therapies are clustered on the left bank, and only hydropathy and osteopathy (with pills) are able to make it to the AMA side of credibility and recognition. The monkey—mechanotherapy—atop the "United We Stand" pole, propped up by propaganda, is attempting to grasp the chain of power: recognition in Congress and money. But the AMA has laid a trap for the ANA. Attempts for recognition and money would ensnare natural healers in political and legal turmoil. From "Cloud of Autocracy," *Nature's Path* 26, no. 1, current no. 435 (Jan. 1931): 6. Courtesy of the National College of Natural Medicine Archives, Portland, Oregon.

other states made their definitions more stringent so that even the simplest care was classed as medicine, whether drugless or drug.[63]

The movement depended on allies, so Lust appeared wherever he could make connections. He spoke before the Vegetarian Society of New York and gave press coverage to the meeting of the New York County Chiropractic Society, a lecture at the League for Health Education, and a large mass meeting of the Medical Liberty League. Lust's travels were chronicled in the *NHH* to get as much mileage out of his speeches and their messages as possible. In 1926 a description of his trip to the convention of the Florida Naturopaths segued into an account of his plans to return to New York for the final battle against the Webb-Loomis Medical Practice Act, expected to be signed by the governor. It would codify drugless healers' inability to use the title "doctor" unless authorized by law to do so. All drugless healers would be forbidden from practicing medicine except under the supervision of AMA-recognized doctors. In response to this bill, the Chiropractic-Naturopathic Defense Society had been formed to establish a closer association between chiropractors and naturopaths and battle common legal issues.[64]

In 1935 Lust was charged with conferring degrees without a license as president of the American School of Naturopathy. Both he and the president of the New York School of Chiropractic were fined five hundred dollars and given a suspended jail sentence. In fact, the 1926 passage of the Webb-Loomis act had invalidated the ASN's charter. Lust claimed that the original New York charter was still valid through its inclusion in the Washington, DC, statute. It did not close its doors until 1942.[65]

Editor Lust also had the luxury of devoting lengthy articles to New York legal cases, while reports of cases in other states would have garnered only a paragraph. One case was that of a naturopath convicted in New York City in 1934. He had graduated from the American School of Naturopathy and later assumed the helm of the electro-hydrotherapy and massage departments of the Middletown Medical Sanitarium in New York. The article about the case detailed his entrapment, his arrest, his nervous breakdown prior to trial, and his being taken to court against his will while sick. The jail sentence was suspended, but the five-hundred-dollar fine was enforced. In heartrending accounts, Lust made the story of one man personal to all naturopaths and New Yorkers, even those who had never been legally harassed. Lust published a litany of allopathic actions to keep the fires burning. A court ruling reported in 1937 was a simple illustration of allopaths out of control: in New York only doctors, and not beauty shops, could perform electrolysis.[66]

Despite the *Naturopath and Herald of Health*'s vast coverage of persecutions and Lust's efforts to champion the cause and expose and resist AMA offenses, discontent was palpable among naturopaths. One disgruntled practitioner, after being jailed for a week without access to an attorney, wrote angrily in 1937 about the inability of the New York–based ANA to offer effective legal assistance: "The Naturopathic Association is your friend, but it is a helpless friend . . . it cannot help you well. Because it is disorganized it is powerless to oppose the enemy [the AMA]." He noted the splits within the ANA and the fact that there were no punishments to curtail the behavior. He astutely observed, "Our movement suffers because of too much anarchistic sentiment in our ranks." The diatribe was on point. Naturopaths turned to the national organization for assistance with legal fees, but as the number of cases mushroomed, even this one aid became difficult to provide, despite legal funds provided by the Lusts for New York practitioners. The splits continued to occur; Lust's attempts at unity were frequently foiled. He noted that some of those who requested legal defense were not even paying members of the ANA.[67]

California naturopaths encountered similar difficulties. Between 1908 and the 1940s numerous California bills were introduced to define—and redefine—the practice and licensure of naturopaths. The legislation reflected naturopaths' desire not only for increased standards but also for increased adoption of allopathic methods. A critical mass moved away from the original ideals of drugless healing and responsibility for personal health and toward the successful model of the AMA, which had been around for almost a century. Standardization and licensure created growing pains for naturopathy, but in California, attempts at new structure and licensure coexisted with traditions in an almost Wild West anarchy.

The Association of Naturopaths of California was created in 1904 to recruit and organize natural practitioners. California was the first state to pass a naturopathic licensing law in 1909, largely owing to the efforts of the ANC and Dr. Carl Schultz, owner of the Naturopathic Institute and Sanatorium of California and creator of the Naturopathic College of California in Los Angeles. Naturopaths had found themselves pitted against osteopaths, allopaths, homeopaths, and eclectics, who were legitimized as viable practitioners under California law. Schultz, as president and secretary of the ANC, tried mightily to persuade both the state board of medical examiners and the governor to legitimize naturopaths as well. Schultz studied law at night for three years and spent thousands of dollars of his own money to fight the allopaths who went after him—to no avail. Finally, the 1909 naturopathic licensing law gave practitioners who were members of the ANC the legal right to practice. But afterward, the law was rewritten to specifically exclude drugless practitioners, instead requiring anyone who had a certificate from the Naturopathic Board to take the allopathic medical state board examination. Anyone who practiced medicine without a valid certificate was guilty of a misdemeanor. Once convicted, a practitioner had to pay a fine of one hundred to six hundred dollars, faced imprisonment for a term of 60 to 120 days, or both. It is little wonder, then, that by 1916 drugless doctors in California united with Oregon colleagues to wage an active campaign for legal recognition. Oregon practitioners were in a strong position to assist because there were roughly five hundred drugless doctors holding state licenses in Oregon and they had a college of chiropractic with a clinic. In 1917 the California Medical Practice Act mandated that every practitioner licensed by the state must register annually with the state board of medical examiners beginning in 1918, and pay a two-dollar fee. Drugless healers earned a victory in 1925 when the state board of health allowed nurses trained in nonallopathic schools to work in California hospitals.[68]

Four schools of healing arts were legally recognized in California: naturopathic, allopathic, homeopathic, and eclectic. The naturopathic curriculum was by far the most inclusive and the most far reaching. Practitioners employed sixteen therapeutics. But the inclusiveness also bred competition, distrust, and turf wars. In 1923 Benedict Lust was disheartened by the discord among various factions of drugless healers in California. "There is fighting, proceedings, actions, quarrels, struggles, etc., which is shameful." Each group was rushing to establish its own superiority and give its own version of events. The old ANC split when some groups formed the Association of Progressive Naturopathic Physicians of California. In the summer of 1935, the *Naturopath and Herald of Health* reported that the new association members were refused licensure by the allopathic board but then secured it through what the article called a "National Bill of California." In a twist of irony, they were allowed to prescribe narcotics and liquors and to perform major and minor operations, all antithetical to core naturopathic principles. One of these men was arrested for murder. Others were found to be in violation of the state narcotic law, giving more bad publicity to naturopathy. Some practitioners licensed under the drugless certificate of the Medical Practice Act were practicing allopathy.[69]

The Medical Board of California cracked down on all those considered quacks during that decade, according to a document it produced in 1995. Using funds from an annual tax on licensed physicians, "the Legal and Investigation Department of the Board grew from four to . . . ten individuals" to track violations and maintain an efficient record system. Originally, when the board's special agent found evidence of a Medical Practice Act violation, it was turned over to law enforcement. In 1939, "legislation made the special agents peace officers themselves and eliminated this dependence. . . . If found guilty, the violator was fined [and] could be given a jail sentence. . . . The Board continually urged stronger penalties for the objectional medical methods practiced by quacks." Lust reminded naturopaths to stay within the parameters of the law, because the state medical board treated drugless healers fairly. He hoped that the sloppy therapeutics and infighting could be remedied by building up membership in the reputable ANA, appointing representatives from each major city, and getting representation on the medical board. Lust liked to use California as an example, for troubles with overreaching practitioners and individuals vying for power were not unique to that state.[70]

In the summer of 1926 Lust visited the state and was hailed as both the national president of the ANA and the founder of the American School of Naturopathy. There were plans for a stellar school of naturopathy with a strong curriculum in Los Angeles, and applications were being taken for the first class. Curiously, this accomplishment was attributed to Lust, when in fact it was equally, if not predominantly, the result of efforts by local organizers. A recent article in the *Los Angeles Sunday Times* had been favorable toward methods of naturopathy, chiropractic, and other drugless practices. For a time at least, Lust and the college created some cohesion and momentum.[71]

A significant blow came in 1935 with the defeat of yet another California bill to license naturopaths. According to a *NHH* report, the leadership had done its duty nobly, but the bill would have "placed parasites under the naturopathic banner." Suggestions for a rewrite included supervision by the governor (even though other states rejected the idea of giving up control this way); limiting licensure to graduates of credible schools (chiropractors, osteopaths, or naturopaths); and a residency requirement of five years. Other suggestions in the *NHH* were to increase the number of curriculum hours to four thousand and to require graduates to take a postgraduate course lasting at least three months. The licensed drugless healers, once properly schooled, should be able to use anesthesia and perform minor surgery. A feverish disdain for the uneducated was a theme that picked up steam in these decades. Among professionalizing naturopaths there was nationwide ambivalence about the lay movement that had created naturopathy. Between 1937 and 1944 three additional bills, with increased details and standards largely in line with those proposed in 1935, were attempted in the state. The difficulty was finding the political leadership to accomplish the goals. A bill meeting naturopaths' wishes did not pass until 2003.[72]

Illinois, like New York and California, pursued legitimacy and the freedom to practice through organizing, organizing, and more organizing. In the early decades the leadership and influence of Dr. Henry Lindlahr was critical. Despite apathy and disorganization, Illinois was revitalized in the 1920s through Lindlahr's efforts. The *NHH* encouraged interested parties to contact Lindlahr, author of the pathbreaking text *Nature Cure* (1914) and founder of a sanitarium in Elmhurst that was highly endorsed by the journal. There were several other naturopathic sanitariums in Illinois in the early decades, and the state was the hub of the American Medical Liberty League, a stronghold for medical and naturopathic rights. The naturopath T. Louise Nedvidek chaired the twenty-first annual meeting of the American Medical Liberty League in Chicago. She was very active nationally, and five women presenters were scheduled. Two items on the league's agenda were injustices regarding licensure and forced vaccination. Chicago passed an ordinance establishing a health board to protect citizens from any kind of compulsory medication. This realized the work of antivaccinationists who had been fighting for years. It also enhanced the credibility of naturopaths, who had collaborated with the antivaccinationists on this cause.[73]

Because the AMA was headquartered in Chicago, allopathic suppression was ever present for Illinois mixer healers. In 1920 allopaths had brought injunctions against fifty or more chiropractors who practiced without a license because they had not been allowed to take the state licensing exam. Also, they could not pay dues to the sectarian organization nor accept money or legal aid from it, lest they be arrested for violation of the Illinois Medical Practice Act. Lust likened these injunctions to the federal government's injunctions against striking miners and steelworkers, saying that in this case the injunctions were being used to eliminate competition. He believed this had been the long-term plan of medicos and osteopaths of Illinois, who had schemed to rob the already licensed drugless practitioners of the right to practice. Lust urged all readers to show their support by joining the Chicago-based American Medical Liberty League. There was power in numbers.[74]

As the decades passed, the nomenclature used by naturopaths in the state signaled a shift to embrace the identity of "physician" and "doctor." Naturopaths had created high standards and were professionalized in the allopathic model despite continued hostility toward allopaths and their political strategies. Increased activism, such as the formation of the Illinois chapter of the American Naturopathic Association, strove to give naturopathy a higher profile. A group calling itself the Drugless Practitioners Association of Illinois held large meetings with the goal of combining the efforts of all naturopaths, naprapaths, chiropractors, and others to pass a statewide bill to legalize drugless healers. In 1940 the unification call went out from members of the Illinois Naturopathic Physicians to all naturopaths, chiropractors, physiotherapists, and other drugless doctors to support a naturopathic bill. It may seem odd that they appealed to these diverse healers to support a bill for naturopaths only, but in fact naturopaths had been supporting the individual efforts of the others for years. Despite increased organizing and cooperation amongst drugless healers, there was little progress. As of 2014 there still was no licensure allowing naturopaths to practice in Illinois.[75]

VICTORIES AND SUCCESS STORIES

Individual states' leadership and gains throughout the 1930s strengthened the movement's credibility with the American public and increased the ranks of its advocates and political allies. For his tirelessness and his frequent lobbying in the states for legal recognition, Lust can rightly be credited as the primary leader, author, inspiration, and supporter of protective laws. Yet his insistent, if at times ambivalent, inclusion of all natural healers hindered the unique chances for naturopathy's legal recognition. His legacy is at times contradictory, but his successes are indisputable. After his death in 1945, legal protection in the Northeast diminished greatly, but it remained intact in congenial political climates. Securing licensure was a remarkably ambitious and often achievable goal, but other parts of Lust's dream, such as the establishment of a national examining board, went unrealized during his lifetime.[76]

By far the largest measure of success nationwide was the amount of organizing and activity that paved the way for later laws. These efforts were accomplished with the help of a large number of allies, whose complex, strained, dedicated, passionate, and irksome agendas were both enriching and enervating within the coalition. There was a variety of fruitful action in the late teens and early 1920s. In Washington the State Board of Naturopathic Drugless Examiners was formed in 1919, supported heartily by women's groups. A 1923 bill passed by Florida legislators was refreshingly clear, regulating the practice of naturopathy and providing for a board of naturopathic examiners that could define its own powers and duties and determine penalties for violations. In 1926 Maine began incorporating its naturopathic association; in 1927 Oregon naturopaths submitted a bill and declared their desire to host the national ANA convention. Also in 1927, the second annual convention of the Indiana Naturopathic Association outlined its mission and procedures; in Pennsylvania, the landmark acquisition of a charter for a naturopathic hospital was announced by the state naturopathic society.[77]

The nationwide appeal and success of naturopathy was repeatedly demonstrated in legislative successes, and Lust made the most of it in the pages of the *Naturopath and Herald of Health*. Three successful bills from the 1920s in Florida, South Carolina, and Oregon were reprinted in 1937 to inspire those laboring in other states. In late 1939 and 1940 another flurry of bills were drafted and passed by Mississippi, New Mexico, and Maryland.[78]

There was continued activism in the 1930s. The Texas State Naturopathic Association was formed in 1931; Rhode Island presented its bill to the state legislature in 1935; and Connecticut made efforts to establish a Natureopathic College in the spring of 1935. Also in 1935 a suit against Lust was terminated, and the ANA of Kansas reported that a suppressive basic science bill had failed in their state. Two years later Idaho naturopaths presented a bill, and another was put before the Pennsylvania state legislature in 1939. That year naturopaths were also buoyed by an invitation from the mayor of Detroit to hold their ANA national meeting in the thriving city that was the center of the US automobile industry. There was no doubt that Americans were seeking naturopaths nationwide, and the practitioners wanted that acknowledged in the law.[79]

By 1939 eight states had introduced bills to allow all practitioners to work in government and nongovernment hospitals. Bills in a number of states were designed to guarantee free choice of healing methods to indigent persons. Though unsuccessful, they served as an index of the increasing interest in protecting individuals' rights to select the healing methods and practitioners of their choice. That same year, laws enacted in South Dakota and Tennessee allowed drugless physicians to receive public funds for medical care; no distinctions were drawn between the legally authorized branches of healing.[80]

The 1940s were a swing decade. In 1940 Texas was granted a charter enabling naturopaths to establish and supervise their own schools. That same year Nevada proposed a naturopathic law. In 1943 Tennessee's Naturopathic Act was passed over the governor's veto, legalizing naturopathy, and Connecticut celebrated an act incorporating the College of Natureopathic Physicians. Alaskans awaited word from the territorial legislature on the proposed Non-Medical Practitioners Act, which covered all the healing arts.[81]

Certainly the news was not all good during these decades. Many states saw their bills defeated or attempts at repeal initiated by their foes. Others saw the creation of virulent antinaturopathic bills. Washington, DC, in 1935 was one such case, a response, no doubt, to the 1931 success. In Indiana a joint bill presented by chiropractors and naturopaths failed. By the spring of 1944 the South Carolina Naturopathic Association faced a bill to repeal its Naturopathic Act. Naturopaths continued to be arrested and brought to trial; state associations dissolved or experienced marked changes in leadership; confusion reigned in the public's eye, and in the law, about what exactly a naturopath was. Formation of examining boards was sometimes thwarted, naturopaths were excluded as eligible physicians in a series of federal laws, and some states stalemated because of inactivity and futile attempts at organization.[82]

Despite these setbacks, by the 1940s naturopathy had achieved full legal status in Florida, Utah, Colorado, Washington, and other states, as well as several
provinces in Canada. By stepping up their standards, naturopaths had stemmed
the bleeding caused by allopathic basic science laws. As Whorton put it, the
"basic science exams became a less effective sieve for separating cultists from
scientists. Beginning in 1967, one state after another repealed its basic science
law, until the last three disappeared in 1979."[83]

The national and local legal battles manifested a variety of deep and frustrat-
ing problems. Despite successes, things were far from harmonious in the world
of natural healing. There were complaints of fence-jumping practitioners who
falsely claimed the title "naturopath," factions with divisive jealousies, outright
sabotage for personal gain, and resentment by college-educated naturopaths toward the unlicensed among them. Along with the battles from within, one more
undermined naturopathy's rise to legitimacy: the usurpation of proven naturopathic theories by allopaths and their affiliates, such as greater emphasis on prevention of disease through nutrition and public health measures.

After Lust's death in 1945, naturopaths across the country believed that they
must unify or be destroyed. A telling move concealed naturopathic dissent from
the public: the State News and National News sections were eliminated from
Nature's Path. Articles deemphasized political and legal topics and focused on
naturopathy itself—food preparation, treatment of specific diseases, the importance of vitamins, women's management of familial health, and much more.
The new front became professionalization.

CHAPTER 9

Professionalizing and Defining
the Nature Cure

Professionalization of naturopathic physicians, schools, and methods became necessary to stop the legal persecutions. A first step was agreeing on a definition of naturopathy, but from the beginning practitioners of a variety of theories and techniques had adopted the label "naturopath," and for some that was enough, regardless of schooling or level of affiliation with the national movement and its goals.

There simply was not a definition that everyone agreed upon. First, there were philosophical disagreements about the cause and treatment of illness. Also, sociopolitical factions hindered any unified understanding within the profession; naturopaths also became entangled in strategies of reform that reached far beyond health; leaders were often naïve about the parameters of nationally shared goals; and individual health reform leaders of different healing systems fractured the profession when they interjected their own programs into naturopathy. Most importantly, the profession's difficulty in reaching a definition of naturopathy stemmed from Lust's insistence on keeping such a wildly diverse and eclectic set of movements united as a political and philosophical force. This sweeping inclusivity reflected two key aims of Benedict Lust's vision: to merge a variety of therapeutics under the banner of naturopathy, insisting that only a multilateral approach to natural healing could correct disease (therapeutic universalism), and to create a radical collective front whose combined forces could upend the medical trust.[1]

The debates about naturopathy as a profession took place in the trade publications, which were numerous. Multiple publications merged with Lust's *Naturopath and Herald of Health* before 1908, including the German *Amerikanische*

Kneipp-Blatter (1896–1901), the *Kneipp Water Cure Monthly* (1900–1901), and the German and English editions of the *Naturopath* (1902–7). The *Liberator of Medical Thought* (1902–8) also brought its eclectic and radical readership with it. The fifth publication joining the *NHH* was Dr. Immanuel Pfeiffer's *Our Home Rights*. Pfeiffer was introduced as a worker for the social masses, an advocate for medical freedom, an antivaccinationist, and a postal reformist. His journal's inclusion reveals the wide net cast by Lust in his cultural reform efforts.[2]

In 1904 Lust wrote glowingly of the meetings of the Naturopathic Society of the United States, which cooperated with several sectarian groups. These included the Vegetarian Society, the Health Culture Club, the Cosmological Club, the Society for Ethical Culture, "and others that work for human progress and for the improvement of the race, physically as well as mentally and spiritually." Lust praised the harmonious exchange of ideas in these meetings.[3]

Later that year the *Naturopath and Herald of Health* acknowledged the influence of the New Thought philosophy on naturopathy. At this time the *NHH*'s masthead read, "Devoted to Naturopathy: The Science of Physical and Mental Regeneration." Descriptive paragraphs below the mastheads of the Lust-published journals by 1907 provide insight into the evolving profession of naturopathy over the years—they are a study in inclusivity and flexibility. That year Lust's naturopathy encompassed popular hygiene, hydrotherapy (Priessnitz, Kneipp, and Just systems), osteopathy, heliotherapy (sun, light, and air cures), diet, and physical and mental culture. The September 1907 issue of the *NHH* has portraits of both Benedict and Louisa Lust on the cover, although Louisa's image disappeared shortly thereafter. This may have been a strategic move, since Benedict possessed more medical degrees and was the movement's front person; it may also have been a capitulation to male professional gender norms.[4]

Beginning in 1902 Edward Earle Purinton periodically served as editor of the *NHH* while Lust pursued his degrees. Although Lust had encouraged a diverse membership, Purinton soon lamented the ever-growing collection of constituents. He was particularly influential, despite his disagreement with Lust on this subject. In 1908 Purinton wrote in his front-page article, "Let me whisper something to you. *Naturopathy will never be a recognized school until we refuse to teach any but licensed physicians*" (emphasis in original). He bluntly cautioned readers to beware of anyone claiming to be an originator, saying that those people were pretenders. And he agreed with Lust's therapeutic universalism, saying that one should not support practitioners with "the Great One-and-Only-Idea." He wrote, "This brother has a food, a magazine or an exerciser, around which he builds an ornate stock company. . . . He's a quack." Purinton critiqued some articles that were too narrow in his view and advertisements touting products too fantastic to be real. He rejected claims that magnetism, hypnotism, and single-food diets were universal remedies. By 1914 fraudulent advertisers had prompted creation of the *NHH*'s Information and News Department, which "encourages honest advertisers and eliminates frauds." This commitment was never really implemented.[5]

When even the two editors of *NHH* did not agree, it was little wonder that at the first annual meeting of the Association of Naturopaths in California, in 1911, one speaker said that if one tried to find the word *naturopath* in the dictionary, "I fear you will have a difficult time." The same could be said when perusing the contents of naturopathic journals. The cause of naturopathy was not helped by the *NHH*'s 1911 incorporation of the *Phrenological Journal*, which had been independent from 1838 to 1910. It was the sixth journal to be incorporated. By 1911, phrenology—a pseudoscience that claimed there were more than thirty distinct organs within the brain and that studied the shape of the skull—had been largely discredited. At times publishers of other magazines wrote Lust asking for his endorsement. He often gave it, but at times even he refused. When the Nebraska editor of *Unfired Food* requested his endorsement in 1914, Lust responded diplomatically that he had not been living the raw-food lifestyle but wished him much success.[6]

In 1914, at its annual convention in Washington, DC, the National Association of Drugless Physicians adopted the *NHH* as its official organ. The journal would be sent to all members and would publish the association's drugless news. It was the eighth system to amalgamate under naturopathy. But the profession as a whole was impeded by its adoption of the term *drugless*, which was interpreted by practitioners and the public in widely different ways. Increasingly, courts held that drugless healers must not use substances applied externally or ingested.[7]

By the spring of 1914, in the well-crafted Naturopathic Legislation Series, written by the New York Naturopathic Association, a definition of *naturopathy* was offered that varied considerably from the one given in 1907. Osteopathy and mental culture had been eliminated, and dynamic breathing, massage, Swedish movements, structural adjustments, and fasting had been added. A broad bill to recognize all of the healing arts of naturopathy was ready for passage in Albany, but as Lust recalled decades later, chiropractors single-handedly killed the bill. The straight chiropractors selfishly believed, he wrote, that they alone should receive the coveted recognition. A regular reader was likely confused, even dismayed, just seven months later when the American Naturopathic Association clarified its components. This time osteopathy was reinserted, and added were mechanical therapy, chiropractic, neuropathy, naprathy, physcultophy, suggestive therapeutics, psychology, spiritual science, and others. Some of these may have been existing systems under new names, but the nomenclature changed with dizzying rapidity. Lust was obviously comfortable with contradictions and tensions, given the trade-off of maintaining allies whose protective laws could benefit multi-credentialed naturopaths.[8]

As the array of practices came together under the naturopathic roof, Lust published the *Universal Naturopathic Encyclopedia, Directory and Buyers' Guide* in 1918, a monumental 1,426-page book that gave clues to naturopathy's identity in its cover graphics. In the corners were what he considered the four cornerstones of drugless therapy—nature cure, manual therapy, food science, and mind cure.[9]

For some, this widely cast net was unsatisfactory. Shortly after the reinstatement of chiropractors by Lust, Leon Bourgonjon, MD, a self-identified "neo-eclectic" physician, expressed his dissatisfaction with the words *drugless* and *naturopathy*. He insisted on the use of two new words, *rational* and *neo-eclectic*. Changes like this were suggested constantly and portrayed a broiling group of radical, if elusive, thinkers. Despite Bourgonjon's suggestion, the first issue of the *NHH* in 1915 bore the subhead "A NEW ERA in the DRUGLESS MOVEMENT."[10]

Purinton continued to demand a firm definition. He asked, "What is Nature Cure?" And he answered with refreshing candor, "I don't know. Furthermore, I don't know anybody who does know." After reading the main journals and magazines of several contemporary movements, he concluded that doctors and professors agreed that nature cure was the sole cure but that no two agreed on its definition or its components, and that's "why the Nature Cure has not been legalized in America."[11]

This diversity was, in fact, Lust's intention. He had earned several medical degrees, believing that each was valuable and increased his credibility. He hoped that a blending of knowledge and experience would help naturopaths adopt all the good in many systems and mix the theories and methodologies gracefully. But he lacked a successful strategy to make that happen. A statement on the front page of the December 1915 issue of the *NHH*, in very small print at the bottom, read, "The Editor does not assume any responsibility for signed articles. It is the desire of the Editor to give the widest possible latitude for discussion to those who may have different ideas concerning drugless matters." This meant publishing articles that differed radically, offering free discussion, and eliciting strong opposing views. He encouraged contributors to express their dissension. It was an antiestablishment, even anarchistic approach; there was no top-down imposition, and no one way was right. Naturopathy shared with other anarchistic groups myriad foci, a diversity of positions that became unmanageable, internal conflicts, and difficulty in identifying goals. It was like herding cats.[12]

Various departments of the *NHH* (such as chiropractic or dietology) continued to promote their own superior therapeutics. Lust in 1915 offered a brief, utopic, and inaccurate definition: "Naturopathy is a collective term for all that is good and rational in any system of healing. . . . Our drugless doctors all know what it means, but the public at large cannot grasp it yet." He announced with much excitement that the magazine would be called the *Herald of Health and Naturopath*, not the *Naturopath and Herald of Health*, but not until readers spoke their minds on the proposed switch; Lust refused to make this decision alone. Placing *Herald of Health* first signified the wide and contradictory views included in the term *natural healing*. The new name was adopted for a period, with readers' approval. The cover page of the January 1918 issue told readers that the journal would still represent naturopathy but that its larger mission was to show the world complementary treatments, to convert the public, and to render both surgery and poisonous apothecary needless.[13]

How did all of this work? The simple answer is, with much promise and great difficulty. One naturopath in 1918 said it was fine to keep all these therapeutics operating at once but that there must at least be agreement on the cause of disease and its removal. If there was no such shared foundation, there would be only temporary relief and sometimes grave harm. Lust seemed clear on this point, even if others were not. In an 1918 editorial, he asserted that chronic disease was cured by increasing elimination and increasing one's power of resistance. These pointed to the causes of disease, but ideas about how to achieve these results continued to range widely. When the Association of Progressive Naturopathic Physicians of California defined the profession in 1920, they said it involved homeopathy (a system Lust was quite ambivalent about) and eclecticism "but also embraces the good in every cult and system." This was followed by a twelve-line list of complementary therapeutics.[14]

In 1924 the journal was renamed *The Naturopath: formerly Herald of Health*, likely in direct response to the growing need to make the system distinguishable. For the first time, it was designated the "OFFICIAL JOURNAL" of the American Naturopathic Association, state societies and branches, the Academy of Healing Arts, the People's Welfare League, and World Reform. It was the journal of the American School of Naturopathy and the American School of Chiropractic (both Lust's). The definition of naturopathy was narrowed considerably to mean natural healing and sane, rational living. This return to basics was juxtaposed with the "Declaration of Medical Liberty Rights" two months later, a call to support all drugless healers so that they could practice their art.[15]

Lust was uncomfortable with these narrowing definitions. Soon the journal no longer referred to itself as formerly the *Herald of Health*, but the list of systems it represented expanded to include eleven. He wavered on rigid beliefs another way, and it was a sign of the times. In the spring of 1926 Lust warned students against too much anxiety in the selection of food, as it might destroy its pleasures. He temporarily eased up on wholesale condemnation of tobacco, alcohol, coffee, and condiments. Previously all of these had been absolutely prohibited. In predictable contradictory fashion, Herbert Shelton wrote a scathing indictment of condiment use the following month.[16]

Lust's desire for change in the way society viewed health consciousness also confused notions about naturopathy. He urged his general readers to call on "the friends of Natural Life and Nature Cure, of Antivaccination, Antivivisection, the movement of medical freedom, of humanity, and true American spirit, to join the movement of Naturisme." Naturisme, also naturism of nudity, was the public and private advocacy of nudism and living in harmony with nature. Naturopathy and the naturism lifestyle had much in common, including natural medicine and emphasis on environment, diet, health, nontoxic agriculture, and the liberty of choice. Yungborn in Butler, New Jersey, was geographically close to the hub of the fledgling movement in the United States. The American popularizers were German immigrants who established two nudist facilities (in 1929 and 1932), both in New Jersey roughly thirty-five miles from Butler. Lust's embrace of the naturism movement was a logical, if controversial, step, since healing sun baths with minimal or no clothes were part of Yungborn's treatments in the Rikli model.[17]

In 1935 Lust brought an abrupt halt to his own therapeutic inclusivity. It was an important turning point. He wrote to his disparate colleagues in the *NHH*, "Let's be NATUROPATHS" and nothing else. He had firmly shifted; he issued a call to naturopaths to join their state and national naturopathic associations, to support the recognized schools, to write laws that benefited naturopaths, and to lobby effectively. He refuted the rumor, begun in California, that his American School of Naturopathy in New York City was closed. He described the rumor as scandalmongering of the worst order, perpetuated by medical scalawags and likely some chiropractors. Internal frictions were bubbling to the surface, and Lust had seemingly had enough.[18]

The profession at this point was more focused. It included physiological, mechanical, and psychological sciences, which rebuilt, purified, and normalized

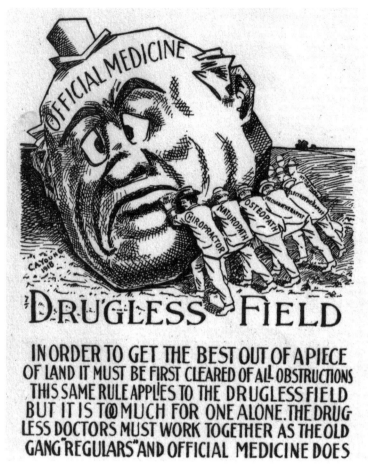

"Drugless Field." Official Medicine is the giant bolder obstructing the success in
the field of natural healing. It can only be pushed out of the way by collaborating
healers determined to improve the public's health—chiropractors, naturopaths,
osteopaths, mechanotherapists, electrotherapists, and hydropaths. From *Herald of
Health and Naturopath* 13, no. 10, whole no. 290 (Oct. 1918): 833. Courtesy of the National College of
Natural Medicine Archives, Portland, Oregon.

the vital forces of the body. With this new focus came more congruous naturopathic definitions in which the word *science* was increasingly invoked. Solid
naturopathic education would distinguish naturopaths from other sectarians
not similarly schooled. Professional naturopathy continued to be more aligned
through March 1937, but in April Lust announced that the American Naturopathic Association's Scientific Assembly would have twenty therapeutic emphases.

The basic sciences increased in importance; the new complementary therapeutics of autotherapy and homo-vibra ray and the scientific disciplines of obstetrics, gynecology, and endocrinology were introduced, and the original core principles of hydrotherapy, herbology, and dietetics remained in place. Cell science appeared later that year.[19]

THE AMERICAN SCHOOL OF NATUROPATHY AND THE AMERICAN SCHOOL OF CHIROPRACTIC

Education was one clear marker of a skilled practitioner and professional. This was vital given the multitude of modalities encompassed within the movement and its far from homogenous standards. From its inception the organized American naturopathic movement created colleges with uneven curricula and widely varying degrees of nature-cure instruction. Louisa and Benedict Lust, with their combined funds, opened the American School of Naturopathy (ASN), sometimes called the American College of Naturopathy, at 124 East Fifty-ninth Street in New York City in 1901. Louisa served as instructor of practical naturopathy. Benedict recalled in his memoir years after Louisa's death that they had planned a large school as a monument to their work, but despite their efforts, promises of additional funding had failed. The ASN offered a comprehensive two-year general curriculum, advanced and postgraduate courses, and summer instruction at New Jersey's Yungborn. The school was chartered in 1905 and legitimized a resident school and correspondence courses leading to graduation and the granting of the ND degree only to those in residence. Oddly, the school is absent from the pages of the *NHH* from 1907 through 1911, and the 1911 *NHH* masthead does not mention the ASN. Nevertheless, by 1914 the school had grown in enrollment and stature. The New York State Society of Naturopaths applied for a new charter to give naturopathic graduates the chance to attend public and professional lectures, demonstrations, and clinics so that they could be schooled in current ideas and methods. The forty-four students enrolled in 1914 included students from all levels, both full and part time, and at least eight were women (some used initials, which were indeterminate). According to Lust, each student had met the school's high standards, set by the dean, the faculty, and the examining board of the American Naturopathic Association.[20]

Although the ASN was a legal entity by state charter, it was repudiated by the medical profession. That did not stop the movement. About a half dozen other naturopathic colleges across the country ran advertisements in the *NHH*, and some sanitariums also provided instruction. The *NHH* announced their registration dates, fees, and application procedures. The problem was that some schools did *not* offer adequate naturopathic instruction during this period. The Physical Culture College of Chicago renamed itself using the term *physicultopathy* or *naturopathy*. This angered Lust, who protested against these comparisons because naturopathy was a complex science. He reminded readers that only members of the ANA could call themselves "naturopaths"; it was not a label to be used by any and all.[21]

Prior to 1919, the American School of Naturopathy drew older students who were largely water-cure practitioners. Texts were Kneipp's books. Then, a high-school education became a requirement for admittance, as it was for AMA-affiliated medical schools. Later, graduates completed a four-year course of study with a passing grade of 75 percent or more, and the ASN's curriculum was updated to include basic sciences, physiotherapy, phytotherapy, geotherapy, electro-therapy, and mechanotherapy.

In 1919 the school was incorporated under the laws of the District of Columbia and then subsequently in California, Illinois, Florida, and other states. The Lusts' building in New York City at first housed the Naturopathic Institute, Clinic and Hospital and the ASN. Four years later they changed the titles to the American School of Naturopathy and the American School of Chiropractic, as Lust had secured a new charter. It offered a four-year curriculum through the two schools.[22]

In 1920 the American Drugless University opened in New York City, stirring controversy. It promised to teach all phases of drugless therapy. One author called it the greatest institution with the most benefits and implied that it had Lust's support. Yet the 1920 bulletin of the ASN stated, "This [the ASN] is the only non-drug college in existence for 25 years. . . . And no other college of any single or several branches of non-drug therapeutics can equal it." So much for Lust's supposed endorsement. In fact an open letter from the secretary of the ANA made a point of informing readers that despite the articles claiming a liaison between the American Drugless University and the ASN, initial thoughts of collaboration had been abandoned when a takeover attempt by the new university was uncovered. Whether for the purpose of increasing its advertising revenue or courting allies, the *NHH* featured advertisements for the American Drugless University's first convention alongside advertisements for the ANA convention in 1920, a decision unpopular with some. Ultimately there would be no collaboration on fundraising. The new school was a defection that interfered with the ANA.[23]

Three years later, to illustrate the official nature of the ASN's credential, full-page photographs of the certificate of attendance and qualification awarded to graduates and the diploma for the ND degree appeared in the *Naturopath*. They, like the journal's new title, reflected the decision to make naturopathy the center of all the healing systems. Lust's dream of a reputable school of naturopathy had been realized. He also envisioned a nonprofit universal naturopathic college and hospital. Attendees at the twenty-first ANA convention were asked to pledge one hundred dollars toward it, but the goal was never realized.[24]

Because of military enlistment during Word War I, enrollment was a challenge in 1918. Schools without endowment funds suffered the most. Lust actively recruited female students, saying that "our profession is a most noble one, and women are most admirably suited for it. The woman is the natural healer. She possesses all the qualifications for the work, and more of her sex should be encouraged to take it up." He rejected the notion that refined and sensitive women should avoid medical work. Naturopathic colleges, he said, had high moral standards and wholesome curricula. In fact, from the late teens on, the numbers of female students, graduates, physicians, authors, and instructors increased, perhaps owing to the growing success of the women's suffrage movement. The increases were undeniably also a result of Louisa's and other female leaders' presence. By 1924, class photographs generally revealed a gender ratio of four men to every woman; in some, a ratio of three to one.[25]

Previous historians have cited a source from this era that claimed that the ASN graduated roughly eight thousand students. This is not possible given the only thirty to forty names listed for each year's class, and the pleading calls for recruits also make it unlikely. The claim of eight thousand graduates originated with Lust and appeared in the *New York Times* in 1924. His motivation was to emphasize the number of skilled naturopaths in the face of a deleterious charge. The ASN had run afoul of the law when a New York State senator sued the school, claiming that it drew undesirables to the upscale residential area and generated unpleasant noises. Lust concluded that the suit was the result of political pressure put on the senator by naturopathy's foes. More likely, it was related to the rabid antiimmigrant sentiment of the 1920s. By 1924 the ASN students hailed from nine countries, a fact the class historian applauded. Lust relocated the school to East Thirty-fifth Street, quipping: "Our new location is among congenial neighbors [away from the] snobbish neighbors, would-be millionaires, and lickspittles of the medical trust." Lust realized a profit from the school of thirty-five thousand dollars, which he added to his Insurance or Protection Fund, also known as the Greater School Fund, hoping these funds would launch a university of naturopathy.[26]

The ASN was a vehicle for professionalization, but it also provided opportunities for public relations and marketing. Lust said in his commencement address to the thirty-six graduates in 1923 that their diplomas symbolized the school's standardized education. The *NHH* heralded the school's nine teachers, who offered postgraduate courses in addition to regular courses.[27]

As usual, there was an inconsistent, ambivalent relationship between naturopathy and chiropractic in *NHH* advertisements for the ASN. Those ads that do not

"A Dignified Profession—Doctor of Naturopathy." At a time when the profession and the school depended on an increase in educated practitioners—and their fees—Lust recruited students by touting "the new marvelous science of Naturopathy." From *Naturopath* 28, no. 12, whole no. 350 (Dec. 1923): 720. Courtesy of the National College of Natural Medicine Archives, Portland, Oregon.

mention the American School of Chiropractic read, for example, "A Dignified Profession—Doctor of Naturopathy." A full-page ad in one issue names both schools and also includes a reassuring portrait of Lust. In 1925, only one month before Louisa's death, Lust claimed before an audience of two thousand that the classes graduated that year were among the schools' largest. It was, Benedict later wrote, the Lusts' biggest triumph and their last together.[28]

It was not until 1925 that the front cover of the *Naturopath: formerly Herald of Health* declared itself the "College Journal of the American School of Naturopathy and American School of Chiropractic," a sure sign that the school was entrenched as a central part of the naturopathic profession. But not for long: this link disappeared shortly after Lust became increasingly wary of straight chiropractors.

Naturopathic schools were signs of the strength of the movement and its legitimacy, so the *NHH* wrote about the plans, openings, and faculties of those schools outside the New York area. The existence of a school correlated directly with the number of professionalized practitioners in a state and their legislative efforts. Rarely were curricula given in detail, except those of schools most closely affiliated with the ASN. Many cropped up in states without licensure or legal status, affiliated with individuals whose names appeared just once in the *NHH*. Plans for others appeared in writing but never came to fruition. Yet the sheer amount of space devoted to schools speaks to the movement's popularity, a growing patient base, and the need for professional training.

One legitimate, highly reputable institution, which produced the most naturopaths in California, Oregon, and Washington, was the California University of Liberal Physicians, opened under the leadership of Dr. Carl Schultz in 1914. He ran it in conjunction with his Naturopathic Institute and Sanitarium. It was approved by the state boards of naturopathic examiners in Utah, South Carolina, and Tennessee, among others. The 1923 business directory and buyers' guide listed five schools: both of Lust's schools, the ASN in Philadelphia, the ASN in Comal, Texas, and the Southern College of Naturopathy, in Tampa, Florida. The next year it listed six. By 1926 the number had increased by five, with one each in Philadelphia; Portland, Maine; and Hartford, Connecticut, and two in Los Angeles. Any given issue carried several small advertisements for sanitariums, cures, and institutes offering classes, although they did not call themselves schools. In the course of one year about one hundred such advertisements appeared. In the 1930s and 1940s, colleges flourished, including the College of Naturopathic Physicians and Surgeons in Philadelphia, which celebrated its eighteenth anniversary in 1940.[29]

The schools garnered mixed reviews. Lust reported at the 1927 annual ANA convention in Minnesota, "We have good schools, of excellent standards." In fact, investigators for the Council on Medical Education and Hospitals of the AMA in 1927 inspected all schools of naturopathy and chiropractic in the country, including the ASN/ASC. The council's findings were condemnatory of the ASN, but their observations must be considered in the context of the ongoing virulent campaign against natural healers. The investigators found twelve schools with fewer than two hundred students. The AMA report stated that the ASN offered night classes only, took up two and a half floors of the building, and had a small demonstration room with an osteopathic table, five chiropractic adjusting tables, and a small chemical laboratory. The inspector was told that twenty students were enrolled in 1927 and that fifteen had graduated in 1926. A four-year degree was awarded after students attended nine months per year and paid $250 in tuition; texts and supplies were available, and discounted, through Lust's store. The report made no mention of a clinic or hospital. Lust, who was in Florida at the time of the inspection, did not meet the reviewer, who wrote, "His school is a very sorry looking affair." He then mocked the dean, Sinai Gershanek, saying that he was deaf, nervous, and thick headed. *JAMA* published the report and had to retract and apologize for the statement about Gershanek. The findings contributed to the AMA's growing stereotype of naturopaths as uneducated and backward, a stereotype increasingly common in the public sector, especially after the availability of sulfa in 1937. The ASN/ASC closed in 1942. By then Lust had established the New York School of Massage and Training School for Physiotherapy, which he ran until his death in 1945.[30]

Legislators had tried to shut down several schools indiscriminately. Lust's memoir states that the New York Webb-Loomis Medical Practice Act, passed in 1926, made the ASN invalid despite its charter. In 1934 the McNaboe Bill axed New York chiropractic schools as well. "However," Lust wrote, "the American School of Naturopathy was also chartered in Maryland, California, Minnesota and Illinois." He proudly recalled that despite legal setbacks, the ASN had never closed all of its doors and had remained the ASN through its charter in Washington, DC. It was able to graduate naturopathic doctors without interruption, despite legal harassment and attempts to close it.[31]

The ASN, like countless American institutions during the Great Depression, struggled to keep going. The Depression caused economic disaster for Lust. Savings accumulated by him and his brother Louis, reported to be $897,000, and Louisa's remaining funds were in a private bank that closed its doors. In a moment all of their savings, accumulated over a quarter century, were gone, he wrote. Plans for the establishment of a naturopathic university ended immediately. Lust kept on, nonetheless, operating the ASN and not allowing himself to become dejected. "I had a job to do and that job was to keep the school going at all costs."[32]

In the 1930s, partnerships and community connections were the lifeblood of the ASN. The school relocated to Twelfth Street, joining the World Reform League and the Benedict Lust Community Center at the same address; the latter offered meetings for members and the public. Graduation classes ranged in size from eleven to eighteen, with students from the contiguous United States and (in 1931 alone) from Italy, Switzerland, the Philippines, Germany, Panama, and Hawaii.[33]

Despite Lust's public optimism, the school's economic difficulties persisted. He made a plea for endowments in November 1935, noting that he and his wife had contributed ten thousand dollars annually for the past forty years. He urged anyone making a will to remember the school and provide an endowment for it.[34]

HOME STUDY, CORRESPONDENCE COURSES, AND DIPLOMA MILLS

Naturopathic credibility was compromised by courses offered through home study and correspondence, raising the specter of diploma mills. But some of those were extension courses for the public rather than for professional training. The home study course offered in the fall of 1920 by the American College of Naturopathy (synonymous with the ASN) did not claim to be a professional course, nor did it grant a diploma. It gave individuals the knowledge to heal themselves, their families, and maintain health. The fee was one hundred dollars; students received the necessary books, including the *Naturopathic Encyclopedia*.[35]

Correspondence courses and diplomas were offered by various nature-cure colleges. Lindlahr College of Natural Therapeutics in Chicago offered postgraduate extension courses. Advertisements detailed the subjects and the wall charts and texts that students received. Content in the correspondence courses correlated with that in the resident courses. The brick-and-mortar school of William F. Havard, ND, the Academy of Healing Arts, also in Chicago, offered correspondence lessons. A serialized version of Havard's curriculum ran for two years in the *NHH*. It focused on scientific issues, among them cell structure, and disease treatment and suppression. It also addressed the physical, moral, and mental aspects of health.[36]

In 1919–20 a lengthy home course in mental science created by Helen Wilmans debuted in the *HHN*. It comprised twenty lessons, and reprinted back issues could be purchased for twenty-five cents each. The course dealt with mental science, New Thought, therapeutic suggestion, psychotherapy, and spiritual and metaphysical research. Each lesson was six to eight pages long and ended with a note on directions for patients.[37]

In the winter of 1926 Lust began his own home study course on naturopathy. The lessons discussed theory, cell structure, nourishment, the body's inner workings, the role of mineral salts, weight and health, and sugar as a toxin. Lust also explored the fat, carbohydrate, nutritive salt, albumen, fiber, and water content of countless foods and their times of digestion; packs and compresses; the use of water; and the importance of good teeth in health.[38]

By 1929 Lust was operating the Preparatory School of Naturopathy out of the Lusts' East Thirty-fifth Street institution, which advertised "Study Naturopathy at Home!" This home extension course of twenty-four lessons covered the hard sciences, specific diseases and their treatments, dietetics, and numerous complementary therapeutics. Lust recruited anyone who wanted to be a health specialist, such as nurses, chiropractors, and osteopaths, to take the postgraduate course, which was also geared toward drugless practitioners.[39]

During the Great Depression home study courses through the journal proliferated. It was a way to educate the public with little overhead, and it made education free to all those seeking enlightenment. Lust's "Post-Graduate Study of Naturopathy for Students and Practitioners" (1935) was a serialized course in food chemistry and allied subjects. Its lessons addressed classification of foods, proteins, acids, chemical agents, and heart action, among other topics. Each detailed lesson was approximately four to six single-spaced pages long, and included definitions, effects and results of treatment, opinions in favor and opposed, and historical beginnings. The lessons continued through 1936. In 1937 and 1939 an examination for this course appeared in the journal. How students submitted their exams and methods of evaluation were not given. A new home-based course with case studies charting treatments and outcomes began in 1936. Wilmans's course, titled "The New Psychology and Brain Building," ran again in the middle of the decade.[40]

In the mid-1940s Lust offered "The Modern Home Study Course in Naturopathy"; the fee for the one hundred lessons and all of the required texts was one hundred dollars. The lack of clarity about the clientele for this course, as well as the "Post-Graduate Study of Naturopathy for Student and Practitioner," and the uses to which the clientele put their diplomas did not help the profession's status. But difficult times, including legal persecutions and economic hardship, made these correspondence courses vital; they kept the values and therapeutics of naturopathy alive in the pre–World War II years and generated funds.[41]

In 1939 Lust spoke glowingly of the successes of the naturopathic schools and how they countered what he saw as the social and health damage created by regular medicine. He reported that despite discrimination against them by the medical trust, not one school had gone out of business. "Our schools may be 'underground' but they are a million times better than these institutions of the medical trust, medical centers and universities, which are operated simply to keep the people ignorant and to maintain a system of class oligarchy, intolerance and dictatorship."[42]

Lust's enthusiastic praise of the schools was tarnished by the close affiliations he had with practitioners whose schools churned out diplomas like a counter-feiter's press. Even allowing for the enmity between natural healers and allo-paths, some evidence greatly discredits these diploma mills. F. W. Collins is a good example. He had twenty or more institutions crowded into one building in which twenty-six students studied. Each graduate received diplomas from two to three of these various schools. Located in Newark, New Jersey, and called the First National University of Naturopathy, it was an amalgam of his Mecca College of Chiropractic, his New Jersey School of Osteopathy, and his United States School of Naturopathy. He also chartered the United States School of Physio-therapy and the American Academy of Medicine and Surgery. The AMA's 1928 evaluative report found about fifteen other such colleges. Tuition for a degree was six hundred dollars. No preparatory education was required, and the report concluded that all of this fooled the public. The findings of the report resulted in the entire Mecca College of Chiropractic being charged with practicing medi-cine without a license in 1926, the year the Webb-Loomis Medical Practice Act was passed. Lust took flack for his association with Collins, and whenever Collins's credentials were attacked, Lust responded that they collaborated for the great cause of medical freedom.[43]

Collins signifies two of Lust's weak spots: his fluctuating relationship with chiropractors and his support of controversial healers. Lust lectured at Collins's Mecca College of Chiropractic and also visited the chiropractic college in Dav-enport, Iowa, run by B. J. Palmer. Palmer despised Lust's liaison with Collins, saying that all of Collins's degrees, save his 1907 diploma from the ASN, had been shabbily acquired. Many naturopaths, like Lust, viewed Collins favorably and considered him an integral part of the movement, so Lust's affiliation with him was not unusual under the circumstances. In addition, they had a common enemy: the AMA viewed both Lust and Collins with derision. So for Lust, "the enemy of my enemy is my friend," and given Lust's belief that AMA leaders lied to destroy irregular practitioners, Lust would have had a hard time believing accusations made by them against Collins.[44]

Lust's opinion about chiropractors swung from embracing the mixers to resenting the political battles with the straights. In his memoir Lust claims that the competition from the new cult of chiropractic was persuading prospective healers to pursue a dangerous one-therapy practice. To combat this, he procured a charter to teach chiropractic and grant DC degrees in 1923. Looking back, he wrote, "This act I look upon as one of my mistakes." He came to believe that single-system healing was flawed. For years Lust had labored endlessly to collaborate with mixer chiropractors, often with great success. At the time, however, Lust expressed no concern that chiropractic needed to be slowed. He openly accepted the system and spoke glowingly of its therapeutic value. His shifting feelings over time explain the on-again, off-again mention in the *NHH* of the American School of Chiropractic with its sister school, the American School of Naturopathy.[45]

F. W. Collins was certainly not alone in the diploma mill business; allopaths could also buy their degrees. In 1923 the *New York Times* reported wide-scale diploma frauds that allowed people to buy both high-school and medical diplomas. One witness named Sachs in Connecticut admitted to selling a thousand fake high-school certificates. He knew of fifteen to twenty thousand doctors practicing unlawfully in the country, two hundred of whom he knew personally. The diploma rings ranged from Boston to San Francisco to Florida. In Illinois, Indiana, and Wisconsin preliminary certificates could be bought for $10, and Arkansas and Connecticut were, he said, the easiest states to work. A medical license in Connecticut could be bought for $800. Sachs said he had routinely taken the supposed graduates of the Kansas City College of Medicine and Surgery (where diplomas cost $200–$250) to Connecticut, where they quickly had Connecticut licenses in their pockets. With greater demand came higher costs. The cost of Dr. Alexander's diplomas from a Kansas City school spiraled from $200 to $500, then to $800. These increases were due in part to the kickbacks given to those who issued the false licenses in Connecticut. When Sachs visited Alexander's school, he saw neither classes nor professors. As a result of the *New York Times* exposé, Connecticut's Waterbury Medical Society passed resolutions to decrease the number of state medical boards from three to one, thus eliminating these scandalous practices and installing one competent, responsible, and honest board.[46]

EDUCATIONAL STANDARDS

Abraham Flexner's AMA-sponsored 1910 study of medical schools led to an increase in national discussions of standards in medical education. As the AMA set higher standards for medical schools, the state boards and legislators demanded higher educational standards for sectarian schools as well. By the 1920s many naturopaths also believed that only consistent training and rigorous educational standards would ensure a professional standing and national legitimacy. But there was debate among naturopaths about those standards. Basic science laws gave naturopaths an incentive to at least agree on the core education. Reaching agreement was not easy, given that they could not even agree on which therapeutics and natural substances should be considered drugless. By the 1930s, naturopaths had joined the national debate about the length of study required for a qualified practitioner. For example, from the School of Drugless Therapy in Nashville a naturopath wrote against the idea of mimicking medicine with a four- to five-year required course of study, as chiropractors had done. But he agreed that a course of merely eighteen months was inadequate. The school decided on a middle ground of 3,661 hours over thirty months.[47]

At the national level standards were set higher. Among trained naturopaths, many favored four years and no fewer than four thousand hours. Hawaii, where naturopathy had been legalized since 1925, was held up as an example. There, a total of 4,520 hours was divided among training in anatomy, histology and embryology, chemistry and toxicology, physiology, bacteriology, hygiene and sanitation, pathology, diagnosis, naturopathic theory and practice, obstetrics and gynecology, clinical practice, biochemistry and dietetics, therapeutics and jurisprudence. Anatomy required the most hours, 650; jurisprudence the fewest, 50.[48]

The common belief was that all allopaths received training that was more rigorous than that in the sectarian schools. Data from the era provide a mixed portrait. In some areas allopaths' requirements were higher, but a comparison of requirements for allopaths, naturopaths, chiropractors, and osteopaths, *depending on the schools*, reveals considerable congruity. Allopaths did their best to exaggerate and promote the belief that their schools were the most rigorous. In 1940 the *NHH* printed a detailed chart, which it claimed was based on the latest available data, that showed the number of hours required for a medical education (3,600), a naturopathic one (4,580), an osteopathic one (4,410), and a chiropractic one (3,528). All required four years of high school or the equivalent. Medical schools required two years of college as well. In response to charges that naturopaths were unschooled or deficiently schooled in the hard sciences, the

State University of New York reported that compared with the medical school requirement of 648 hours in anatomy, the requirement for naturopaths was 850 hours. Medical schools required more than two hundred hours in pharmacology and therapeutics and another 612 hours in surgery; naturopaths had no course offerings in these categories since the practices were antithetical to their beliefs and it was illegal for them to pursue them. These requirements applied to seventy-three medical schools in the United States.[49]

THE AMERICAN NATUROPATHIC ASSOCIATION: "DOCTORS, WAKE UP!"

The Naturopathic Society of the United States was founded in 1902 after Lust dropped his primary identification with Kneipp. Yet in his memoir and in all of his writings Lust claimed that it was created in 1896 so that he could assert its establishment as part of the nineteenth-century natural health movement. He also wanted to show that there had been an organized and cohesive plan of action from one date, as early as possible, to prove stability. In fact, of course, any national naturopathic endeavor was far from cohesive or stable, but branding was key to creating a profession. In 1907 the masthead of the *NHH* read, "The Official Organ of the Naturopathic Society of America."

The organization changed its name to the American Naturopathic Association in 1919 due to the economic woes associated with the Naturopathic Society, and over time it added eighteen state organizations. Lust counseled practitioners to form a state society or a branch in their vicinity in order to enjoy the advantages of a local and a national organization. In 1922 he described the ANA as "a 'union for the mutual advancement of all healers who abjured drug therapies,' an organization 'under [whose] wings all practitioners, all schools, that use no drugs can find shelter.'" To facilitate the transition from the Naturopathic Society to the ANA, Lust published the first *Universal Naturopathic Encyclopedia Directory* in 1918. Annual additions to the *Directory* appeared in the *NHH*. This 1,426-page book had the motto "In Unity There Is Strength." Lust carefully collected the names of practitioners of the "Rational School of Medicine" in America, Canada, and the British Isles. He claimed that there were forty thousand practitioners of naturopathy and that all could benefit from the *Directory*. He called it a vital tool of the trade and reached out to all, stating that he wanted every professor of drugless therapy to become his friend and coworker in this great cause. Articles in the *Directory* were written by the top names in the field, all of them NDs, and a few were MDs as well. Lust opened the volume by explaining the principles, aims, and program of the nature-cure system. The *Directory* provided lists of books, complementary publications, and book reviews. It contained advertisements for sanitariums, schools, and individual doctors. Photos of buildings provided proof of institutionalization. Drawings and photos of prominent founders, along with those of the current leaders and authors, gave a visceral sense of the constant growth and progress in the profession.[50]

With this mammoth text as his primary tool and proof of a professionalized discipline, Lust traveled the country promoting the ANA. He claimed that he was licensed in nearly every state in which naturopathy had been legalized, beginning with California in 1907. Through the ANA Lust led nationwide legislative, organizational, and publication efforts. His travels were legendary. On one trip to the West Coast he spoke at naturopathic institutions in Philadelphia; Chicago; Kansas City, Missouri; El Paso, Texas; Los Angeles; San Francisco; and Portland, Oregon. Lust was elected ANA president every year until 1921, when he was elected president for life, in part to quell internal factionalism. He held that position until his death in 1945.[51]

In the *NHH*, most aspects of professionalization were linked with the ANA. In 1915 templates of business cards were displayed to encourage practitioners to use them in a standardized fashion to signify their professionalism. Special rates for the *NHH* were extended as a benefit to ANA members. In 1918 Lust stressed that a membership of only one hundred deeply committed members would make the movement an unstoppable force. The paltry number of members prompted Lust to call his colleagues slackers, and he urged them to look into the future and realize ANA's value.[52]

In 1923 Lust designed tactics to increase a sense of collective professional identity. One was an ANA color button for practitioners to wear, along with pins and brooches for women, fob chains, cufflinks, and scarf pins. He published an addendum to the *Universal Naturopathic Encyclopedia, Directory and Buyers' Guide* that listed accredited practitioners, schools, institutions, societies, publications, food outlets, and publishing houses by state. Foreign affiliates were noted in Mexico; Vancouver, Montreal, and Toronto in Canada; Cuba; Kingston, Jamaica, West Indies; and Sydney, Australia. Three hundred thirty practitioners were listed in the states of California, Florida, New Jersey, New York, and Pennsylvania, of whom sixty-two were female, with another sixty-five using non-gender-specific initials (many women used only initials professionally). A significant moment in professional development occurred in 1926, when the Spanish Naturopathic Society was created.[53]

The front page of the *Naturopath*, the official journal in the mid-twenties, listed affiliated state societies. The societies deemed well organized in 1923

included those in New York, New Jersey, Illinois, California, and Florida. Because bolstering the ranks of the ANA was a constant goal, Lust counseled every member to work to increase its numbers. He advocated recruitment of all
drugless practitioners to increase membership, explaining his insistence on
inclusivity.[54]

The drafting of a constitution for the ANA in 1929 was another step toward professionalization, but just who it represented and governed is difficult
to ascertain. Presented at the thirty-third annual convention in Philadelphia,
it formalized the organization's name, objectives, component societies, fel-
lows and officers, board of trustees, meetings, funds and expenses, and official seal. However, this document may have been created by the new western
American Naturopathic Association, rather than by the original group, since
it declared that officers should serve for one year or until their successors were
elected or appointed (whereas, as stated above, Lust was granted a lifetime
tenure in 1921). Because of the competing agendas of groups having the same
name, the only thing certain is that years later Dr. Robert V. Carroll, chairman of the executive committee of the newer western ANA, noted at the 1935
annual convention in San Diego that a new constitution and bylaws had been
created.[55]

Lust was just one of several leaders in the movement who made all efforts to
personally professionalize in the paradigm of the day and advocated for consistent standards for naturopathy. Among the many was Jesse Mercer Gehman,
ND, a close associate of the Lusts'. Gehman founded and served as president of
the American Vegetarian Union and was the first president of the ANA after
Lust's death. Another naturopath central to professional development was
Dr. Theresa M. Schippell, secretary-treasurer of the ANA, who later served as
president and resumed publication of the *Naturopath* as a professional journal.
She had been key in getting naturopathy recognized in the District of Columbia. Sinai Gershanek, who served as dean of the ASN in the 1920s, also played a
valuable role in the journey to legitimacy.[56]

The growth of naturopathy in the 1930s and 1940s was evidenced in the focus
on sanitariums and hospitals. Connecticut's New Haven Community Hospital
was open to physicians of all schools, regardless of affiliation, so that patients
could be treated by their own physicians. On staff were naturopaths, registered
nurses, trained attendants, and a trained physiotherapist. The Ka'ōnohi Naturopathic Clinic opened in Hawaii under the direction of the phytotherapist and
naturopath Alexander K. Ka'ōnohi. Regional demand for naturopathic physicians grew, speaking to the system's popularity; in New Mexico trained practitio-

ners were recruited. State associations more often collaborated with the national office. Among them were associations in Michigan, Washington State, Colorado, and Oklahoma.[57]

Liaisons between naturopaths and other drugless healers continued to expand and contract, as always, in the attempt to secure legal rights. The ANA strengthened its ties with the National Association of Naturopathic Herbalists in California, it joined its educational efforts with those of the Universal Health League and the World Naturopathic League, and New Jersey's naturopathic association joined with all drugless healers in the state in an attempt to pass a drugless bill.[58]

Two other milestones solidified professionalization efforts. Lust's ASN in Cuba had long signaled solidarity with Spanish-speaking people; the curriculum was taught in Spanish, which prompted Lust to publish Spanish tracts. Outreach in 1926 resulted in the Spanish Naturopathic Society, bringing in more practitioners and patients. Lust was the dean of the school, and the nationally respected A. L. Lopez served as secretary and on-site leader. The Tampa Latin Naturopathic Medical Group and the West Coast (of Florida) Naturopathic Physicians Society, facilitated by the publication of a Spanish-language journal, met in July 1943. They maintained close and productive ties with Cuban naturopaths.[59] The second milestone was a shift in favor of scholarly and laboratory-based research. Harry Riley Spitlers's edited volume *Basic Naturopathy* (1948) is evidence of this, as is A. W. Kuts-Cheraux's *Naturae Medicina* (1953), which chronicled botanical modalities, published under the auspices of the American Naturopathic Physicians and Surgeons Association.[60]

CONVENTIONS

The ANA conventions were the primary means of bringing practitioners together, sharing ideas and goals, and reporting on legal issues. These national meetings also modeled process for state conventions. Policies were adopted; state representatives gathered, most with an eye toward building community; priorities were set; internal schisms were addressed; and most importantly, therapeutic and philosophical ideas were shared. Women's involvement shows how central they were as leaders, healers, and organizers. Planning and attendance were dominated by people of varied Euro-American ethnicities, but there were also Latinos, South Asian Indians, and occasional African Americans. Invitees were drugless practitioners of any system or school and the journal's readers. There were lectures, discussions, demonstrations, and exhibitions, some of which took place in neighboring cities. The convention in 1917 was a three-day event in which sessions were held in drugless colleges in New York and in Newark and Butler, New Jersey. "Don't fail to be there!" Lust exhorted.[61]

Lust saw the conventions as ideal public relations and recruitment tools. They were referred to variously as conventions, congresses, and international congresses. The 1923 convention in Chicago was the first one at which presenters were from every sect, including allopathy. Others represented old-time nature cure, hydrotherapy and diet, naprapathy, osteopathy, chiropractic, neuropathy, and physiotherapy.[62]

Photographs in the *NHH* depicted full banquet halls with hundreds of attendees dressed to the nines. Lust claimed with marketing hyperbole that there were ten thousand attendees at the Los Angeles meeting in 1924; in reality the number was likely much lower, if the several hundred people shown in the dinner photos are any indication. Reports meticulously listed delegates from the various states, validating (and advertising) members' contributions and their buying into the processes and the cause. In 1923 the delegates came from twelve states and Canada. The halls were festooned with patriotic banners, and women were ever present, although it is difficult to know who were practitioners, spouses, or both (the same can be said of the male attendees). At the 1935 San Diego convention there were fifty-two national vice presidents and state representatives listed from forty-eight states, as well as Hawaii, Panama, the British West Indies, and the Canal Zone.[63]

Lust used the *Naturopath and Herald of Health* to place professional action and ideology in the public eye. The *NHH* advertised convention agendas on a second front page preceding the journal's index. Statements of purpose and the full program details confirmed conventions' importance. Calls for attendance were actually public relations for participation in the movement: "You will leave loaded with enthusiasm, your courage renewed to continue the fight for freedom"; "Great things are up for consideration and some very important decisions will be made." Tourist sites in the host cities encouraged attendees to make the most of their visit. Welcome addresses were delivered by the who's who of the field. In 1923 Henry Lindlahr, MD, pledged legal action and drugless harmony. The main speakers that same year included Bernarr McFadden, Dr. F. W. Collins, and Dr. Candrian, a famous health counselor from Wisconsin.[64]

There were full reports during the months following conventions. Of the 1918 Cleveland event a reporter wrote that all who had attended had learned a great deal and had their hazy conceptions cleared up. A plan of action was adopted for the work to be done. The mayor sent his greetings, and the issues tackled included the ANA's problematic alliances with drugless physicians and legal issues.

The reports were extremely detailed. The text for the New York City convention in 1920 covered seven single-spaced pages, and that did not include reprints of the papers presented; the report was indicative of the importance of the conventions. Lust gave welcoming greetings and stressed the need for internal harmony and solidarity. Speakers were named, their professions indicating the breadth of the movement and the effectiveness of conventions as movement-makers. Attendees were practitioners, directors of schools, authors, editors of journals of other movements, and founders and leaders of complementary movements, such as the Anti-Vivisection Society. Papers addressed yoga, mind power, hydropathy, drugless freedom, suggestive therapeutics, mechanical actions, vegetarian philosophy, diagnoses of the eye, causes of disease, and chiropractic and chromotherapy, among other topics.[65]

There were also brief reports on the Spanish and German sections of the meetings, revealing, for example, that Dr. Carlos Brandt was publisher of *Cultura*, a Spanish naturopathic magazine. The business section included reports from delegates and officers. The *NHH* even provided the meal menus, and musical performers were thanked and praised. Most reports ended with comments on the good fellowship that prevailed, although some conventions were overshadowed by recitations of legal problems. Even that publicity served to foment resistance and solidarity. Lust said, "Let us do more damage," meaning intentionally run afoul of the law in the name of progress.[66]

Like all trade conventions, the ANA's featured exhibitions and demonstrations with charts and graphs; displays of food and other products; photographs from individual sanitariums; equipment and other paraphernalia; and the distribution of health newsletters, journals, magazines, and books. The Battle Creek Food Company was a regular vendor, reflecting John Harvey Kellogg's commitment to vegetarianism and hydrotherapy.[67]

Minutes of the executive sessions showed immense progress in professionalization. In 1926 they addressed the *Naturopathic Directory*, letters from practitioners in various states, the need to place the ANA on a solid business footing, the imperative to organize state associations, problems with collecting dues, and the appointment of a national organizer to increase memberships. To be on par with other health professionals, by the mid-1920s convention papers and exhibitions were to be scientific and advance the education of physicians and their clients. By using the term *physician* during this period of scientific institutionalization, they signaled to allopaths that their laws would not stop a growing field.[68]

The conventions were particularly effective in making the public aware of the ANA during the twenties and thirties, when natural health was part of popular culture. One convention report advertised Dr. Jesse Mercer Gehman's daily radio talks over the bigger metropolitan stations. People liked to hear naturopathic truths, Lust said, and they were eager for real health information. In 1922 President Warren G. Harding, the first president to recognize naturopathy, greeted each Washington, DC, convention attendee at the White House with a hearty hand shake. It was a high-profile coup, followed by sadness when Harding died the following year. Naturopaths blamed allopaths for mishandling his treatment. Other public nods came in 1923 when the convention was recognized by the health commissioner of Chicago and the *New York Times* gave positive coverage to Lust's talk on the benefits of shower baths and fasting to "stop old age."[69]

THE ENEMY WITHIN

The purpose of the conventions, or congresses, was to provide unity and direction. Yet they frequently provided a forum for contention, personal and political agendas, and attempted coups. In his presidential address to open the convention in 1920, Lust told his listeners, "We are assembled here together for the purpose of bringing greater solidarity in our ranks, for settling our differences with all factions of the art of drugless healing, [and] to spread the gospel of right living." But by 1925 Lust had become aware of the degree to which dissension had risen. The "enemy that sprang from our own ranks" with Louisa's death that year was made up of those who wanted power diverted away from the centralized national leadership. Anticipating Lust's weakness following Louisa's death, nine conspirators thought they could push him aside and run the ANA and the school, as well as take over the two publications and the two Yungborns. He lambasted other kinds of traitors, men who joined the movement and made divisive moves, such as inserting an *e* into *naturopathy* (*natureopathy*) and claiming they had founded it. Most of these individuals failed, but later Lust recalled that to the extent that they succeeded with "new groups, new associations, new magazines and new schools they have weakened the universal proposition on which Naturopathy was founded, developed and recognized. They were our worst enemies." His belief in the need for a solid front against allopaths was so strong that he said their deeds were more damaging than medical basic science laws, the conniving of the medical trust and their guardians, and public health officials. He consistently refused to name names, which he believed would fuel the war of words, damage public credibility, alienate fence-sitters, and give recognition to his—and the movement's—enemies.[70]

At the same time, there was carefully crafted publicity about harmonious mutual support. E. K. Stretch, DO and ND, reflected this cooperative spirit when

Benedict Lust and twin nieces Helen and Lauren Lust, 1926, one year after Louisa's death. Their companionship brought him comfort. Reprinted by permission of Lauren Lust Proctor.

he was offered the job of editing the Osteopathy Department in the *NHH* in 1915. He wrote that he would edit without prejudice against any faction or school of drugless practice. Sects' relationships with naturopathy were guided by their leaders' personalities and individual leanings. At times entire schools, such as the Atlas College of Osteopathic Medicine and the Connecticut College of Chiropractic, were deemed friendly to drugless practitioners. Even when the *NHH* lambasted chiropractors, one might also find a notice proudly announcing chiropractic training at the American School of Naturopathy.[71]

Despite Lust's claims in his memoir of harmony in the early years, there is evidence to the contrary. Purinton, who served as interim editor of the *NHH* in 1915, applauded Lust's campaign of organized cooperation to benefit everyone. He recalled attempts to do so as early as 1902, but he believed that drugless leaders had been too blind and stubborn to achieve this. Perhaps, he added, naturopaths had also been too rash and inexperienced. Certainly Lust was not always the ideal diplomat. Late in the second decade of the twentieth century, speakers at the ANA convention had been severely criticized for their attacks on straight chiropractors. Lust commented that the criticism had not been aimed at all chiropractors, only those narrow-minded ones who claimed that spinal adjustment cured all without correcting the disease-breeding habits that had caused the disorder. In 1920 the Lucas affair reflected a lapse in Lust's judgment in an attempt to be inclusive; the result was very awkward and fueled discord. Dr. Alzamon Ira Lucas complained angrily in an open letter to B. J. Palmer that Lust had invited members of the American Drugless Association to join the naturopathic convention. By Lucas's telling, Lust had then changed the speaking time of a prominent chiropractor, who could not present at the new time. Lucas claimed that Lust had given the chiropractor such a hard time that the president of the American Drugless Association had had to intervene. Chiropractors had been highly offended, said Lucas, and he warned Palmer to be wary of Lust and not presume him to be a friend.[72]

Name-calling and subterfuge often peaked at ANA conventions. At the annual convention in Chicago in the fall of 1923, Henry Lindlahr gave the welcoming address. To curb the rising tempers, he chanted a series of affirmations encouraging all attendees to work as one. The affirmations, each one beginning with the words *"We will,"* included the following: "achieve unity, present a solid front, unite our best efforts, sink our petty differences, and further the cause of Naturopathy" (emphasis in original).[73]

As the ranks descended into discord, the public viewed all drugless healers as generally one mass, as exemplified in the unflattering 1926 *New York Herald Tribune* article title "Drugless Cults Organize to Fight for Recognition." That year the strife was ugly and public. Some naturopathic leaders created their own organizations, schools, and state societies in direct opposition to Lust's centralized leadership. For example, Washington's Dr. John Lyden, a graduate of Lust's school, created the term *sanipractic* to distinguish himself from naturopaths. Whether he did this for political expediency, public relations, his divergent philosophy, or personal gain is unclear. The sanipractors' own internal conflict then escalated so dramatically that the *Naturopath* condemned them for creating an unfavorable impression and sullying the name of sanipractic.[74]

In 1927 a takeover attempt resulted in a new publication, the *Journal of the American Naturopathic Association*. This marked the first of several attempts by combatants adopting ANA titles and committee names to which they had no credible claim. The more embattled that Lust or the national headquarters and the old-time naturopaths became, the more his supporters rallied around him, extolled his virtues, and increased efforts to make shows of strength. The *NHH* described the 1931 national convention as "an epochal event in the history of Naturopathy characterized by the spirit of progressiveness, harmony and devotion to the Great leader of naturopathy, Benedict Lust."[75]

By the mid-1930s factionalism threatened to destroy the ANA. Professional cohesion required an undisputed leadership, but Lust was constantly challenged. A 1935 announcement noted the creation of *The Journal*, the official organ of the United States Naturopathic Association, Ltd., of California. There was also an announcement of a National Society of Naturopaths, whose affiliation with the original ANA was vague at best. To set the record straight (for the umpteenth time), in September 1935 the *NHH* listed the members of the National Legislative Committee and the Membership Committee of the original ANA. At the convention that year these committees discussed professionalism and unification. A Kansas practitioner laid out once again the stakes for a unified front: unification meant progress, cooperation, promotion, education, legislation, and protection.[76]

By the mid-1930s the confusion and mayhem were so pronounced that calls to naturopaths to professionalize and organize were in direct response to these public disagreements. The ongoing need for a national identity and legal recognition was more important than ever. In 1935 the Arizona division of the ANA filed trade names and went to court for use of the words *naturopathy* and *natureopathic* (with an *e*). One *NHH* author complained that "the A.N.A. is being used by every upstart, every claim staker, every crack pot for the enhancement of their own little ego." Also in 1935, B. J. Palmer and the straight chiropractors sold out and helped defeat the naturopathic law in California. Lust that year railed against the International Non-Medical Alliance, or INMA, proposed by Washington and Oregon, saying that it discredited naturopathy; he described it as another new organization, like so many before, that would weaken the ANA. "They get us nowhere. They are moves for our ultimate defeat and annihilation." What would the legislature and the public think "when they discover we are a disorganized mob with no one organization at the head?" But Lust had not said all he had to say on the topic: after the usual clashes at the annual ANA convention that year, he said, "We are clearing our ranks of . . . opportunists, fake naturopaths, criminals, pretenders, deceptionists, double-crossers, traitors, [and] false friends." While he did not single out individuals, he focused on groups or laws that undercut the eastern branch's leadership, the ANA, or naturopathic interests.[77]

Lust also turned his frustration and sense of helplessness toward Dr. Robert Carroll and Dr. W. Alfred Budden in Washington State, founders of a separate American Naturopathic Association in the western United States. Carroll served as that organization's first president. The loyal Lust camp charged the group with pseudomedicalism or mimicking the AMA. Lust said that the original ANA was

absolutely opposed to the formation of other organizations that would usurp programs and prerogatives of the ANA. He did not publicly find fault with Carroll or
Budden individually—a calculated political move—but condemned the upstart
organization's splintering of the movement.[78]

In 1937 things got worse. The Connecticut National Society of Natureopathic
Physicians held its fifth meeting in Hartford. Whether in direct response to this
or in response to cumulative issues, Lust condemned doctors who had hijacked
naturopathy's identity. While there were fewer of them than there were medicos,
"we despise them," he wrote, adding that "they have wrecked the standing and
honor of our profession." The next month, another naturopath put it bluntly:
naturopathy's primary enemy was the naturopath, not the medico-politician.
The irony of saying this in 1937 was that it was a watershed year for the medicos
and scientific medicine. Sulfa drugs came onto the market, giving scientists the
ability to cure through prescriptions. These medical advances were met with
hostility by most naturopaths, but they signaled the end of the height of naturopaths' popularity.[79]

Lust's biggest battles were now internal. Self-proclaimed naturopaths without
a standardized education laid claim to the term and argued that a diploma from
a reputable school—which, it turns out, some did *not* possess—and a certificate
of good moral standing should suffice for a license. Several competing societies
proclaimed themselves naturopathic associations and produced publications and
hazy educational standards. It was nearly impossible to know who were competent renegades and who were self-serving quacks.[80]

Problematic relations with the national office, even among those who supported it, continued. State secretaries were slow to send in news, which made
professional communication and cohesion difficult. The plea to follow through
appeared on the cover of the *NHH* monthly. The nagging problem of dues and
paid membership continued; many loosely affiliated with the national ANA de-
faulted on dues that would have supported the movement's efforts. This was frus-
trating and debilitating, to say the least, and it was happening at the state level as
well. Add to this the countless wildly eclectic practitioners of every stripe who
opposed *any* professional affiliations. J. C. Thie was a naturopath, chiropractor,
and physiotherapist who wrote against those dissenting voices. Many saw no
need for a national organization or thought that state membership was sufficient,
he wrote. Worse yet, many shunned membership in any organization. It was piti-
ful. Thie called them narrow-minded and said that they used naturopathy for
their own goals. The movement's previous inclusivity had been based on the
premise that drugless healers had more in common than not, and that their dif-

ferences were overshadowed by their common loathing of the medicos. But as different schools sought legal recognition, each was faced with a dilemma: to go it alone or form utilitarian bonds. This was further complicated by strong personalities, each seeking dominance in the field. In Lust's numerous writings he had always listed his degrees following his name—ND, DO, DC, MD—signaling his relationship to all branches. But increasingly these liaisons were deleterious.[81]

With so much work and so many battles, Lust was overburdened and haggard. In the fall of 1937 he was tired of the damage caused by legislative defeats when naturopaths merged their fate with all drugless practitioners. The horrible infighting among sectarians and naturopaths had taken a toll. In his remarks before officials at the annual ANA congress he cut to the chase: "FIRST, WHOEVER USES THE WORD 'Naturopathy' professionally must be a living member and a living part of our organization." Those persons had to be qualified and standardized as set by Washington, DC's congressional naturopathic acts of 1929 and 1931, and they had to meet requirements set by recognized college and naturopathic boards assisted and protected by the ANA.[82]

Lust's belief in an academic exchange of methodologies went hand in hand with his philosophy that if only the professionals would stick together, they would get the legal standing they deserved. He said in 1937 that the misunderstanding, jealousy, and mistrust that had existed between single schools of naturopathic methods (a list of seven-plus) had given way to a closer relationship for the purposes of self-protection and legislative success. Lust could not possibly have believed that such cohesion already existed; rather, it is likely that he was again attempting to nurture a *sense* of cohesiveness by advocating open, democratic access to the naturopathic institutions. Lust's wavering leadership toward and away from the other sects fueled breaks away from him and naturopathy, however. He pleaded with mixer osteopaths to stay loyal to the more progressive and accessible naturopathy and not to adopt the style of medical dictatorship. He also pleaded with chiropractors not to forsake their naturopathic colleagues. The time to unify as one common front had arrived. Perhaps most importantly, he called the upcoming congress in Pittsburgh a *naturopathic* convention, not a drugless one.[83]

Lust had made the transition. He had clarified in his own mind, if not in the minds of his contemporaries, the need for legitimate *professional* naturopathic allies and the need to shun those outliers who embraced questionable methods, called themselves drugless practitioners, and were harmful to the movement. Those were now his boundaries.[84] However, Lust's greatest impediment as a leader was his tendency to vacillate, stemming from his academic desire to expand, embrace, and share the best therapeutics, on the one hand, and his need to zero in on professional legitimacy—without leaving any supporter out—on the other. There is no better example of the clash in these ideologies than his Christmas message in 1939. Lust asserted that naturopaths believe in the laying on of hands—a new and incongruous claim (and one quite different from therapeutic touch). This came within two years of his emphatic boundaries of professional naturopathy. It is difficult to calculate just how much Lust's own vacillation cost the movement in cohesiveness or whether it reflected last-ditch efforts to hold it together. Clearly he had helped lead naturopathic methods, laws, schools, and professionalization; but he had greatly overestimated his ability to balance the organization's anarchist inclusivity, members' individual self-interests, and the need for naturopaths to be, as he put it, qualified and standardized.[85]

Everyone was sick of the infighting. The Connecticut Society of Natureopathic Physicians submitted to the *NHH* a sarcastic list titled "How to Kill Your Society" in 1937. It was right on point: don't attend meetings or arrive late; find fault with the officers, members, and the way things are being done; do not work, just criticize; if on a committee, get sore and don't attend; tell the president you have no opinion on an issue, then after the meeting tell everyone how things should be done; don't do actual work yourself, but when other members are hard at work on a project, accuse the organization of being run by a clique; don't pay dues; and don't bother recruiting new members, let someone else do it. These timeless ideas are familiar to anyone who has worked in an organization, but for naturopaths seeking credibility and unity, the behavior placed their profession on the line.[86]

More ugly infighting provided negative publicity again in 1938. One Arizona practitioner filed suit demanding that the Arizona State Board of Naturopathic Examiners issue him a license, which he had been denied despite possessing a partial license. He attacked the legality of having as a board member someone who had not practiced in Arizona within the five-year window required under Arizona law; he asked the court to remove the board member from his position as secretary-treasurer and revoke his license. He named two other naturopathic defendants in his suit. Eight months later a joint bill drafted by Pennsylvania naturopaths and chiropractors was defeated at the eleventh hour because a group of straight chiros submitted their own bill.[87]

All of this took its toll on Lust, and his supporters tried to bolster him. A friend wrote him a public letter in 1939 offering very personal words of encouragement. "They would be lost without you," he wrote. "It is too bad that there are so many who must become personal. The work is so great! . . . I really do not believe they appreciate [your] fatherly guidance. . . . My hope is that you do not grieve, for that would age you. We need you!" Acknowledging the depth of the divisiveness, he concluded, "I shall always stand behind you—and if not in this in an organization to come." Lust continued to publish letters from his supporters. In March 1940, in a gesture of friendship and professional esteem, the president of the National Association of Naturopathic Herbalists, Dr. Arthur Schramm, presented Lust with an honorary membership certificate. The New York State Society of Medical Masseurs heaped praise on Lust after he lectured at their forum. The president of the Territorial Board of Examiners in Naturopathy in Honolulu, Hawaii, asked that all naturopaths give Lust their unequivocal support. Finally, something that surely brought a knowing smile to Lust was that the AMA voted out osteopathy at their 1940 convention, fulfilling his and Andrew Taylor Still's 1915 predictions.[88]

In the meantime, the bad news continued to roll in. Dr. Robert V. Carroll had written a new constitution and bylaws for the ANA at Lust's request, and the ANA board of directors had approved them. Afterward, a convention of the competing ANA in the West, led by Carroll, was held in Chicago, while the original ANA, headed by Lust and Schippell, met in Atlantic City. Both claimed to be the original ANA and went to court over the naming rights. As one contemporary recalled, the judge, disgusted, threw the case out. Despite efforts at reconciliation, the split persisted. Not long after, in 1940, it was reported in the national news that chiropractors and osteopaths had submitted a joint bill to the Seventy-sixth Congress that would include them among practitioners who could be paid through workers' compensation benefits when treating someone hurt on the job. Naturopaths, who had been mixers' allies for years, were blatantly excluded.[89]

In the early 1940s new support came from unexpected—and familiar—places. The number of anti-AMA health rights organizations had been thriving, a sign that the popular health movement endured in the mid-twentieth century. The United Front for Freedom in Healing in Chicago pushed for recognition for all healers, in all departments of the government, including military service. Mrs. George T. Vickers and the New Jersey Anti-Medical Trust Federation put forward a naturopath and mixer chiropractor to run for office. The Progressive Alliance of the Medical Arts, claiming to represent forty thousand drugless practitioners from all states who treated 80 million patients, presented a resolution to the National Democratic Convention urging the recognition of all drugless healers. Other staunch Lust allies were the International Association of Liberal Physicians; the Loyal Liberty Legion, an Indianapolis based group begun in 1925

that was devoted to wresting control away from the AMA; and the National Association of Naturopathic Herbalists, supportive of the national office. Over
time, beginning in 1920 with the collaboration of Swami Yogananda, another
complementary modality, yoga, was embraced and advocated by naturopathy.
Through this bond an appreciation for South Asian Indian culture and lifestyle
was forged.[90]

Despite this significant support, the credibility of the movement continued to
be severely compromised by those calling themselves naturopaths, epitomized
by one Dr. Compere of New Mexico in 1943. Compere had previously been discredited in New York when he failed in his first takeover attempt of the ANA in
1921. In 1943 Compere was a practitioner in Albuquerque, where he had risen to
local leadership, then exploited his position by becoming, according to the report
in *NHH*, a dictator who ruled with an iron fist in multiple venues. He misappro-
priated funds and, in anticipation of pending protective legislation for sectarian
practitioners, he sold diplomas and licenses to people from San Diego to Vancouver. He illegally served as president, vice president, secretary, and treasurer in
all of his endeavors. He advertised a bogus cancer cure that so enraged the AMA
that it branded all drugless practitioners as charlatans. He also misrepresented
himself as a representative of the national ANA while trying to reorganize state
societies. His goal was to take over the ANA and enthrone himself as the undisputed king. Other bogus claims included Per Nelson's calling himself the
chairman of the Committee on Education of the ANA, which elicited wrath
from E. W. Cordingley, who legitimately held that position. In the wake of this
negative publicity, more supportive testimonials for Lust poured in to the *NHH*.[91]

As the ANA convention approached in 1943, Lust was tired and sickly from his
exposure to the fire at his Florida Yungborn. The fire destroyed the establishment, which was never rebuilt. He was exposed to the smoke and severely burned,
and he was treated with sulfa drugs. Friends and family believed his declining
health over the next two years was due to the sulfa. He was emotionally drained
from this constant turmoil and spoke openly and passionately about the enemy
within. The raging infighting made naturopathy a "house divided against itself
[and] easy prey for enemies without." But naturopathy was born amidst conflict
and would survive this too. He made an open plea for convention attendees'
moral support to meet the trials ahead.[92]

Lust, in another impractical move during this tumult, embraced the controversial Reverend Major Jealous Divine (1875 or 1880–1965). Divine, the son of a
former slave, was a charismatic spiritual leader and a self-named minister who
created his own eclectic religious doctrine and likened himself to God. His fol-

lowers believed he embodied the Second Coming of Christ. Divine advocated economic independence and racial equality, which appealed to the impoverished. He also founded the International Peace Mission Movement. Divine ran afoul of local ministers and the law, was institutionalized in a mental asylum briefly, and many Americans were troubled by his liaisons with the American Communist Party. Some called him a charlatan and a cult leader; Lust admired his values. Lust was aware of the controversy surrounding Divine but in published correspondence between them expressed his admiration and good wishes for what he believed was Divine's wonderful work. The impact of Lust's support for Father Divine is difficult to measure. It was typical for Lust to weigh in on the side of a champion of peace and of the poor, but his endorsement added to the drama of a movement already reeling from dysfunction and divisiveness.[93]

The tensions within the naturopathic movement were still at a fever pitch when Dr. Theresa Schippell reported on the 1944 convention in Philadelphia. She lambasted former members of the ANA who acted on the "Trojan Horse principle of betrayal from within." She accused them of attempting to destroy the association's ability to defend itself and undermining the morale of members. They had failed, she said, to garner support through scare tactics aimed at creating chaos and crushing unity, and they had not upended the ANA leadership structure. Like Lust, Schippell portrayed the infighting as resolved, or at least greatly diminished—a common strategy in the ANA's efforts to present a united front.[94]

Lust's intellect and belief in a free exchange of ideas led him to embrace an ever-growing circle of practitioners who he hoped would become allies. But his on-again, off-again relations with osteopaths, chiropractors, and charismatic healers caused shifting loyalties. He believed he had been chosen because of his history and his institutional legitimacy to keep the movement afloat, no matter how nasty the schisms. His calls for unity were juxtaposed to his own controversial decisions.

Lust died in September 1945. The official cause of death was coronary thrombosis. The obituary in the *New York Times* identified his accomplishments and said he and the ANA "do not believe in vaccination, drug treatments, medicinal remedies or vivisection." The headline identified him as an advocate of water cure; things had come full circle. This description of the ANA was to the point and far more focused than the movement's history reveals. Upon his death the divisiveness escalated and the schisms deepened. Six separate organizations emerged as World War II and federal policies undercut naturopathy. To make things worse, the AMA attacked naturopathy and chiropractic anew. The painful years up to the mid-1940s foreshadowed ongoing problems for naturopathy: how to distinguish schooled, professional healers from self-described healers who were incompetent and how to develop a national organized profession amidst internal dissension, wildly diverse approaches, struggles over leadership, and philosophical differences.[95]

CHAPTER 10

Deepening Divides, 1945–1969

Until the death of Benedict Lust in 1945, the East Coast organizations were central to naturopathy's development in America. But nationwide, other groups had developed their own regional—and even national—ideologies. After World War II they vied to take over the national leadership, to define naturopathy, and to chart its future. Fierce arguments that had been kept to a low boil under Lust spilled over after his death and led to entrenched factionalism. In addition, the AMA increased its control over national health discussions and had the power to define what was best for Americans in the name of science. There had been solid medical advancements during World War II, and its growing membership had increased the AMA's power and political clout. The government gave greater powers to the AMA through public health initiatives such as vaccination.[1]

Beyond internal schisms, several cultural trends contributed to naturopathy's decline as a recognized healing system by midcentury. The 1910 Flexner Report did not doom naturopathic, chiropractic, or osteopathic schools, but it did ensure the domination of the biomedical profession. The American School of Naturopathy and other schools struggled to obtain funding and enrollment, which were assured for accredited allopathic medical schools. The rise of hospital care, from which naturopaths were largely excluded, also negatively impacted the profession. Naturopathic hospitals had long been a goal, but only one opened, and for just one year.[2]

Medical freedom was dealt a hard blow when the AMA succeeded in defeating compulsory health insurance in 1935 and, just as important to naturopaths, "any lay control of medical benefits in relief agencies." Perhaps the strongest one-two punch naturopaths suffered was the 1937 ruling by the US Supreme Court that

Congress could create broad regulatory and interstate laws to establish Social Security and Medicare, excluding naturopaths. That year sulfa drugs also became available, changing forever the relationship between Americans and pharmaceuticals.[3]

Before 1937, natural medicine had enjoyed nearly a century of acceptance and success within its niche despite persecution by allopaths. Even though medical doctors were gaining power and authority for their methodology, they left a lot to be desired when it came to providing verifiable cures for scores of diseases. But the invention of sulfa, the use of antibiotics during the war, and finally Salk's polio vaccine in 1955 were pivotal in turning the public away from naturopaths. In the immediate postwar years, courts viewed naturopaths, who were unwilling and unable to prescribe pharmaceuticals, as outsiders and unprofessional compared with allopaths. New laws also held that drugs could include anything that one could eat, drink, or apply to the skin, which meant that naturopathic nutritional advice could be viewed as an illegal use of drugging. State supreme court rulings in Washington and Arizona severely restricted naturopaths by calling them "drugless healers." This was a profound irony, since for decades the naturopaths had alternately welcomed and distanced themselves from self-described drugless healers—practitioners who used any combination of scores of recognized and unrecognized modalities. Some were characterized as quacks by naturopaths and allopaths alike. Now, trained naturopaths became immutably merged with them in the eyes of the law, and this contributed to naturopathy's decline.[4]

Allopathic medicine achieved a pinnacle of success during World War II. To execute the war, the government called upon the entire nation, and the size of the federal budget increased to ten times what it had been during the New Deal. With the advent of sulfa and penicillin, medical science received a boost through its partnership with the federal government. Popular magazines of the day constantly announced the newest breakthroughs and the marvels of wonder drugs. Battlefield injuries could be treated with morphine and antibiotics, saving lives and limbs in numbers never before experienced in war. This resulted in the quick development of high-technology medical approaches and very visible successes. The American public readily accepted pharmaceutical innovations. While vaccination remained controversial and abhorrent to some, Americans came to welcome its ability to curtail diseases like polio.[5]

Science was the new religion in the postwar decades. Lauded not only for improvements in health, science also provided other life-changing innovations, such as radar, the atom bomb, and the insecticide DDT. Venereal disease went on the decline thanks to new drugs, further adding to the authority of medical science. Yet it was also true that improved hygiene had reduced the prevalence of cholera, dysentery, materno-fetal mortality, and childhood infections in the nineteenth and early twentieth centuries, for which naturopaths and their followers, as advocates of hygiene and healthful living, had received no credit.[6]

In the postwar years almost all sectors of society agreed that *everyone* could benefit from medical science. In the Cold War milieu of hostile political polarities, medical science seemed to be a unifying force. In fact, one scholar noted that in the twentieth century Americans came to credit scientific technology with reducing the astoundingly high death rates seen in previous centuries and effectively treating most disease and suffering. The rhetoric of scientific progress ignored or downplayed improved food supplies and nutrition (especially with electrical refrigeration), environmental sanitation, improved standards of living—particularly for the white middle and upper classes in postwar America—and preventive and pretechnological medical measures. Natural healing's emphasis on hygiene and diet should have been credited for many of these advances. The broad adoption of and trust in medical science also came as a result of the culture of conformity in the 1950s. As the Cold War escalated, Americans saw a right way and a wrong way to do things. People were either for us or against us. Anyone stepping outside the norm was regarded with suspicion.[7]

Naturopaths critiqued and mostly rejected the cultural demigods of scientific expertise and medical "breakthroughs," as well as processed foods, pesticides and fertilizers, the acceptability of cigarette smoking, atomic energy, and more. They were out of step. It was difficult, if not impossible, to sell naturopathy. It demanded time-consuming personal responsibility for health, when magic bullets were now available. It deemphasized the omnipotent physician at a time when MDs were more than ever regarded as authoritative experts. It advocated natural methods at a time when new synthetic innovations were hitting the market every day. Naturopathy seemed to be downright unscientific, less dramatically interventionist, and to many backwards.

The dissent among naturopaths, especially between those in the eastern states, where the movement had begun, and a newer western coalition, almost sealed naturopaths' fate as has-beens. However, two new organizations developed to bridge the East-West divide and merge the best of old naturopathy with new scientific ideologies and practices. Thus, even as the number of naturopaths and legal locations diminished, because of a small core of consistent, unified thought naturopathy survived beyond midcentury.

COMPETING VOICES: OSTEOPATHY, CHIROPRACTIC, AND HOMEOPATHY

Naturopaths' long and contentious relationship with chiropractors and osteopaths also changed after World War II. Straight osteopaths had credibility among allopaths, who recognized value in osteopathic modalities and found them complementary to their own. Most significantly, osteopathy carved out a place in medical science by abandoning its drugless healing and treatments, effectively negating any lingering alliances between them and naturopaths.[8]

At the same time, and continuing through the mid-1960s, chiropractors, like naturopaths, were challenged by the AMA's unrelenting attack on their system. Those who practiced naturopathy with a chiropractic license were targeted even more. In 1932 Dr. Morris Fishbein, editor of the *Journal of the American Medical Association*, or *JAMA*, charged that naturopathy had developed to give chiropractors additional revenue beyond spinal adjustments. He noted that several chiropractic schools taught naturopathy and speculated that 50 percent of naturopaths had begun as chiropractors. He claimed, inaccurately, that any chiropractor could become a naturopath by simply completing a three-month postgraduate course in a naturopathic school. While this accusation was ahistorical and had no basis in fact, it merged the two systems in the public eye, to the chagrin of both sects' practitioners.[9]

Chiropractors were under periodic fire from state medical licensing boards, which damaged naturopaths as well. In 1953 New York's Seelye-Morgan Chiropractor Bill would have licensed chiropractors had it passed. The AMA opposed it, calling chiropractors dangerous, cultists, unqualified, insufficiently trained, and scientifically unsound.[10]

Despite the opposition, after the war chiropractic schools were successful because chiropractic physical manipulation techniques differed dramatically from innovations used in scientific practice during the war. Chiropractors also grew in ranks thanks to the legitimacy granted to their schools by the GI Bill; returning soldiers were subsidized to train in chiropractic. Chiropractors had both the funds and the numbers to legally challenge the AMA's medical monopoly. In 1973 an international conference provided formal scientific recognition of the benefits of chiropractic, which was renamed at that meeting "spinal and manipulative therapies," in part to distance it from its contentious past.[11]

In the 1940s and 1950s, chiropractic mixer schools, longtime allies of naturopathy, dropped their ND diplomas. Some of these schools had offered dual degrees, but the cut was made because states increasingly extended the scope of chiropractic practice under medical licensing laws. Turnover in leadership allowed the transition. Under Dr. William A. Budden's tenure, 1929–54, Western States Chiropractic College graduates had been able to study another two years and earn a naturopathic diploma. But in 1955 the National Chiropractic Association, wanting to cut ties with naturopathy, announced that it would only accredit schools offering DC degrees. When Budden's successor resigned in 1956, the college dropped its naturopathic program, which was one of the factors contributing to the founding of the National College of Naturopathic Medicine (NCNM).[12]

The last justification for chiropractors to disassociate from naturopaths was the discovery that naturopathy would not be included under Medicare in 1968. As the historian Walter Wardwell noted, chiropractic offered something that naturopathy did not: a focus and identity that emphasized correcting subluxations "to the exclusion of most other therapeutic modalities."[13]

In the 1950s, naturopathy's relationship with homeopathy became a strong part of naturopathic identity, but it also created further problems. The National College of Naturopathic Medicine had in its curriculum a third-year course in homeopathy and offered electives in homeopathy to third- and fourth-year students. In 1978 the newly formed Bastyr College of Naturopathic Medicine's curriculum required a minimum of forty-four hours in homeopathy, with additional coursework and clinical instruction optional. This raised concern in two camps. It raised more AMA suspicion about claims of drugless practice, for what were homeopathic remedies if not drugs? Some homeopaths were distressed at what appeared to be self-taught naturopaths claiming to be homeopaths.[14]

FACTIONS WITHIN NATUROPATHY

Whatever cohesion existed in the naturopathic movement at the national level had all but disappeared by the outbreak of World War II. After Lust's death in 1945, factions grew; six different organizations emerged from the American Naturopathic Association at the same time that chiropractic colleges stopped offering ND degrees. The role of the Carroll family was crucial in these schisms. Otis G. Carroll, who sought a melding of the eastern and western camps, was the most respected drugless practitioner in Washington State during the 1920s. In 1926 Lust nominated him to the ANA's board of directors, and he was the state's ANA representative from Washington. He was Lust's stalwart ally. Otis's brother Robert was the first president of the Seattle-based splinter group that had also named itself the American Naturopathic Association, and Robert's leadership made it a reputable organization. Robert became a lightning rod when he led the move to break away from Lust's ANA and form a separate organization. The old leaders, loyal to Lust, charged Robert with pseudomedicalism because he supported certain allopathic principles and methods, and they said that he had betrayed naturopathic principles. Lust had always tried to highlight similarities between the eastern and western groups, and in 1935, at Lust's request, Robert V. Carroll wrote a new constitution and bylaws for the ANA that were signed and endorsed at the ANA convention in Omaha that year.[15]

Divisions between naturopaths in the West, who adopted more scientific methods, and those in the East, who stayed true to natural healing practices, escalated after 1937, so that in 1942 the rebels (the western group, led by Drs. Robert Carroll, Dugdale, and Henry Schlichting) met in Chicago at their own convention. They claimed to be the original ANA, even as the older, traditional ANA met in Atlantic City led by Lust and Theresa Schippell. The competing conventions tore the movement apart. Otis Carroll continued to work for unity while aligning himself with both groups. In 1947 again there were two separate conventions, in Detroit and New York. Otis attended both. There were personal conflicts between the two camps in addition to the philosophical disagreements over the naturopathic approach to healing. The most significant quarrel was over the degree to which medical science should be incorporated into naturopathy.[16]

Nineteen forty-seven was also the year of the Golden Jubilee Congress. It had been planned for years by Lust and his followers. Held in New York City, it was a week-long event that the officers and trustees of the ANA encouraged every follower of natural living to attend. Lust put out the call to nature healers of every stripe, despite the deleterious effects of associating and identifying with all drugless practitioners. The conflicting agendas and capricious alliances, not the least of which were those in the Carroll family, were demonstrated by Otis's situation. He made an impassioned plea to the old-school naturopaths to understand his membership in both the old and new associations—and of the dangers ahead. He explained that he had joined the new group in Washington State when it was the only group in the region for naturopaths. In the meantime, his brother, the group's president, had persuaded others to accept Otis's presence, however distrustfully. The relationship between the brothers was tenuous at best. At the New York Golden Jubilee Congress, Otis lambasted the pseudomedical naturopaths over which his brother reigned supreme, calling Robert a "high pressure salesman" determined to redefine naturopathy: "He has a sales group of men around him . . . and they are going to overcome you. Now what are you going to do about it?" Otis mocked the western ANA's adoption of medical science procedures, "hydrochloric acid application and oxygen under the skin, etc.," saying, "That belongs to the medical profession. That doesn't belong to our group at all." Despite

Otis's distrust of the western ANA, he tried to merge the two groups in 1948 and again in 1951 as vice president of the western group.[17]

The national split and the question of Lust's successor worried the leader in his last years. A few key players who were committed to naturopathy for decades deserve full biographies. Dr. Jesse Mercer Gehman, of New Jersey, was not on Lust's list of possible successors, but he had been a loyal Lust apostle and played a role in the debates during the Jubilee, heading the Golden Jubilee Congress committee. He became the first post-Lust president of the ANA. Prior to that, he had been the founding president of the American Vegetarian Union. At the Jubilee Gehman emphasized the eastern ANA's commitment to remain loyal to the principles of basic natural healing. He believed in teaching the works of the old nature-cure masters and spoke against acceptance of germ theory and other scientific advances, saying, "Our work is not based on a warped and decadent pathology, bacteriology, or biology."[18]

Six months before his death, Lust tagged F. W. Collins as his chosen successor. Lust wrote that Collins had fought tirelessly for naturopathy and was his stalwart ally. If Collins declined, Lust's second choice was Dr. Paul Wendel, an advocate of Kneipp's health teachings. Lust's loyalty to Wendel extended back to Paul's father, Dr. Hugo Wendel. He also tried to keep Dr. Theresa M. Schippell in the top leadership, as she too had proven herself on countless occasions. Collins had put his name forward for the presidency of the eastern association prior to the Golden Jubilee Congress. But Collins, caught creating faux diplomas, was a controversial figure. His run for president was packed with self-aggrandizement and hyperbole. In a full-page advertisement in *Nature's Path* promoting his candidacy, he wrote that he was running for president of the original ANA, saying he was a good choice in an atmosphere of distrust among naturopaths. He identified as an MD, but he listed his full credentials, a degree from the ASN in 1907 and five other degrees he said he had earned, adding that he had done postgraduate work at many other schools as well. By identifying as an MD only (Philadelphia College of Physicians and Surgeons, 1909), he extended his appeal to those drawn to scientific naturopathy. He claimed that he graduated more than four thousand students through his schools (all housed under one roof), and he highlighted—and exaggerated—his other professional deeds. Lust's promotion of Collins failed, signaling deepening divides and the naturopaths' choice to disassociate from an accused charlatan. Paul Wendel became president of the New York–based ANA after Gehmann, and Theresa Schippell later became president as well.[19]

The split widened. Gehman and his followers rejected the western group's incorporation of biomedical philosophies and procedures; Robert Carroll said in

1948 that it was necessary to critique old methods and embrace new ones. The eastern and western groups printed textbooks highlighting their differences. Paul Wendel, in his 1951 principal text for the eastern group, advocated no drugs but continued to advocate the use of water, massage, heat, light, and air. The western group's primary text was written in 1948 by Harry Riley Spitler, who served as the secretary of the Lay Publications Committee of the western ANA. Spitler wrote that the value of drugs was in how they were applied and that the term *drugless healing* was a misnomer, because even the simplest substances were drugs. The very word *drugless* conjured images of unprofessional methods and the antithesis of a medical doctor for Spitler. To raise themselves to the level of respect accorded to MDs, the western group increasingly used the word *medicine*, which riled the traditionalists.[20]

NATURE'S PATH IN THE EAST

The publications of the late 1940s and beyond demonstrate the continuation of a naturopathic culture, as well as the ongoing conflicts, differing views, and warring personalities. However, *Nature's Path* contained less evidence of rancorous infighting, regional divides, and unrealized plans for unity than did the other naturopathic publications. As a voice for the naturopathic lifestyle, it was the one most clearly written for the lay public. It contained articles by many lay authors, frequent contributions by chiropractors (but few, if any, by osteopaths), and numerous pictures of smiling, happy, and contented people.[21]

Nature's Path merged with the *Naturopath* between 1925 and 1927, but then reemerged as a separate publication in 1939. It was a vibrant publication with a voluminous advertising base. By 1947 it averaged sixty to sixty-five pages in length with about thirty articles and original reports and fifty-five to sixty ads. There are no records of the number of subscribers. This publication remained aligned with the older eastern ANA and at all times exalted Lust's memory, works, leadership, and vision. The index–cover page of each issue rallied readers to the cause with the statement, "This publication was founded by Dr. Benedict Lust in 1896. You can help carry on this great work by telling your friends." In the late 1940s *Nature's Path* continued to cover the same themes and concerns that had been covered in the earlier *Naturopath and Herald of Health* and the 1920s *Naturopath*. To keep Lust's centralizing presence alive, it reprinted earlier Dr. Lust Speaking columns without giving the original publication dates and with no editorial comment. At times these reprints seemed strangely out of place. In March 1947, for example, the column read, "Fear, malice, jealousy, hatred will squander vitality to such an extent as to be quickly seen in bilious attacks, with accompanying headaches and coated tongue." What had been a plausible theory in the early twentieth century was simplistic forty years later without additional comment. At other times his reprinted words rang prophetic and anticipated ideas and values that did not come to pass until the late twentieth century. A column reprinted in May 1947 spoke of hereditary influences on health, the power of will and thought, and spiritual power, which informed mind and body. The journal also featured other early trailblazers, praising them in the new column Personalities. This emphasis on charismatic leaders in the profession reflected one of the biggest problems in naturopathy during this era, the competition between leaders of factions and the struggle to be the next Benedict Lust.[22]

Other topics in *Nature's Path* were typical: disease-specific treatment, rejection of germ theory as the universal cause of disease, ANA conventions, a commitment to professionalization and unity, women's and children's health, instances of sects bonding or fighting, shifting philosophies, constipation and bowel evacuation as systemic detoxifiers, the necessity of exercise, legal persecutions, and legitimate naturopathic education. Newer themes included a greater inclusion of lay authors, organic gardening, distrust of consumerism, the importance of vitamins, and how to combat aging. This last topic had particular appeal as the movement's leadership aged. It also reflected the burgeoning youth culture of the 1950s. Virulent arguments against tobacco and sugar were pervasive in the journal as more Hollywood stars blew lazy streams of smoke on screen and Americans joyously consumed newly concocted sweets in the aftermath of wartime rationing.[23]

The prevalence of cancer and its connections to meat eating and unhealthy living were constant themes, along with the dangers of toxins in everyday life. Naturopaths denounced radioactivity, nuclear energy, and garden fertilizers. While there was a proliferation of chemical use in the postwar years, naturopaths' warnings about them were not new. Back in 1928 *Nature's Path* had been one of the first American publications to tout Dutch studies showing the correlation between cancer and exposure to petrochemicals and chemical pesticides.[24]

Naturopaths' long-held beliefs in food dangers continued at midcentury. They warned against eating exclusively processed foods and criticized fluoridation of public water, which they called "chemical poisoning"; their distrust of fluoride, chlorine, and impurities created a market for filters for home faucets— something that saw a resurgence at the turn of the twenty-first century. Companies advertising these products provided a steady stream of publishing revenue. A three-part article by Barbara Lust titled "Poisons Formed by Aluminum Cookware" discussed aluminum's corrosion by hard water and stated that it was a possible carcinogen, resulting in the destruction of foods' vitamins and a gaseous stomach and contributing to constipation, colitis, and ulcers. While there was a stronger emphasis on toxins in the 1940s and 1950s, much of naturopathy remained consistent with earlier years. Critiques of allopathic (and other) doctors went hand in hand with condemnation of electric convulsive therapy, drugs, prescriptions, and tobacco. Diet, manipulation, and exercise were the therapeutics advocated through the years for varicose veins, constipation, stomach ulcers, high blood pressure, coughs, aging, and bursitis. And of course there was the continuing advice to stay away from allopathic methods.[25]

In the mid-1950s and early 1960s Nature's Path was subtitled Health through Rational Living. During this period it was edited by Benedict's nephew John B. Lust. John had published two books, in 1953 and 1967, through the New York City publishing house founded by Benedict. Nature's Path reflected the consumerism of the times in the abundant products advertised for sale, including organically grown produce, salt-free bakery products, wheat germ oil, cider vinegar, laxative vegetable tablets, herbal teas, vitamins, and goat's milk. The journal praised the usefulness of yoga, the importance of foot health, and products for chronic problems. Advertisements plugged natural health clinics; nature-cure establishments; home study courses; literature from countless healers, with an increased emphasis on East Indian healing theories; foods and teas; vitamins and herbs; individual physicians' services and claims; and various forms of equipment, ranging from weight reduction machines to juicers.[26]

Nature's Path was clearly geared toward women readers. Female NDs penned recipes and addressed readers' concerns in a question-and-answer format. Virginia S. Lust, daughter-in-law of Benedict's brother Louis, and Dr. Alice Chase were prime among them. Virginia Lust and others provided specific recipes. The accompanying images showed aproned middle-class women preparing meals in their modern kitchens. Dr. Alice Chase's columns addressed women's health and the healthful properties of individual fruits and vegetables. Many of her correspondents sought advice on gynecological problems. Not surprisingly, from the late 1940s through the 1950s the diminished number of women active in the movement were largely relegated to the Ladies' Auxiliary. Most of the articles geared toward women emphasized the exaggerated, stereotypical gender roles that characterized those conservative postwar years. An emphasis on beauty by women authors was introduced under the guise of health, reinforcing the idea that a woman's value was in her appearance and sexual desirability to men.[27]

THE INTERNATIONAL SOCIETY OF NATUROPATHIC
PHYSICIANS AND THE *JOURNAL*
OF NATUROPATHIC MEDICINE

The group that came closest to bridging the East-West naturopathic schism at midcentury and also contributed to a consistent, professional presence during the declining years was the International Society of Naturopathic Physicians. The ISNP was founded in 1938 by a group of eight California physicians who had first formed a research group in phytotherapy (herbal, or plant, medicines) in Los Angeles. The ISNP became the world's largest naturopathic association, with members hailing from forty-six countries. The *Journal of Naturopathic Medicine*, the society's official publication, was published in several languages. The society also produced the lay publication *Nature's Way to Health* for distribution to patients throughout the second half of the twentieth century.[28]

The *Journal of Naturopathic Medicine* gradually included more medical science over time, but in the early years it emphasized the use of herbal remedies and botanical preparations, pain relief, obesity cures, and other methods common to all American-trained naturopathic practitioners. Articles about homeopathic remedies and liaisons with homeopaths appeared regularly, and within a short time other fields began to be included. By 1952 the journal claimed that the ISNP had more than two thousand members. At the fifteenth-anniversary celebration in 1953 Dr. Arthur Schramm, cofounder of the ISNP and its first president, was praised for birthing the society and for remaining a steadfast presence in uncertain times.[29]

Mario T. Campanella, the society's outspoken next president, edited the *Journal of Naturopathic Medicine* beginning in 1950 from his offices in Graham, Florida. Graham also became the administrative headquarters of the ISNP. Having its roots in both the eastern and western United States, the society tried, but failed, to unite the discordant organizations, claiming that they were the real center of naturopathic professionalism after Lust's death. In the early 1950s Campanella repeatedly urged feuding naturopaths throughout the country to "end the intolerance, bigotry and rivalry that had shattered the profession." "We are our own worst enemies," he said, "and until those conflicts are laid aside no good will come to the profession." While Campanella said he refused to take sides with any group, the journal's articles written by US and international authors clearly embraced scientificism, a philosophy associated more with the western ANA. In fact, the West's National Association of Naturopathic Physicians (NANP) openly affiliated with the ISNP.[30]

The themes of the *Journal of Naturopathic Medicine* were quite similar to the *Herald of Health and Naturopath*, the *Naturopath and Herald of Health*, and *Nature's Path*. Authors condemned drug medications, vaccination, and the power and control of the AMA and touted the importance of eliminating toxins through diet and evacuative therapeutics (enemas, colonics, etc.). The success of the *Journal of Naturopathic Medicine* among the eastern American Naturopathic Association practitioners shows that with the passage of time and insight, Lust's most stalwart allies, including Drs. Theresa M. Schippell and Paul Wendel, came to accept the language and at times the methods of scientific naturopathy. This swelled the journal's and the ISNP's influence. Wendel, defender of naturopathy's roots, came to define naturopathy as a scientific system of natural healing *and* an art of natural healing. In short, the rhetoric and beliefs of the groups were similar, despite the severe schism. Wendel was a perfect bridge in the East between the old and the new. When the American School of Naturopathy's alumni association elected him as its custodian of records in 1952, they were maintaining ties to Wendel's friend Benedict Lust.[31]

The ISNP conventions, such as the 1950 gathering in Wyoming, were held separately from those of the eastern and western ANAs. Reports from the president and the secretary underscored the organization's continuity and the need for professionalism to take precedence over personalities. The number of attendees is unknown, but President Emeritus Schramm thanked the many organizations who officially participated. The ISNP emphasized educational standards and provided reports of state legal activity and news and obituaries of luminaries. The never-ending hope for unity was articulated by Schramm, who wrote, "We have so many points on which we can agree, there are so many good points in the other fellow, there is no use trying to search for reasons to disagree." To illustrate this the journal in 1950 advertised the "Eastern and Western ANA Convention" as a joint convention of the two groups, and it diligently advertised both groups' conventions and state boards of examinations.[32]

The philosophy of the ISNP blended old-school methods and scientific theories, and members' articles addressing specific diseases and therapeutics reflected this. Indeed, a few articles that originally appeared in the *Journal of Naturopathic Medicine* in 1950 were reprinted verbatim in the western's group's *Journal of Naturopathy* in 1957. Further evidence of the blended approach were the *Journal of Naturopathic Medicine*'s articles offering research news, medical illustrations, cellular explanations, along with articles about acupuncture, physical therapy, and the importance of dental health, the latter long advocated in the *NHH*.[33]

The *Journal of Naturopathic Medicine*'s continued alliance with traditional natural healing and the ISNP's roots in botanicals were evident in the ISNP presidency of Arno Koegler, a homeopathic specialist who served between 1956 and 1972. Of German origin, Koegler was a prominent Canadian naturopath who was also active in the United States. He was a board member on provincial and national naturopathic associations and was the first president of the Ontario College of Naturopathic Medicine, the first of its kind in Canada. Koegler's binational leadership strengthened the ISNP. He had become well known for reducing the number of homeopathic remedies to sixty through twenty-five years of experimentation and streamlining their use for naturopaths.[34]

William Turska, ND, also played an important role in the ISNP while concurrently serving as a staff contributor to the western ANA's *Journal of Naturopathy*. In 1953 he wrote a catechism of naturopathy that articulated the natural methods of living and treatment. These included age-old methods as well as notable scientific adaptations, among them autogenous vaccines, clinical education, referrals to specialists, surgical intervention when necessary, and a nod to germ theory. This article was so well received that he was asked—and agreed—to head the ISNP's Council on Naturopathic Philosophy.[35]

In 1956 Turska wrote several articles that appeared in the *Journal of the American Association of Naturopathic Physicians* (whose contributors hailed from both the eastern and western ANAs) and the western group's *Journal of Naturopathy*. His goal was to clearly define a blended approach to naturopathic healing, stressing the compatibility between natural laws and science. Naturopaths lacking solid scientific training damaged the profession, he wrote, but this problem was being remedied with the required 4,200 to 5,200 classroom and clinical hours necessary for graduation from naturopathic colleges. He also advocated patient referrals to specialists when necessary.[36]

By 1956 Turska's earlier rejection of metaphysical systems of healing had shifted. He asserted that health was tied to interwoven energetic fields and forces and must be treated as a dynamic process. His ideas reflected the tensions between natural therapeutics and science when it came to theories of energy and healing. His explanations were metaphysical (and convoluted).[37] Turska's elaborate and at times ethereal explanations reflected the wide range of opinions and explanations given by naturopaths at the time; articles ranged in their ideas from quite modern to distressingly outdated ones. In 1952 Henry Bodewein, vice president of the ISNP, suggested treating arthritis, rheumatism, and high blood pressure with bloodletting and sweating (advice that was idiosyncratic and alarming). Other naturopaths focused exclusively on the theory of vital force or advocated suggestive mind techniques and hypnotism as primary therapeutic methods—limited approaches that Lust would have found unacceptable. Yet in the same issue, Dr. Ellen Schramm's sophisticated article "Obstetrics and Pre-Natal Care" offered sound advice that mixed the best of the old and new knowledge and techniques.[38]

As a blended entity, the ISNP exemplifies the ways in which naturopathy was adapting to survive in the postwar years. The membership agreed on the importance of monitoring naturopathic professional aids, and in late fall 1953 the ISNP created the International Council of Naturopathic Medicine to evaluate the efficacy of naturopathic medicines advertised and sold to the profession. A number of naturopaths, the council noted, had been duped "by firms that sold them instruments, supplies and medicines which were worthless."[39]

When Arno Koegler became president of the ISNP in 1956, he urged naturopaths to perform and publish research. Many individuals had already been doing so, he wrote in 1959, but little of the information had been made available to the profession at large. "Let us not depend on what the medical men find out for us; let us do our own research and see if we can develop better ways and means to help suffering humanity."[40]

THE WESTERN AMERICAN NATUROPATHIC ASSOCIATION AND THE *JOURNAL OF THE AMERICAN NATUROPATHIC ASSOCIATION*

Beginning in January 1948 the western American Naturopathic Association, under the presidency of Dr. Robert V. Carroll of Washington State, began publishing the *Journal of the American Naturopathic Association* (*JANA*), with the occasional second title *Magazine of Naturopathic Research*. All of its authors held either the ND or the PhD degree. In 1948 the western convention was held in Salt Lake City, which became the western ANA's headquarters. Its board of trustees came from eight states, a sign of naturopathy's continued national appeal even in an era of decline. In 1949 the association's secretary, Henry R. Schlichting, of Midland, Texas, became president. Schlichting's editorials in *JANA* throughout 1950 and 1951 pleaded for a vibrant membership, signaling a lack thereof. He spoke of the need to monitor and control articles and news items prepared by well-meaning physicians that nevertheless harmed the profession. All material would be edited, screened, and corrected before publication first at the state level and then at the national level by members of the Public Education Committee. This was a profound change from old-school naturopathy's willingness to print unfiltered articles written by a variety of practitioners who claimed an affiliation with natural medicine. A peer-review process was introduced, a practice that became vital to ensuring quality research. But it also undercut the value of empirically derived knowledge that had not been systematically documented. The policy alienated many drugless physicians whom the ANA had, indeed, sought to exclude. It may also have impeded membership and increases to the subscription and advertising base. In its first issues the publication was called both the ANA *Journal* and *JANA*, and it had shrunk in length to 22–24 pages; by 1951 it was only 14–23 pages long, with only eight to ten advertisers, reflecting naturopathy's struggling economic and membership base.[41]

The clout of international connections is evident in *JANA* through the occasional appearance of a Department of International Relations column, by the prestigious Schramm, who was also chair of the International Relations Committee. Global links had been evident for decades in Lust's publications, and now links with Australia, Canada, England, Denmark, and other nations could be seen in *JANA*.[42]

JANA's practitioners were in the process of rebranding themselves to emulate the successful model of allopaths. By the spring of 1950 the logo on the journal's cover page was a caduceus with the letters *ND* at its center, a clear sign of identification with allopathic medicine. From 1948 to 1951 *JANA* contained both old and new features. Carried over from the eastern ANA's *Naturopath and Herald of Health* was the State News section, although it had far fewer entries (at times only two states wrote in), and over time none from eastern states. *Health for You*, a new lay magazine spin-off by Dr. Robert Carroll giving healthful and practical advice, competed with the eastern group's *Nature's Path*. To accompany *Health for You*, the western ANA created the lay organization Nature's Way Public Health Association and encouraged readers to join.[43]

In *JANA*, unlike in the eastern group's publication, few to none of the columns were written by women. This compromised its popularity, its readership, and its subscription and advertising bases as well. The few pieces by women emphasized women's beauty. The Ladies' Auxiliary of the western ANA referred to its members only as the wives of practitioners. This invisibility and passivity surely would have made Louisa Lust and the eastern ANA's female leadership, authors, and practitioners cringe, and it signaled a decline in women's centrality—and public intellectual value—in the western movement.[44]

The western group also formed the American Naturopathic Sanitarium Hospitals, Inc., described as a western ANA–sponsored hospital foundation. A naturopathic hospital had been a longtime goal of Lust's. A facility opened in Salt Lake City after a privately owned sanitarium was purchased from a naturopath.

Its obstetrical department offered modern facilities and conveniences for giving birth naturally. Nonresident staff members needed either a state licensing board credential or certification from the National Board of Naturopathic Examiners, as well as membership in the Sanitarium Foundation and recommendations from two trustworthy naturopathic physicians. The purpose of this final criterion was to exclude self-defined nature doctors who lacked credible credentials. Despite the unproven claim of the Membership Committee in 1948 that more than one hundred naturopathic physicians had applied for nonresident staff membership, the institution was short-lived. In January 1950, less than a year after its opening, the naturopath who had sold his sanitarium to the hospital concluded that naturopathic physicians were not interested in it. His own former patients occupied more than 95 percent of the beds; fewer than 5 percent were occupied by patients of other naturopathic physicians. None were patients of allopaths, osteopaths, or chiropractors, all of whom declined to use the hospital. The failure could have been due to several factors: other sects wanted no part in naturopathic hospitals; the quality of facilities or services may have been subpar; or practitioners with private practices saw no economic benefit to joining its staff. Another, more likely explanation is that naturopaths relied on ongoing healer-patient relationships built through trust over time, which the impersonal nature of a hospital diminished. It also did not serve the needs of those engaged in home care. Perhaps its most unavoidable flaw was that it replicated the AMA model, which was still rejected by most naturopaths. In short, what use would naturopaths have for hospital-based care?[45]

The western group approved three Class A schools: Central States College in Eaton, Ohio; Western States College in Portland, Oregon (dropped from the list in 1949 with no explanation) and the Arizona College of Naturopathic Medicine in Bisbee, Arizona. Two others were approved but did not receive a Class A rating. Not surprisingly, no eastern schools were mentioned. In July 1949 resolutions were passed at the Houston convention stipulating that no school would be approved that offered or advertised correspondence instruction, stressed diplomas earned rather than courses of instruction in its advertisements, granted more than one degree or gave overlapping credits for the same degree, or granted advanced standing based on transcripts from schools not recognized by the ANA. These professional standards excluded countless self-identified naturopaths, who were still accepted by the eastern ANA, and demonstrated the western ANA's focus on scientific expertise and its alignment with biomedical standards.[46]

In May 1950 the western ANA's headquarters relocated to Des Moines, Iowa. Schlichting said that it did so to consolidate and centralize resources. More likely, this neutral territory of the Midwest held promise for greater support. *JANA* would be published from there as well, under a new editor, Dr. A. R. Hedges of Portland, Oregon. That same spring a new definition of *naturopathy* appeared. In fact it was not new but more detailed than the cryptic definition that had appeared in *Blakiston's New Gould's Medical Dictionary* in 1949. Its articulation of more inclusive therapeutics may have been an attempt to gain more support and a more solid operating base. Naturopathy now included the use of physical, mechanical, chemical, and psychological methods, and it still excluded major surgery, x-ray, drugs, and radium for therapeutic purposes. There was one proviso about drugs: naturopaths did utilize health-giving substances that the body could assimilate or that already existed within bodily tissues, such as vitamins. The old school did indeed coexist with the new: discussion of antibody theory, cancer, physiotherapy, and asthma recommended treatments similar to old-school therapeutics.[47]

The popular press and newspapers responded favorably to the western naturopathic activities because they replicated those that gave the AMA legitimacy. An obstetrician who wrote on childbirth in *JANA* had his articles reprinted in two issues of *Reader's Digest*, signaling their widespread appeal; he did not recommend technological approaches, but emphasized natural methods. And coverage by the *Houston Star Telegram* shows how publicity savvy the western ANA could be. Film stars such as Hollywood's Pat O'Brien and Dennis O'Keffe attended the 1949 convention in Houston, and Dorothy Lamour's radio show was broadcast live from the convention. There were performances by Russ Morgan and his band, Dorothy Shay the "Park Avenue Hillbilly," and eight orchestras. Convention representatives reported that the convention was at 100 percent capacity, although no specifics were given.[48]

Perhaps the most significant contribution of the western ANA, in addition to its emphasis on science, was H. Riley Spitler, ND's announcement in the winter of 1953 that a publication five years in the making was complete. Dr. A. W. Kuts-Cheraux was the general editor of a naturopathic reference book declared the "Magna Charta" of the profession. *Naturae Medicina and Naturopathic Dispensatory* was a collaborative effort. The book included information on natural healing remedies, including methods of preparation, chemistry and dynamic actions, plants and their naturopathic usage, the healing properties of vitamins and minerals, and the effects of endocrine secretions (chemical and physiological) and their role in controlling bodily functions in health and disease.[49]

In the summer of 1954 *JANA* underwent a name change when attendees at the annual convention of the western ANA voted to call it the *Journal of the American Association of Naturopathic Physicians*, or *JAANP*. The western ANA was by this time highly organized internally, professionalized, and aligned more so than ever with scientific methods and allopathic structures. It had standing committees on ethics, membership, hospitals and sanitariums, scientific research, advertising, fund-raising, legal and legislative concerns, malpractice insurance, physiotherapy, public education, public health, and the patient publication *Health For You*. There was also a new Council on Education, as well as state boards of naturopathic examiners and a National Board of Naturopathic Examiners. Some of these elements did not exist or were less central in the East. The western group, through the new journals, was committed to a middle-of-the-road course presenting both sides of controversial issues. Things had seemingly come full circle: the eastern ANA, for decades a progressive and radical coalition, was now deemed the old guard.[50]

FRAGMENTATION AND ATTEMPTS AT UNIFICATION

Despite philosophical differences and competing personalities, the eastern and western naturopaths both urged unity, but neither was willing to compromise. In 1949 Henry Schlichting, of the western ANA, wrote that individualism was self-defeating; success lay with cooperative organization. Unity, he said, "is a *must* for our survival. . . . In Unity is strength!" He noted that the eastern ANA's Chicago convention had appointed a committee to meet with a committee from the western group. Not surprisingly, the proposed unification convention never took place. Gehman said, rather unconvincingly, that it had not taken place because his contingent had refused to approve the suggested dates. In April 1948 an anonymous author had reported that the eastern ANA had voted 7 to 3 in favor of the convention. This revealed disagreement among the easterners, and the cancellation of the proposed convention was likely a result of division within the eastern group as well as the East-West split.[51]

All actions belied the lip service paid to unification. The western ANA's 1949 convention was held in Houston; its primary emphases were the elimination of charlatan naturopaths and making naturopathy credible. As always, there was also praise for Lust. The continued verbal and written nod to the leader by both groups was likely one reason why some found a tiny glimmer of hope for unification despite indications to the contrary. But the two camps continued to create their own textbooks, hospital foundations, therapeutic regimens, radio programs, and so on. Both formed unification committees. Dr. Jesse Mercer Gehman, representing the eastern ANA, negotiated with Drs. Whiting and Gramm for two days in Chicago. Gehman joined with the western ANA's Schlichting to plan a unification convention at which the officers of both groups would resign and a new leadership would be chosen. In 1949 the western group's Houston convention created a new organizational structure for itself with an eye toward the planned national convention. Yet none of those changes signaled any movement toward reconciliation; in fact they embedded the western group's agenda more fully. The westerners designed a new insignia that differed greatly from the easterners', generating further ire before the scheduled unification convention.[52]

The year 1950 was a replay of the previous one. An ANA unification convention was announced in JANA, followed by two pages detailing the luxurious St. Louis hotel accommodations. No direct mention was made of the eastern ANA; it was to be a gathering of American naturopathic physicians under the auspices of the western ANA. In fact, misunderstandings between the two factions were immense and so convoluted that they baffled even the journal's writers. In September 1950 JANA reported Schlichting's claim that the convention in St. Louis had resulted in amalgamation of both groups. Yet Dr. Paul Wendel, president of the eastern ANA, denied that amalgamation had occurred. Before the convention, Dr. Gehman had declared adamantly to the profession "that he was still president of the ANA (Parent Body), and that the 'Parent Body' was not holding their convention in St. Louis." This led the writer to comment, "All this is very confusing, to say the least. According to our summation, it seems that we still have three ANA's"—the eastern and western groups and apparently a parent body.[53]

In the late 1940s, while the primary difference between the two groups was allegedly the western group's emphasis on scientific naturopathy, there were actually very few articles on disease, often only one, in each month's JANA. Both groups still condemned many allopathic modalities. Advertisements also contradicted the western ANA's adoption of scientific expertise. One advertisement for "The Zodiac and the Salts of Salvation" was available from the Carey-Perry School of the Chemistry of Life in Los Angeles. The need to generate revenue trumped philosophical purity.[54]

Throughout the late 1940s and the 1950s, as calls for unification continued, so did the splintering. The western ANA in 1950 became the American Naturopathic Physicians and Surgeons Association, and in 1956 it took the name Association of Naturopathic Physicians. With the books by Ohio's Harry Riley

NATURE'S PATH

Spitler and Indiana's A. W. Kuts-Cheraux, the western group had by then a significant body of scholarly work. In the East, the seminal text was Paul Wendel's *Standardized Naturopathy* (1951). One author has described the difference between the Spitler and Wendel texts as "the somewhat opposing perspectives of the more science-based, or 'green allopathic' and the purist 'nature cure' camps."[55]

THE NATIONAL ASSOCIATION OF NATUROPATHIC PHYSICIANS (NANP) AND THE *JOURNAL OF NATUROPATHY*

The closest thing to a truce between the eastern and western ANAs during legal crackdowns, decline in naturopathic institutions, and the massive confusion of 1950s organizational shifts came through the National Association of Naturopathic Physicians. It was aligned with the International Society of Naturopathic Physicians. These two organizations, through their journals and conventions, created a middle ground for professionalizing naturopaths and carried the movement through the dark days. Leaders were active in both organizations, with the NANP on a national scale and the ISNP working both nationally and internationally. Rising cultural and legal pressure led to a Joint Naturopathic Convention, finally, in 1956. According to the pamphlet reporting on the convention, the American Association of Naturopathic Physicians, the American Naturopathic Association, and the Naturopathic Physicians, Inc. merged into one single national association, which they named the National Association of Naturopathic Physicians. Genuine excitement and comradeship jump from the pages of the report, which states that the merger "has proven that when men with a common cause and an equally common destiny are brought together—they will work in harmony for the common good." The usual considerable differences existed about the principles of medical science, but it was clear that something had to change if naturopathic methods and way of life were to survive. Among other strategies, the NANP had a Council of Education, based in Washington, DC, that kept tabs on legislation and naturopathic strategies and continually argued for standardized naturopathy.[56]

A tool of the new organization was its *Journal of Naturopathy* (*JN*). Its originators reflected both a national and an international coalition. Dr. W. Martin Bleything, from the Northwest, was the national editor; the ISNP cofounder Arthur Schramm, of California, was the international editor; the ISNP president, Mario Campanella, of Florida, was the business administrator; and Dr. Robert Spears, of Texas, served as secretary. The *JN* proclaimed itself the official organ of the NANP and the ISNP. This was a curious claim, given that the *Journal of*

Naturopathic Medicine also deemed itself the official organ of the ISNP. The two publications existed simultaneously, and it may be that the *JN*'s truly national (as opposed to eastern or western) viewpoint was responsible for its loyal clientele. A few months after it began, the *JN* deemed itself the official journal of organized and representative naturopathy.[57]

There was good reason to pull together in the late 1950s. All of the work to create standards, coalitions, viable professional organizations, educational institutions, and texts had taken place against a backdrop of vicious infighting and legal battles. The results were destructive: in 1953 the attorney general of Texas ruled licensure of naturopaths unconstitutional; successful lobbying by the AMA caused the Georgia legislature to eliminate the naturopathic board of examiners in 1956; and Florida ended its licensure in 1959. By 1958 only five states—Arizona, Virginia, Utah, Colorado, and Connecticut—licensed naturopaths, and a few others licensed them under laws for drugless practitioners. In 1964 California stopped issuing new licenses, although it allowed those in practice to continue. One reason why naturopaths were denied Medicare reimbursement in 1968 was that there were so few—only 553—practicing in the country at the time, and the number were falling.[58]

The falling numbers reflected the "desolate state" of the schools. A "Survey of Naturopathic Schools" was prepared for the Utah State Medical Association in 1958. Its purpose, the authors stated, was not to authorize or censure any school, person, or group but to present facts uncovered by competent investigators in many states. In 1957 the Utah legislature had passed a bill that reestablished naturopaths' right to use drugs and surgery and postponed the issuance of any new licenses for two years. The governor had vetoed the bill and called for more research. Of the twenty-six US schools investigated, only nine were still in existence by the fall of 1958. Of the top five US naturopathic schools and colleges, three were in decline and only two reported that they were accepting new students and offering training and degrees. The survey report disregarded those two schools, concluding that "the total number of schools has declined to a point where there are virtually none in existence." One of the schools deemed insignificant was the new National College of Naturopathic Medicine, today known as the National College of Natural Medicine. The dismissive attitude of the report belied its authors' so-called disinterestedness.[59]

According to the survey, Lust's American School of Naturopathy was no longer in business. In fact it had closed in 1942. Lust had only operated the New York School of Massage and Training School for Physiotherapy until his death in 1945. Originally chartered by the state in 1905, the naturopathic school was

dissolved by proclamation of the governor in 1926. Lust said that the incorporation out of Washington, DC, gave him the right to operate a school and issue
diplomas. It was one of the reasons why the acts of Congress in 1929–31 that defined naturopathy and protected it in Washington, DC, were so important for his
own work as well as for naturopaths' collective credibility.[60]

The loss of legitimacy over these years led to the formation of the NANP. In
January 1957 even Dr. Campanella admitted that the NANP was facing adjustment problems but that it had a solid foundation. Nine states had affiliated with
it, and Campanella claimed that many others had requested affiliations. Critiques of allopathy appeared in nearly every issue, but it was clear that there was
no united front, since the pleas for unity in the ranks were constant.[61]

Like the ISNP, the NANP garnered support through a blend of old and new
healing theories and methods. The NANP was far more attuned to and welcoming
of advances in medical science than its predecessors. The 1960 "NANP Summer
Report," written by President Harry F. Bonelle, outlined a professional curriculum
with a two-year premedical and/or liberal arts college prerequisite and the pan-
theon of original naturopathic therapeutics, along with anatomy, chemistry, pathology, bacteriology, hematology, psychology, x-ray technique, obstetrics, minor
surgery, orthopedics, and gynecology.[62]

Dr. Bonelle was motivated to focus on professionalism and educational standards. Not only was he writing in a largely hostile sociolegal climate for his profession but earlier that year a sensationalized incident had received national
front-page coverage. Some of the worst negative press naturopaths had ever
endured emanated from serious accusations against two prominent NDs. In January 1960 Robert Vernon Spears, a Dallas naturopath, was secretary of the NANP.
According to the New York Times, he had a long criminal record, although no
specifics were given and it was possible that his "criminality" was practicing medicine without a license. Spears was accused of taking dynamite caps from a road
construction site onto a plane flying from Tampa, Florida, to New Orleans that
crashed in the Gulf of Mexico, killing forty people. When the FBI took over
the case, it was discovered that Spears had never boarded the plane; another
man had flown under his name. Spears had been presumed dead, but in fact he
had hidden out for two months with the well-known Dr. Turska in Arizona.
Allegedly, dynamite and blasting caps were found at Turska's desert home near
Phoenix. When the FBI searched Spears's home, they found many books on hypnotism, which they speculated he had used to get his longtime friend William
Allen Taylor to take his place on the flight. Several prominent articles in the New
York Times alleged that Spears had hoped to fake his own death so that his wife

could benefit from his life insurance policy. No charges were ever brought against Turska. Spears was ultimately charged with transporting Taylor's car from Tampa to Phoenix without the owner's permission. In short, both men were exonerated in the bombing accusations, but Spears's so-called tainted past and alleged attempted insurance fraud (his wife had not pursued a claim) painted all naturopaths as disreputable criminals unworthy of trust.[63]

In this climate, naturopaths fought an uphill battle for legitimacy; emphasizing scientific methods and adopting a structure of expertise was one way to do that. The antiauthoritarian philosophy of naturopaths had long been abandoned by formally trained practitioners. Authors stressed the value of observation and diagnosis, clinical studies, referrals to those with greater expertise, complementarity with allopathic therapeutics, avoidance of environmental toxins, therapeutic management, and use of vitamins and minerals. Practitioners wrote about the importance of psychotherapy referrals—since general naturopaths were not sufficiently trained—and said that general referrals should be to trained naturopaths. The *JN* made a point of specifying which articles were based on original research. Medical articles were accompanied by detailed illustrations, graphs, charts, and figures. In a book guide for the naturopathic profession, a *Handbook of Pediatrics* had entries written by physicians from the Yale, University of California, and Stanford medical schools. But natural therapeutics were not abandoned. J. H. Tilden's 1935 *Toxemia Explained* was still a standard.[64]

While naturopaths combined an eclectic mix of allopathic remedies, natural methods, and Eastern medicine, they rejected fundamental biomedical treatments such as vaccination against polio. They were also unconvinced about nonemergency surgeries such as tonsillectomies. They wrote about allergies, food adulteration, dangerous pharmaceutical drugs, and uterine and other pathology, along with other ailments and therapeutic techniques. They valued research, referring to controlled clinical studies, new and old physical therapy techniques, and the dangers of artificial fertilizers. Obesity and its treatments were addressed constantly. Product endorsements served to protect patients from bogus sales claims and generated advertising revenues.[65]

While the NANP resounded with the western ANA's ideology, reflecting the eastern ANA philosophy was that women were present as both practitioners and recruits for the Ladies' Auxiliary. Dr. Lorna Mae Murray took over the presidency of the NANP in February 1957. She had graduated from Blumer College of Natureopathy in Hartford, Connecticut, and practiced in Miami in the mid-twenties. She had become president of her state organization and led the counterattack when the state legislature abolished the state naturopathic board. She persuaded lawmakers to permit those still practicing to continue to do so. Her qualifications and knowledge were impeccable. Her first article as president stressed the need for unity and cooperation. At the international level, meetings between American naturopaths and Drs. Marie Marchesseau and M. L. Seller of Paris were reported in the *Journal of Naturopathic Medicine.* These international liaisons resulted in the National and International Naturopathic Convention in Winnipeg, Manitoba, in 1961.[66]

The members of the Ladies' Auxiliary of the NANP were to raise funds to assist worthy students enrolled in approved colleges, hold meetings with entertaining programs at the local and national levels, support naturopathy 100 percent, and encourage their husbands' attendance at the NANP conventions. A member of the Ladies' Auxiliary made it clear that this latter task was not easy, as evidenced by the continued journal calls for membership and attendance.[67]

Meanwhile, fierce attacks against naturopathy continued in various state legislatures. Under assault were naturopathic rights in Florida, Utah, Texas, Wyoming, Missouri, and Maryland. Washington State remained a significant bright spot; there naturopathic licensure still operated under the provisions of the 1919 Drugless Physicians Act.[68]

THE NATIONAL COLLEGE OF NATUROPATHIC MEDICINE

The work that went into creating standardized scientific naturopathic training between 1956 and 1960 belied Utah's 1958 "Survey of Naturopathic Schools." Contrary to the supposed death of naturopathic training described in the survey report, there was in reality a flurry of activity to pioneer enduring educational institutions. In the aftermath of the watershed 1956 ISNP Colorado convention, President Arno Koegler urged doctors to concentrate on founding naturopathic schools and to participate in a public relations campaign. The founding of the National College of Naturopathic Medicine in Portland, Oregon, that year was considered by many to be the NANP's most significant accomplishment. This was the school discounted by the Utah report. The articles of incorporation were executed by Charles Stone, W. Martin Bleything, and Frank Spaulding. Bobbie Carroll, Robert's son and Otis's nephew, was also instrumental, as was Joseph A. Boucher. For a year Boucher traveled from Vancouver, British Columbia, to Portland every Friday to teach at the NCNM. For twenty-seven years, not only did he teach there but he was a member and chair of the board of trustees. John Bastyr (1912–1995) was the individual who most linked old-fashioned naturopathy with the modern vision. At the NCNM he taught obstetrics and gynecology and was the executive director, a board member, and eventually president. Bastyr worked at the Portland, Oregon, campus as well as at the institution's branch campus in Seattle. He lobbied the Washington State legislature to recognize natural medicine and served two terms on the Naturopathic Advisory Committee for the Washington State Department of Health. He was an honorary member of the Advisory Committee until his death.[69]

In early 1957 the National College of Naturopathic Medicine had some sixteen thousand dollars in assets, with an excellent clinic, laboratory, and classrooms in a three-unit store building. The Naturopathic Foundation for Health Research, based in Chicago, had donated more than ten thousand dollars to the NCNM. The foundation's forty thousand dollars in assets in late 1957 meant that the NCNM began with a firm financial base and significant support across the country. Bleything was named dean of the college, and an infrastructure of committees and individuals was quickly assembled to manage business affairs, the clinic, faculty hiring, and college administration. Bastyr's colleague Dr. Joe Boucher argued that rigid school requirements were vital for naturopaths. The college's curriculum, he said, must adequately prepare students to treat their patients skillfully and carefully and to exercise judgment that embodied naturopathic therapeutics and philosophy. At the NCNM, the 4,647 required class hours were based on Kuts-Cheraux's *Naturae Medicina*.[70]

In 1959 the NCNM moved to Seattle. Other colleges were founded shortly thereafter. Seattle's College of Naturopathic Missionary Medicine, Portland's College of Naturopathic Medicine, and Los Angeles's Sierra States University were approved by the NANP in 1959. In the spring of 1960 the *Naturopathic Magazette* became the official publication of Sierra States University, edited and published by Dr. George Floden. It boasted about the school's curriculum of physical therapy and massage (Swedish massage, passive and resistive, and medical gymnastics), colon therapy, psychology, naturopathy, and hydrotherapy.[71]

NATUROPATHIC CULTURE IN THE 1960S:
THE NATUROPATH

The *Journal of Naturopathy* discontinued publication in 1962. It was supplanted in 1963 by *The Naturopath*, which had a format that was friendly to a general readership. *The Naturopath* was "Dedicated to Natural Health and the Preservation of Organic Life in Our Soils." It was the official publication of the Oregon State Association of Naturopathic Physicians of the NANP; it did not replace the *Journal of Naturopathy* as the organ of the ISNP.[72]

The Naturopath was a modest publication, seven to twelve pages in length, that reflected the largely regional base of naturopathy in the Northwest. There were very few ads, and there was a small directory of practitioners from the West and a smattering from elsewhere. The covers frequently pictured smiling children outdoors and panoramic scenery. Many of the articles had a homey tone, such as "Did Grandma Practice Medicine Without a License?" and, for rural readers, "How to Treat a Cow." Authors also emphasized toxins found in everyday life and naturopathic means of combating them. This was not a new theme, but the journal created new—and clear—messaging with studies as proofs and featured public discussions that reaffirmed decades-long naturopathic contentions, as well as practical ways to avoid dangers. Rural and suburban readers were advised to grow unadulterated organic foods. The urban dweller was directed to healthful food products that could be purchased from advertisers listed in each issue. They offered organic honey, vitamins, dried fruits and vegetables, natural juices, and dozens of other healthy products. The publication's subscription price was $2 a year, or $15.55 in 2014 dollars.[73]

Naturopaths' emphasis on environmental toxins in food production was connected to the need for natural foods, but maintaining a natural kitchen was an increasingly uphill battle by midcentury. American cookery had undergone dramatic changes in the early twentieth century. The availability of electricity, cookbooks, and prepared foods reflected the home-economics movement as well as the widespread use of technology. Decades earlier, in the 1930s and 1940s, advanced methods of food processing technologies had yielded canned, frozen, and packaged food items with increased shelf life and greater availability. This also led to "enriching" products with supplements. From the 1950s on, red meat and poultry were widely available. Food ads reflected the convenience, price, and reliable consistency of the product. The distribution, marketing, and technologies for preserving and preparing food had begun to change before the war. Sugar-based treats like Twinkies (1939) and highly popular sodas increased the national consumption of sugar. Packaged products such as Bisquick (1931), Spam (1937), Ragu spaghetti sauce (1937), and Kraft macaroni and cheese (1937) promised to lighten the housewife's workload. A plethora of snack foods were introduced in the 1930s. The gradual replacement of iceboxes with refrigerators made keeping foods easier. And the ubiquitous fast-food chain restaurants, beginning in 1921 with the first White Castle hamburgers in Wichita, Kansas, contributed to the explosion of the fifties auto culture. What better outing than to drive to a restaurant and eat a popular burger! It all went against naturopathic advice.[74]

By the 1950s, food preparation and etiquette increasingly reflected social class. Middle-class women were expected to run their households efficiently and economically with new products, demonstrating that they could afford certain luxuries. After the war, upwardly mobile women who left the wage labor market returned home and were encouraged by cookbooks and women's magazines to create meals that took hours to prepare as a show of love for their husbands and children. Any shortcuts that packaged products provided were appreciated by them. At the same time, demands on working-class women also contributed to the popularity of new products and methods. Women had been forced to relinquish their well-paid manufacturing jobs to the returning male soldiers, but many had to remain in the work force, but in low-paying, sex-segregated jobs such as caretaking, retail, and domestic service. They valued the ability to produce quick and inexpensive food as well.[75]

Also after the war, synthetic chemical additives—four hundred new ones during the 1950s alone—exacerbated the trend toward using more convenient food. This was when injecting chickens with antibiotics and growth hormones began. The unfettered use of additives led US Representative James Delaney to call for restrictions on their use in 1952, which eventually led to the 1958 passage of a legislative amendment bearing his name that banned additives causing cancer in animals. Rachel Carson's *Silent Spring*, published in 1962, described the deleterious effects of DDT and other chemical sprays on bird populations, putting food processors and the agrichemical industry on the defensive.[76]

Families adhering to a naturopathic way of life had a radically different relationship to food. Processed foods were rejected; the healthfulness of the food, not the time it took to prepare it, was what was important. They had to produce their food themselves or buy it locally from those who did. This explains why so many articles in *The Naturopath* targeted female readers. Men also engaged in planting and growing, but the preparation of food and bringing it to table continued to be women's work. The regular column Organic Gardening the Natural Way, by the biochemist Dewey Pleak, urged readers to garden consciously, improve the soil, and create mineral-rich composting pens. Pleak cautioned readers not to be deceived by the billion-dollar companies that push chemical fertilizers. He stressed the health benefits of individual vegetables and juices and the bodily organs they benefited. He touted the joy, satisfaction, and rewards of gardening.[77]

The Naturopath worked in concert with other media that exposed the dangers of toxins. Chemical poisons posed a potent threat to man, and the journal cited the CBS television program that concurred with Rachel Carson's dire predictions in *Silent Spring*. But it seemed that few in mainstream America knew how serious the pesticide danger actually was. Appearing alongside the CBS story in *The Naturopath* was a terrible tale of a pesticide that killed rodents, insects, and children that *finally* aroused health authorities and parents. Thallium sulfate, an inexpensive salt of a metal similar to lead, killed nine children in Texas between 1954 and 1959 and left another twenty-six with lasting brain damage. The article quoted a study by the AMA showing that the chemical was deadly for children. In 1957 the Texas legislature cut the allowable dose to a third of what it had been, but it took the US Department of Agriculture until 1960 to do so.[78]

While denouncing toxins found in everyday life, naturopaths praised the use of herbs and healthful recipes and emphasized the necessity of preserving forests because of their importance for the wild plant and animal species and air quality. There were frequent suggestions to emulate Native American practices. "Old Doc Cherokee" had a monthly column. Whether or not this author was actually Cherokee, synchronicity with native ways of life was encouraged. Herbs were recommended, but the *Naturopath* authors did not prescribe or offer direct advice nearly as often as earlier naturopathic publications had done. The authors were usually NDs who described treatments for numerous diseases and occasionally told their readers to consult their doctors (presumed to be NDs). One column, The Question Box, reporting experts' findings in nutrition, health, and related subjects, stated clearly that its author did not prescribe. This approach reflected both a growing reliance on expertise and harsher legal constraints against naturopaths.[79]

Authors were compelled to report on the rigid criteria applied to naturopaths and to provide information about conventions; little was said of legislative activities. The medical monopoly, as usual, was ridiculed and denounced. One author mocked the age of miracles that caused side effects and made cures more deadly than the maladies. Publicized instances shocked the American public into taking a second look at wonder drugs. There were allergic reactions to penicillin that caused skin eruptions, asthma attacks, aching joints, and permanent deafness. Dr. John W. Noble, president of the NANP, spoke at the Congress on Medical Monopoly in April 1965, held to coincide with the AMA's Congress on Quackery. He wanted to keep foremost in Americans' minds the monopoly of orthodox medicine and harassment natural healers suffered from it.[80]

Accusations of collusion between the medical monopoly and industry also continued. In 1963 an author hinted at collusion between allopaths and the tobacco industry when it noted that the AMA had dropped a tobacco study. The AMA claimed that it had discontinued its study of the effect of tobacco on various diseases because it could not find appropriate physicians to serve on an investigative committee. It would not have occurred to the AMA to consult natural healers, who had long opposed the habit. The announcement, the author deadpanned, "caused tobacco stocks on the N.Y. Exchange to spurt upward." And when Joe Boucher wrote on the subject of poor public relations for naturopaths, he stated bluntly and with scorn that political medicine crushed all who opposed it and all competition.[81]

The midcentury organizations and publications demonstrate a movement racked with divisiveness, conflicts based on region and personality, differing philosophical approaches, and wavering credibility. Yet they also offered a growing cohesiveness during naturopathy's era of decline. Against all odds, educational institutions were rebuilt and naturopathic ways of life were bolstered by publications for professionals and lay followers. There was still a passionate commitment to natural ways of living for an evolving American culture, and some embraced scientific expertise. Just as in the earliest years, there was a shared belief in the dangers of scientific power and authority and a desire, albeit unrealized, for professional unity.

CHAPTER 11

The 1970s and Beyond

Cultural Critique and Holistic Health

Despite the difficulties and decline of naturopathy in the mid-1950s, there were some promising signs. In 1955 the businessman Fred J. Hart founded the National Health Federation (NHF), a nonprofit organization emphasizing consumers' rights and medical freedom. It opposed an allopathic medical monopoly, and its membership included, among others, vitamin manufacturers, food supplement companies, natural health store owners, naturopaths, chiropractors, natural hygienists, and MDs who incorporated natural health methods. It opposed compulsory pasteurization, fluoridation, and vaccination and provided legal aid to those who violated restrictive health laws. The federation served many natural healers and natural hygienists (also called straight nature curers). Natural hygienists and formally trained naturopaths often overlapped in their work, and the NHF created something of a coalition.[1]

Several prominent authors in the 1950s, such as Cash Asher and Herbert Shelton, paved the way for the natural health rebirth in the 1960s and 1970s. They asserted that poor diet determined the severity of communicable diseases and that vaccines made billions of dollars in profits for medical industries; they stressed the importance of hygiene and fitness and the success of holistic healing and said that many allergies were caused by foods that individuals could identify and eliminate. These texts and the naturopaths' decades of cultural critiques and alternative methods set the scene for the tsunami-like changes to come.[2]

Despite the positive steps and the promising coming together of naturopathic societies in the late 1950s, they fought an uphill battle. The pharmaceutical industry was growing and benefited from the broad acceptance of the idea that drugs could eradicate diseases. By the 1970s, the state of the new branch of the

National College of Naturopathic Medicine, which had opened in Seattle in 1969, reflected the hard times, and it closed after its 1976 class. At that time even the main campus struggled mightily with too few students and economic problems.[3]

THE 1960S AND 1970S

On a national scale, sociopolitical critiques during the 1950s led the way for a counterculture revolution that swept the nation in the 1960s and 1970s. In the wake of 1950s conformity and conservatism, Americans, pushed by the largest young adult generation in US history, condemned the lack of civil rights for people of color, women, and gays and lesbians; the Vietnam War; the military industrial complex; educational institutions; racism and segregation; familial conformity; and law enforcement and government. The American love affair with expertise was challenged by an antiestablishment or anti-status-quo under-current in these movements. Women's self-help groups and the holistic health movement contested the biases and shortcomings of organized medicine and the social class privileges and norms it upheld. They denounced the slow progress in curtailing diseases that debilitated and killed most Americans. They also spoke out against the excessive, often unnecessary drugging of, and surgeries on, women, such as the prescription of tranquilizers and hysterectomies. There were exposés of unconscionable medical experimentation, such as the Tuskegee experiment of the 1940s and the forced sterilization of poor women, many of whom were women of color. These issues pointed to the ethical shortcomings and therapeutic limitations of biomedicine. As one author noted, "The high-tech medicine that came out of the world wars had been centered around crises. . . . It was pretty flashy stuff. But people started to see that it had shortcomings when it came to health maintenance and prevention of chronic degenerative disease."[4]

Joining this wave of cultural dissent and change, more alternative health books appeared in the 1960s. They emphasized therapeutic fasting and foods as medicines and provided family guides to nature cure. These books enhanced the appeal of alternative medicines, as did the counterculture desire to get back to nature. Caring for and understanding one's body and health became a statement of self-determination. For some it was an outright rejection of institutional authority harkening back to the popular health movement a century earlier. In the late sixties, this desire for independent health knowledge was exemplified by the consciousness-raising of the women's health movement and then the publication of *Our Bodies, Ourselves* by the Boston Women's Health Book Collective in 1971. The book's call for personal knowledge and responsibility for health struck a nerve. By 2012, after multiple editions, the book had been translated into twenty-nine languages, and it was chosen by the Library of Congress as one of eighty-eight books that helped shape America. Alternative journals and magazines proliferated throughout the 1960s and 1970s, and naturopathy benefited from mass-released publications. They focused on practical natural healing and its diverse applications. Some books by MDs, psychologists, and others encouraged patients to pursue a mind-set that would allow optimum healing to occur—a strategy employed by naturopaths for a century. Extremely popular texts were written by Mark Bricklin (1976), Norman Cousins (1979), Ken Pelletier (1977 and 1979), Bernie Siegal (1986), Dolores Krieger (1979), Andrew Weil (1972 and 1983), and Paavo Airola (1970 and 1974). All had an immense impact in recasting cultural attitudes.[5]

Many people adopted self-help modalities while continuing with regular medicine. Practices in this era knocked from its pedestal, at least for a while, the once entrenched demigod of professionalized medicine. As one naturopath noted in retrospect, the public was realizing that how one lived greatly impacted chronic diseases, and allopathic medicine and pharmaceuticals barely affected them. A prime example was the overuse of antibiotics and bacteria's resistance to them. Furthermore, the 1970s natural foods movement opposed chemicals, additives, and preservatives. All these efforts increased questions about prescription drugs as the go-to remedy. Today, the rise of neighborhood farmers' markets and the exhortation to "buy local" fresh produce are expanding that movement.[6]

Some have argued that the promotion of holistic healing and natural medicine was helped by the popularity of Edgar Cayce (1877–1945), who was called the "Sleeping Prophet" and the "Father of holistic medicine." The most documented psychic of the twentieth century, he had a worldwide following. Naturopaths agreed with his emphasis on bodily manipulation and nutrition, yet affiliation with him added to naturopathy's outsider status. For more than forty years of his adult life, Cayce gave psychic "readings" to thousands of seekers while they were in an unconscious state, diagnosing illnesses and revealing lives lived in the past and prophecies yet to come. By the 1990s a select number of new naturopathic physicians could study and apply Cayce's methods at his Association for Research and Enlightenment, or ARE, in Virginia Beach, Virginia, founded in 1931. They studied his therapies and concepts. Two naturopathic authors noted that the goal of ARE was to increase the number of Cayce-inspired physicians nationwide.[7]

The New Age and holistic movements were theoretically separate, but they shared certain key features with each other and with traditional naturopathy. In keeping with the ethos of self-determination, each person was seen as responsible for her or his own life and was encouraged to seek out ways to achieve a better quality of life. The medical anthropologist Hans Baer wrote that the New Age movement fostered a cosmology that valued "tranquility, wellness, harmony, unity, self-realization, self-actualization, and the attainments of a higher level of consciousness."[8]

Yet this did not mean that there was harmony and agreement among natural healers. Key differences continued in the 1970s between natural hygienists, also called straight nature curers, and those identifying as naturopathic physicians. They flared in the first issue of the *International Journal of Alternative and Complementary Medicine* in 1993. The journal published a scathing piece by Keki Sidhwa accusing practitioners of alternative medicine of cure-mongering. Sidhwa, a graduate of the Edinburgh School of Natural Therapeutics and the British College of Naturopathy and Osteopathy and a self-defined natural hygienist, practiced straight nature cure, rejecting scientific remedies of all sorts, which he said natural living could make obsolete. At the Third World Congress of Alternative Medicine he asserted that enervation and toxemia caused ill health and result from daily living habits. Originators of nature cure, he said, "would turn in their graves to see modern naturopathy turn into second-hand imitations of doctors' prescriptions with emphasis on the modification of patients' symptoms." It was, as Friedhelm Kirchfeld noted, the old controversy between naturopaths and natural hygienists, the fundamental disagreement about the extent to which there should be a blending of therapeutics.[9]

THE REBIRTH FLOURISHES

Both alternative and mainstream publications heralded the advantages of natural medicine. During the 1980s and early 1990s they noted that toxic drugs and dangerous surgical complications scared growing numbers of patients away from mainstream MDs. Popular, alternative, and newspaper publications detailed natural healing's successes. An article in the *Connecticut Valley News* in 1992 chronicled the success of a naturopath in New Hampshire who treated a young boy for chronic asthma, ear infections, and tonsil and adenoid removal. The boy had been on a steady dose of six medications (antibiotics, antihistamines, steroids, and others). After treatment by Dr. Pamela Crocker, the boy recovered remarkably. The mother testified before the New Hampshire House Executive Departments and Administration Committee (which oversaw naturopathic licensing) that she had found a solution through naturopathic care. These kinds of testimonies, which echoed healing testimonies of seventy-five years earlier, boosted the profession immeasurably.[10]

During this positive coverage the shift toward standardized education and medically trained naturopathic physicians continued. Yet unlicensed naturopathic colleges offered degrees without uniformly regulating curriculum or practice. The fight for legitimacy was also complicated by insurance companies' policies. At this time expenses for only 25 percent of natural healing modalities were reimbursable.[11]

John Weeks, the executive director of the AANP, said in 1992 that there were 1,002 naturopathic doctors in thirty states in the United States and another 250 in Canada. The number of physicians was growing, he reported, but was nothing compared with the 10,000 practitioners in the 1920s. His comments fueled concerns of natural hygienists, or nature curers, whom he characterized as yesterday's old healers, utilizing herbs and old family remedies; the short-lived coalition from the 1950s was gone. Weeks said that they had been replaced by a new generation of NDs whose schooling was comparable to traditional medicine's but emphasized nature's healing methods. The naturopathic colleges had helped build confidence in their methods. Looking back from 2012, Bastyr University's cofounder and president Joseph Pizzorno said that the naturopathic colleges had shown that science-based natural medicine was attainable and efficacious. Bastyr was a leading force in resurrecting public interest in natural medicine. Its graduates were physicians, proficient authors, and trustworthy lecturers who Pizzorno said demonstrated the great value of natural medicine to the world.[12]

There was palpable hope and excitement in the naturopathic profession by the 1990s. Michael Murray, ND, author and faculty member at Bastyr, published with Pizzorno the seminal *Encyclopedia of Natural Medicine* in 1991. They demonstrated the clinical alignment with biomedical standards, but with the principles and personalized health attention of traditional naturopathy. The typical first visit to a naturopathic doctor was one hour long, involving a medical history and physical exam, lab tests, radiology, and other standard diagnostic procedures. Conversation with the patient included lifestyle, diet, stress, environment, and exercise. The doctor and patient together established a course of treatment and a health program involving nutrition, botanical medicine, homeopathy, acupuncture, and physical medicine.[13]

Friedhelm Kirchfeld, the founding library director (1978–2006) at the National College of Naturopathic Medicine, demonstrated this hopefulness when he sent out a letter to "130 old-timers" on the American Association of Naturopathic Physicians list urging them to submit their documents to the 1993 "National

Treasures: Homecoming '93 Celebration" and then to consider donating them permanently to the NCNM library. But understanding the history of the profession, he said to this author dryly, "Let's see what will come out of it."[14]

HOLISTIC MEDICINE AND THE AMA

As naturopaths incorporated more elements of biomedicine, some mainstream physicians adopted some of naturopathy's long-held modalities. This signaled the success of naturopathy's long struggle for legitimacy. MDs were responding to the growing published evidence of the effectiveness of the modalities and to their patients, who were embracing the claims and care of other providers. Medical doctors capitulated in part because of the consumer influence on health care. The combination of these factors led to medicine's gradual incorporation of complementary and alternative medicine (CAM).

This shift was reflected in writings that appeared in progressive as well as some traditionally conservative professional outlets. One MD stated in 1990 that various studies showed that 10–50 percent of cancer patients were using alternative therapies that varied from bolstering health through exercise and diet to quackery. He added derisively that alternative therapies gained in popularity in an antiestablishment, anti-intellectual climate. A professor at the University of Colorado School of Pharmacy added a different perspective. Health professionals' tunnel-vision approach and dismissal of natural medicine as antiquated and useless reflected an ahistorical ethnocentric view. To be a good doctor, physicians must understand language, cultural notions of disease, and perceptual differences. Doctors must not ignore patients' cultural backgrounds.[15]

By the 1990s, MDs were observing that unorthodox medical practices of all kinds persuasively challenged allopathic medicine's ability to meet patient needs. They acknowledged that while patients might rely on physicians for medical care, they were also trying many alternative therapies. The words of early-twentieth-century naturopaths found new life in the *Journal of Medical Ethics* in 1992. According to one author, there was a gap between the "interests of medicine as a social system and the patient's need for comfort and support." Another author argued that journalists should stop reflecting dominant ideologies that negatively compared alternative medicine with mainstream allopathic medicine. Yesterday's alternative would become today's mainstream. It was a prophetic statement about regular medicine's impending cooptation of successful alternative techniques.[16]

NATURAL MEDICINE: POPULARITY, PROFIT, AND INSURANCE

In 1990 Cathy Rogers, the president of the American Association of Naturopathic Physicians, wrote that there was a rising demand for naturopathic doctors across the United States. She noted the number of calls coming into the national office from people looking for naturopathic doctors in states where they could not be licensed. She suggested that moving licensed naturopaths into these states offered tremendous potential for growth. Callers wanted competently trained doctors who offered alternatives to conventional approaches.[17]

The growth in naturopathy's popularity in mainstream society was reflected in a favorable article about naturopathic medicine in *Good Housekeeping* in 1994. As a result of that article, the national office of the AANP received about sixty requests for referrals per day. That brought the year's total number of patient referrals to nearly six hundred per month, up about 75 percent from the previous year.[18]

This increased interest parlayed into more profitable work for naturopaths and other natural healers. Sid Kemp, writing in the *Nieman Reports* in 1993, raised an important point: the designation "alternative" was often applied to those systems that were not recognized by insurance companies. Because insurance companies both shaped and validated what counted as legitimate care, their stamp of approval greatly impacted naturopathy's stature and accessibility. Americans spent tremendous amounts of money on complementary and alternative medicines—how much depends on which study you believe. According to one study in 1994, Americans spent about $13.7 billion on alternative medicine, only $3.4 billion of which was covered by insurance. According to another estimate, expenditures on alternative medicine were $14.6 billion in 1990 and $21.2 billion in 1997. Either way, alternative healers, and naturopaths in particular, had seen increased business. This increase coincided with criticism of "the high cost, bureaucratization, specialization, and reductionism of biomedicine."[19]

At the same time, insurance companies urged caution. The Travelers Insurance Company in a 1993 pamphlet cautioned people to "*beware*" (emphasis in original). The pamphlet stated that according to the *New England Journal of Medicine* one in three people used alternative medicine. Spotting fakes was imperative, and the pamphlet listed five signs. Readers were encouraged to ask questions and consult legitimate agencies for information.[20]

Insurance coverage increasingly became a real possibility. A naturopath in Oregon encouraged patients to seek medical insurance that covered naturopathy, to contact employers or insurance agents when it did not, and if these measures failed, to switch to insurers who covered licensed naturopaths. This was more plausible in Oregon, where there was a historically favorable climate for naturopaths. One researcher in 1997 found that individual insurance companies would cover alternative therapies on a case-by-case basis, and some health maintenance organizations (HMOs) and insurance companies arranged contracts with alternative healers. One insurer that specialized in coverage of alternative therapies was Alternative Health Services, in Thousand Oaks, California. Other insurance companies allowed medical doctors to refer patients to them. But most insurance policies, as well as Medicare and Medicaid, did not cover holistic health services. The result was that those patients with viable incomes (predominantly white) made up the bulk of patients of CAM practitioners. Some researchers also argued that short-term higher costs for alternative therapies were justified by CAM's long-term cost effectiveness.[21]

HOLISTIC HEALTH
AND INSTITUTIONALIZING NATUROPATHY

Hans Baer noted that the holistic health movement included a "diverse cast of characters . . . lay alternative practitioners, psychic or spiritual healers, New Agers, holistic M.D.s, and at least some osteopathic physicians, chiropractors, naturopaths and acupuncturists." The movement overlapped with the hippie counterculture, humanistic psychology, and some aspects of the New Age movement. These eclectic visions were critical of technological, depersonalized Western medicine and society and emphasized self-determination through self-help. It was from these combined ideologies and practices in the late 1970s that the concept of mindbodyspirit wellness arose. The merged words signified that all aspects of a person's well-being were interrelated. A number of authors exchanged these ideas. The senior editor of *Alternative Therapies* said that healing treatments must speak to patients' self-defined needs and embrace pluralism, and thus there was no place for practitioners' arrogance. But how could practitioners in the holistic movement, who had a wide variety of skills and philosophies, organize more effectively for the sake of patients? It was an old question in the history of naturopathy, and the answer channeled the philosophy of Benedict Lust: naturopathy could be a baseline and fundamental framework within which other natural therapies could develop and expand.[22]

Naturopathy had been a precursor to, and was clearly in sync with, the holistic health movement of the 1970s and was poised to flourish. Naturopathy was an

age-old medicine for the New Age. Or as another author put it, naturopathic medicine has been at the forefront of the paradigm shift occurring in medicine. Its
techniques are being incorporated by conventional medical organizations.[23]

Naturopathy had always emphasized self-care; the physician was regarded
more as a teacher than as an irreplaceable sage. Its history and outsider status
made it congruent with the other wide-sweeping cultural critiques. It did not,
however, embrace all of the multitudinous aspects of holistic health practices
from the 1970s on. Many insisted that naturopathy should remain loyal to therapeutic methods discussed throughout this text. Yet the naturopathic, holistic,
and New Age movements were complementary; they all fed off the same cultural
distrust of authority and the same desire to return to nature in a postindustrial
society. Naturopathy for decades had been viewed as a medical catchall for a variety of nonconventional medical practices—even when that was not true. Then
the influences of the holistic movement swept naturopathy back into the public
eye alongside holistic health.[24]

By 1978 *institutional* tides favored naturopathy. This was a result of decades of
deliberate—and contentious—work. At that time there was only one naturopathic
school, the National College of Naturopathic Medicine, in Portland, Oregon. In
1980 three NCNM graduates—Joseph E. Pizzorno Jr., Lester E. Griffith, and William Mitchell, all former students of John Bastyr—established the John Bastyr
College of Naturopathic Medicine, now Bastyr University, in Seattle. The fourth
cofounder was Sheila Quinn, previously a University of Washington administrator, who served as Bastyr's sole administrator for more than ten years, stabilizing
and expanding the institution. Pizzorno became its first president and served in
that post for twenty-two years. The founding of the new college was a strategic
move to prevent Washington State from eliminating naturopathic licensing. Bastyr College would protect licensing, inspire the naturopathic field, and operate
on science-based precepts. It graduated its first class in 1982, and the next year it
was accredited by the Northwest Association of Schools and Colleges—a first for
the naturopathic profession. Since then, the university has added degree programs in the Schools of Naturopathic Medicine, Traditional World Medicines,
and Natural Health Arts and Sciences; and just some of the advanced degrees,
besides the ND, are in nutrition, counseling psychology, midwifery, acupuncture, ayurvedic sciences, and public health. It met scientific standards while contributing to the public's demand for natural medicine.[25]

But these important advances were offset by the paltry state of naturopathic
licensure and recognition nationwide. In 1980 an effort began to create a new
national organization that eventually became the American Association of Natu-

ropathic Physicians in 1985. As part of that effort, the director of student services at the National College of Naturopathic Medicine in Oregon, Sarah Davies, sent inquiries to all the other states about the status of naturopathy. The responses from seven states (Pennsylvania, Rhode Island, South Dakota, Texas, Vermont, Virginia, and West Virginia), now in the NCNM archives, show that none of them had licensing provisions for naturopaths. Either naturopaths were not recognized at all or they were prohibited from practicing, with little hope of doing so in the foreseeable future. Responders attached copies of statutes or case law that chronicled unsuccessful licensure attempts.[26]

The responses showed that the AANP was necessary and that standardized educational advances were a priority. In 1993 the third four-year naturopathic school, the Southwest College of Naturopathic Medicine, opened in Arizona. The University of Bridgeport in Connecticut, chartered in 1947, created a College of Chiropractic in 1991 and the College of Naturopathic Medicine in 1996. The nation's first university-related chiropractic college, it is an example of the peculiar relationships that result from economic need. On the verge of bankruptcy, the university was saved by donations of about $100 million from the Professors World Peace Academy, funded by the Unification Church, over a period of years; the group maintained control of the board of trustees, and the university even awarded an honorary doctorate to the Reverend Sun Myung Moon, founder of the highly profitable Unification Church and self-proclaimed messiah, in 1995.[27]

By the 1990s a number of cross-over practitioners were contributing to the *Journal of Naturopathic Medicine*, reflecting scientific interest in naturopathy. One volume contained articles by naturopaths, public health officials, PhDs, and pharmacists. Articles stressed a move from anecdotal to evidentiary recordkeeping, alternative treatments for brain tumors and cancer, and postsurgical treatment. The philosophy of naturopathy was also publicized on the website of the Naturopathic Medicine Network, an organization that advanced naturopathy worldwide. Naturopathy was "the treatment of disease through the stimulation, enhancement, and support of the *inherent healing capacity of the person*" (emphasis in original). Naturopaths, according to the website, use the healing power of nature, identify and treat the cause of the illness, subscribe to the principle to first do no harm, treat the whole person, become the physician as teacher, and believe that prevention is the best cure.[28]

In 2001 the American Association of Naturopathic Medical Colleges formed to energize and encourage the naturopathic medical profession by supporting accredited and recognized naturopathic schools. The standards rose: by 2014 the typical entering ND had a bachelor of science degree and three years of premedical training. Seventeen states legally recognized and regulated naturopathy.[29] As of 2014, the American Association of Naturopathic Medical Colleges recognized seven institutions, five in the United States: Bastyr University (Kenmore, WA, with a California campus in San Diego); Southwest College of Naturopathic Medicine (Tempe, AZ); the National University of Health Sciences (Lombard, IL); the University of Bridgeport (Bridgeport, CT); and the National College of Natural Medicine (Portland, OR), formerly the National College of Naturopathic Medicine, which had changed its name in 2006 to reflect studies in traditional world medicines and any changes to come in the future. The change was a significant reclaiming of a broader base of healing systems.[30]

The recognized Canadian schools are the Boucher Institute of Naturopathic Medicine, in Vancouver, British Columbia (2000), and the Canadian College of Naturopathic Medicine, in Ontario (1978). Graduates of both institutions can practice in the United States in states that regulate the practice of naturopathic medicine. All of these institutions are accredited by the Council on Naturopathic Medical Education, based in Great Barrington, Massachusetts, as well as by the US Department of Education agencies. The Council on Naturopathic Medical Education works to foster and protect high-quality naturopathic medical education in the United States and Canada.[31]

FEMALE PRACTITIONERS

Since the late 1980s the ranks of licensed naturopaths have swelled, and female practitioners and leaders have become common again. The large number of women graduating from the John Bastyr College of Naturopathic Medicine prompted the creation of Women in Naturopathy (WIN) in 1987. By the mid-1990s, women made up more than half of the graduates from naturopathic colleges, prompting two authors to write that "naturopathic history will be as much a history of women as of men." Indeed, women have been naturopathy's cocreators for more than a century. The history of women and naturopathy in the modern era deserves a detailed study of its own. In a nutshell, since the 1960s and the second wave of the women's movement, opportunities for, and self-perceptions of, girls and women have changed dramatically. Legislation such as Title IX in 1973, known mostly for its sports equity provisions, guaranteed girls and women equal access to educational resources. The exponential rise in divorce rates and the desire for meaningful work outside the home led young women to think more realistically about finding fulfilling work to pursue throughout their lifetimes. In addition, the holistic health movement was in many ways a women's movement. Women of all races, ethnicities, social classes, sexual orientations, and gender identities began to see control of their physical health and reproduction as a primary right.[32]

Since 1980, women have increasingly outnumbered men in college enrollments nationwide. It makes sense, then, that the number of women in naturopathy has increased, since jobs in health care (as nurturers and caretakers) had traditionally been deemed culturally acceptable for women. The tenets of naturopathic medicine—a healthy and balanced environment, natural living and healing methods, and a healthy skepticism of biomedicine—have also been important issues for many women in recent years. Modern-day naturopathy thrives on women as officers, organizers, researchers, practicing physicians, thinkers, and advocates of the profession. It is an environment in which women can flourish; its ethos is inclusive. Cathy Rogers had been a licensed midwife, graduating from the National College of Naturopathic Medicine in 1976. In 1989, as AANP president, she offered seminars on women and work, as well as renewal retreats, in her native Seattle. She urged licensed naturopaths to pursue distinct strategies for naturopathy's growth and survival. It was good advice. Strategies included forging alliances "with the self-care movement; with active environmentalists; with people who are disabled or disenfranchised by conventional medicine's narrow focus of disease, drugs and surgery." Rogers was succeeded as president by Jamison Starbuck, JD and ND, another woman who envisioned a future for naturopathy that was creative, intelligent, and unified. Her leadership lasted only one year, but she was credited with a lasting and positive legacy. Through the next decades women were instrumental in stabilizing and institutionalizing naturopathy. They were convention managers, college administrators, institutional stalwarts, faculty, advocates in Washington, DC, and much more. A 2008 survey revealed that 73 percent of AANP members were women and that 77 percent of nonmember naturopathic physicians were women. Emblematic of this trend, in 2014 all board directors and committee chairs of the California Naturopathic Doctors Association were women.[33]

Articles by women in naturopathic journals have covered a range of general therapeutics, as well as food and family health. Journal profiles of individuals have often focused on women. For example, in 1991 Dr. Jeanne Albin, a member of a Coast Salish tribe, was the first Native American naturopathic physician to have graduated from the National College of Naturopathic Medicine and integrated native and naturopathic healing methods. In 1997 the *Journal of Naturopathic Medicine* devoted an entire issue to women's health care, signifying not only the importance of naturopathy to women's health but the diversity of subjects women's health encompassed. The issue contained research reviews, reports on original research and case studies, and theoretical articles. The foci ranged from donor egg conception for lesbian and heterosexual women to cancers of the breast and ovaries to alcoholism, osteoporosis, adolescent pregnancies, infertility, battered women, and eating disorders. Authors continued to contribute to the journal in areas of women's health, which was still understudied in biomedical science. Alternative publications, such as *EastWest*'s thirteen-page 1990 feature "50 Ways to Better Women's Health," also utilized the expertise of female naturopaths when discussing women's health.[34]

COMPLEMENTARY AND ALTERNATIVE MEDICINE (CAM)

The growing professionalization, acceptance, and influence of naturopathy and other alternative practices were recognized and reaffirmed through the opening of the National Institutes of Health Office of Alternative Medicine (OAM) in 1992. It was given more autonomy in 1998, when it was elevated to the status of an NIH center, becoming the National Center for Complementary and Alternative Medicine (NCCAM). Congress urged its creation in response to concerns about biomedical healing costs. NCCAM's mission is to determine which therapies work. Inclusion in the NCCAM is evidence that naturopathy gained significant legitimacy and reflects the influence of consumers on health care. It is also the case that this kind of institutionalization brings alternative medicine within the control and purview of allopathic medicine. As one scholar quipped, it has, ironically, fostered "biomedical co-optation of the holistic health movement."[35]

In 1993 the AANP was invited to testify at a hearing on alternative medicine at the NIH Office of Alternative Medicine. It was the first time in modern history that Congress recognized the profession's expertise. Naturopathic credibility got another boost in 1994, when the OAM granted Bastyr University a grant in excess of $900,000 to conduct research on alternative therapeutics for HIV and AIDS, and two naturopaths served on the NCCAM advisory council. Since the OAM's founding, naturopathy has been a part of its mission to explore therapeutics for mind and body wellness, dietary supplements, probiotics, and botanical remedies. In a 2012 study, the NCCAM found that 38 percent of American adults relied on complementary and alternative medicines. Yet the entire budget of the NCCAM has been small. It was granted $2 million at its creation, and in 2013 it had a budget of nearly $120.8 million—representing a marked increase but small compared with the full NIH budget of $30.85 billion that year. By way of comparison, the National Cancer Institute, which receives the highest funding among the NIH centers, was allotted just over $5 billion; out of the twenty-five funded centers, the NCCAM ranked twenty-fourth, with only the Fogarty International

Center receiving less ($69.8 million). As of 2001, thirteen specialty research centers had received funding from the NCCAM, all but two of them allopathic; Bastyr and the Palmer Center for Chiropractic Research in Davenport, Iowa, were the only two nonbiomedical centers funded. Labeling legally recognized practitioners of CAM as quacks is no longer appropriate, although if money talks, then the funding granted to allopaths for work traditionally performed by naturopaths speaks to the work yet to be done to achieve parity.[36]

Change comes incrementally, and as the historian James Whorton has pointed out, placing the word *complementary* before *alternative* to create the popular acronym CAM was a game changer. By making alternative therapeutics also *complementary*, the acronym "subtly transform[ed] the image of alternative from dubious option pursued as a last resort to plausible treatment elected in situations where allopathy is insufficient." When the NIH created a center whose name included those words, the concept was institutionalized. Despite the credibility that these recognitions award, medical science has continued to ridicule the NCCAM's regulatory effectiveness because of the widely eclectic troops it oversees. This is a legitimate concern because there are still self-defined alternative healers whose methods are dubious, even deleterious. The sheer number of services offered in a given region makes regulating practice and measuring efficacy virtually impossible.[37]

In this resurgence of alternative medicine, attitudes from chiropractic, osteopathy, biomedicine, and nursing ranged from support and deep commitment to ambivalence and hostility. In part this is because of CAM's eclecticism, a trait that has always represented both the strength and the weakness of naturopathy. To be housed under the big tent once again meant implied association with diverse modalities (some not even endorsed by naturopaths) and a hazy professional identity. Osteopaths had repeatedly aligned in earlier decades with, and sought recognition from, allopathic medicine—more so than had chiropractors. But they had their own uphill battles to fight for public acceptance. For instance, in 1962 osteopaths cut a deal with MDs to merge; the California public also voted in a referendum to prohibit new DO licenses in the state. As a result, the College of Osteopathic Physicians and Surgeons in Los Angeles converted to an allopathic medical college with the blessing of both the California Medical Association and the California Osteopathic Association. Roughly fifteen hundred DOs received MD degrees as a result, and while they could continue to practice, they could no longer identify as osteopaths. Osteopaths challenged the law with a court battle that did not end until 1974, when the state supreme court ruled that licensing of DOs could be resumed.[38]

Many of the legal stumbling blocks naturopaths continued to face did not exist for osteopaths, because they aligned with biomedicine. Legal recognition allowed osteopaths to admit patients to MD hospitals, testify as expert witnesses in court, sign birth and death certificates, and enjoy all the other privileges of licensed MDs. In addition, osteopathy had been rejuvenated as a result of biomedicine's specialization, the need for general practitioners, the unequal distribution of health care, and the limits of allopathic medicine.[39]

But even though osteopaths benefited from the rise of holistic medicine philosophies, some hesitated to align themselves with the movement. One osteopath in 1988 argued that applying the term *holistic* to osteopathy put it back into the cult category, along with naturopathy and other systems regarded as quackery, occultism, or superstition.[40]

Other osteopathic authors moved toward a holistic approach. In a 1991 open conversation in the *Journal of the American Osteopathic Association*, one argued that osteopathy should at least *consider* holistic medicine because it so clearly met a consumer demand. Another said that osteopathic research needed a paradigm shift to incorporate holistic principles and ideas. Leery of too much inclusiveness, others argued that a clearer articulation of osteopathy's *own* theory and practice was needed to distinguish it from alternative medicine.[41]

Chiropractic's liaisons with naturopathy, although much diminished by the mid-twentieth century, continued. Decades of disassociating from other sectarian healers had given chiropractors credibility. By 1960 only two states failed to recognize chiropractic: New York, which capitulated in 1963, and Massachusetts, which added it in 1966. Chiropractors were included in workers' compensation laws, and tax deductions were allowed for their services. At midcentury there were 25,000 chiropractors with 35 million patients—one sixth of the US population. Despite these achievements, the medical profession did not let chiropractors forget they were the enemy. This proved advantageous in the long run. Yes, their treatments took longer, but they treated ailments for which biomedical science tended to use drugs or invasive surgery, and their treatments were cheaper. Chiropractors fit into the holistic movement because their interpretation of disease origins was more inclusive of causalities than allopathy and because they made their diagnoses based on interaction and rapport with patients rather than primarily on machine testing.[42]

At the same time, chiropractors were ambivalent about being associated with holism. On the one hand, they were emboldened by winning the US Court of Appeals case *Wilk v. AMA* (1990), which found "that the AMA illegally conspired to destroy the chiropractic profession" through the AMA's Committee on Quackery. But success came at a high price. Legal recognition also meant being bound to licensed acts rather than a range of holistic treatments that centralized the doctor-patient relationship. Although chiropractic gained official and public recognition in Australia, Britain, and the United States, it did so by disassociating from naturopathy, which stayed true to its broad-spectrum holism. One health economist expressed similar concerns. Chiropractic's future, he said, would be best served by building alliances with naturopaths, acupuncturists, and other holistic medical practitioners, as well as preferred provider organizations and health maintenance organizations. He expressed concern about chiropractic's future as charges of quackery and fraud came from *within* the profession.[43]

The shifting landscape of medical alliances created unexpected liaisons. In 1978, 220 biomedical physicians founded the American Holistic Medical Association (AHMA). As of January 2013, membership in the association required certification by the American Board of Integrative Holistic Medicine, instituted in 1996. Training at one of five US schools or one Canadian naturopathic school was recommended, yet holding a degree from one of them did not guarantee certification as a holistic healer.[44]

The American Holistic Nurses' Association was founded in 1981, and like the American Holistic Medical Association, it emerged in response to the shifting medical landscape. Holistic nurses also laid claim to traditional naturopathic methods without formally acknowledging the link to its practitioners. The association produces the *Journal of Holistic Nursing*, has its own newsletter, and as of January 2013 had 5,400 credentialed members. There is no naturopathic nursing degree per se, but curiously, web searches for *naturopathic nursing* take one to the association site. Hyperlinks to "Naturopathic Definition and Philosophy" are inoperative. Only four schools are listed as offering holistic degrees. Since its inception in 1981, the organization has offered a revised approach to biomedical nursing and incorporates CAM, but it developed its own distinct identity, which includes naturopathy by name but without recognition of its theories or modalities. Aspects of care that authors and practitioners of holistic nursing emphasize include compassion, length of time with patient, relationship between care and religion, psychotherapy, biofeedback, psychosocial needs of patients, stress reduction through breathing and affirmations, soothing environments and environmental harmony, reflexology, and therapeutic touch. They advocate for affordable health care and a balance between body, emotions, mind, and spirit—all tenets of naturopathy for a century. Authors writing in the holistic nursing journals are RNs, MSNs, MSs, EdDs, PhDs in public health, public health nurses, licensed home care practitioners, and registered midwives.[45]

Integrative medicine was making its mark. When the November 1998 issue of the *Journal of the American Medical Association* focused exclusively on alternative medicine, it was a clear example of biomedical physicians' increasing acceptance of CAM and the legitimacy of naturopathic methods. It was also a savvy and timely nod to public skepticism about the limits of scientific expertise. But it spawned a collective concern among naturopaths that biomedicine's use of CAM would draw away their patients. Historians and practitioners alike have noted that at least some biomedical physicians integrated aspects of holism to ride a profitable wave. Given the history of naturopathic-AMA relations, the cynicism was not unwarranted.[46]

In 2005 a study by Terri A. Winnick traced the AMA's changing view of CAM, from quackery to complementary medicine. Winnick examined the coverage of CAM in five prestigious medical journals from 1965 to 1999. She discovered that the shifting views were driven by changes in the medical landscape, including relaxed licensing, the rise of managed care, increased consumerism, and establishment of the OAM. Winnick divided the years examined into three periods: the late 1960s to early 1970s, which she labeled the condemnation period, when authors mocked and overemphasized the risks and urged legal control of CAM; the mid-1970s to 1990s, a time of reassessment, when greater numbers of consumers utilized CAM, implicitly critiquing regular medicine and the needs it did not meet; and the 1990s, the integration phase, when legal hounding of CAM practitioners mostly ceased and biomedical physicians began to utilize CAM themselves. Regular medicine also tried to control CAM in the 1990s through scientific scrutiny.[47]

Of course, national and international criticisms came from biomedical professionals who did *not* incorporate CAM into their practice. In 1990 an article in *Quality Assurance Health Care* asserted that patients with diseases who were helped through conventional medicine could be harmed by CAM. The article highlighted a Swedish study about the deaths, medications withheld, and crises brought on through CAM. One Australian author that year praised medical school teaching about nutrition but disparaged that of naturopaths as unsound. The author asserted that only dieticians, nurses, general practitioners, pharmacists, and dental therapists could legitimately dispense this information. This was an interesting fiat, since many of those deemed "skilled" based their knowledge on long-avowed naturopathic nutrition principles. Once again, naturopaths schooled in scholarly colleges and universities, were lumped together with self-proclaimed naturopaths, herbalists, and self-styled nutritionists—all deemed unqualified.[48]

Two alternative therapies, acupuncture and herbalism, have gained considerable acceptance amidst some continuing skepticism and outright hostility. Both are now central elements of standardized naturopathy. In the case of herbalism, lack of quality assurance and discredited claims continue to undercut its credibility. A modality once considered suspect, acupuncture has emerged as a recognized and valuable contender, gaining a patient following and insurance reimbursement status. Acupuncture's acceptance coincided with naturopathy's, and acupuncture was included in naturopathic curricula by the 1980s. Like naturopathy, it increasingly incorporated biomedicine's social organization and theory. It was also met with the same range of acceptance, disdain, and gradual inclusion. Naturopathy's inclusion of acupuncture brought both new allies and biomedical criticism.[49]

Positive portrayals of acupuncture in the press came with its adoption within traditional medicine. A 1988 Associated Press story reported on the relief and satisfaction it brought to patients. The story quoted registered nurses, internists, and psychologists, who praised its benefits to patients. Another complimentary newspaper article praised acupuncture for its success in treating addictions and phobias. The licensing of acupuncture reflects the general, gradual integration of—and ambivalence toward—CAM. States vary in their treatment of acupuncturists, from demanding that they also be licensed MDs, to requiring medical supervision, to licensing and registering acupuncturists independently from MDs.[50]

Acupuncture was accepted in part because of its growing adoption of scientific methods. In 1989 the Columbia University School of International Affairs held its Fifth Annual International Symposium on Acupuncture and Electro-Therapeutics. Scholarly papers were presented that examined acupuncture's use as an analgesic and its impact on the immune system, circulation, low back pain, phantom limb pain in amputees, and cessation of smoking.[51]

Acupuncturists have their share of detractors as well. In 1991 the National Council Against Health Fraud denounced acupuncture's value outright in an official position paper, stating that physicians offering it to patients should tell them that its use was experimental. The council sent legislators a plea to abolish licensing of acupuncturists. A journalist who surveyed physicians found them ambivalent or hostile. One physician said that acupuncture was like a placebo. It might help a person if he or she thought it would; "otherwise, the idea of inserting needles into the body to help the mind is unproven." As recently as 2011 an article in the *Atlantic* traced MDs' growing acceptance of acupuncture, along with homeopathy and other CAMs. Yet some individual physicians said that acupuncture and CAM were merely cleverly marketed, dangerous forms of quackery. Steven Salzburg, biology researcher at the University of Maryland, College Park, and according to the *Atlantic* "one of the angriest voices attacking [CAM]," proclaimed acupuncture analogous to voodoo, saying, "There's only one type of medicine, and that's medicine whose treatments have been proven to work." Ambivalence aside, growing acceptance of acupuncture prompted Hans Baer to note that "acupuncture, with its more geographically far-flung network of schools, may already have come to supersede naturopathy as a professionalized heterodox medical system."[52]

Medicinal herbalism, which is more diffuse and harder to pin down, also became mainstreamed. Herbs are not by nature benign; some have powerful, dangerous, or fatal effects. This field is also plagued by intense and at times misleading marketing. A clear example is the scandal around the marketing of HerbaLife as a Ponzi scheme. There is no quality assurance nor oversight of claims made by producers of supplements. There are self-identified herbalists and naturopathic licensed curricula that use the word *botanics*. Botanics have historical precedent and may, to a discerning consumer, signal a separation from the field of herbalism. Naturopaths' use of herbs created a market for them long before they became popular. The holistic movement, however, both popularized herbalism and divorced it from its historical and naturopathic roots. This is not to say that naturopaths owned this branch of healing, but its commodification, beginning in the 1970s, and its proliferation came with little reference to naturopathy. In 1988, Jane Jones traced the history of the Weleda pharmaceutical company, founded in 1923, which provided pesticide-free and chemical-free herbs to the medical and veterinary professions. She likened this work to homeopathy, overlooked naturopathy, and noted that natural medicines were an expanding industry and a common sight in supermarkets. Similarly, a seventy-two-page booklet detailed herbs' benefits without acknowledging the long naturopathic tradition of herbal use for cleansing, normalizing body function, nutrition, raising energy levels, and stimulating the immune system. It also offered principles for better health and living that echoed three of naturopathy's own: diet, will, and exercise. These publications present a naïve sense that these are recently discovered remedies, are at long last available, and have limitless safety and value. By failing to acknowledge naturopaths or longstanding natural healing systems, herbalists miss an opportunity for collaboration, but they can also taint naturopathy's judicious use of botanical remedies.[53]

Newspapers also contributed to the idea that scientists were now suddenly discovering plant-based cures. In 1990 an article in the *San Diego Union* covered the National Cancer Institute as it quested for green medicines that could reveal the secrets of nature's healing properties and be of value in AIDS treatment. Scientists would unlock this knowledge. The "discovery" of rainforest botanicals also flourished during the rebirth of holistic healing. Publications in this genre were written by biomedical doctors, traditional Chinese medical practitioners, herbal manufacturers, pharmaceutical companies, and self-publishing enthusiasts.[54]

During this era of "discovery and invention" naturopathy was acknowledged by *Medical Herbalism*, a clinical newsletter for the herbal practitioner. Some of its authors were naturopaths. Gaia Herb's symposium in 1994 focused on naturopathic healing wisdom. Most of the symposium speakers were naturopaths, and academic credit was given to attendees through National University and Bastyr University, with continuing-education credits for NDs. In this case, acknowledgment and collaboration were productive.[55]

EDUCATION

At the seven accredited US and Canadian schools now offering four-year ND degrees, students pursue the same curricula in basic and clinical sciences that MD and DO students do. The ND curriculum combines scientific advances with natural therapeutics and disease prevention. Besides the basic sciences, naturopaths' training includes clinical nutrition, acupuncture, botanical medicine, homeopathic medicine, physical medicine, and counseling. Clinical training during the final two years takes place in supervised settings with licensed professionals.[56]

Disagreement about what constitutes a naturopathic education continues among scientifically trained naturopaths, those taught through correspondence (or online) courses earning "degrees and titles," and those self-identified natural healers (often self-taught). All believe they know the methods and means of preserving health and treating disease. To some lay, self-identified naturopaths, any adherence to scientific expertise betrays naturopathic principles. One writer identified another type of practitioner beyond those schooled or self-identified: the historical naturopath, who follows time-honored traditions of natural healing and folk medicine. The author called for a screening out of unqualified people and allowing naturopathic physicians and historical naturopaths to practice. But this suggestion overlooks a key problem: how are unqualified people to be screened out, when so many refuse to be acknowledged or to be bound by uniform standards? With the exception of graduates from the accredited schools, the field is amorphous, scattered, largely unregulated, and elusive. Thus, the old discussion continues about similarities and differences between schooled and self-taught naturopaths.[57]

As alternative medicine gained credibility in the mainstream, many formally trained, professionalized naturopaths set out to scientifically prove their system's efficacy in chronic disease, postdisease treatment, and rehabilitation. But for the many thousands of practitioners not affiliated with institutions or practicing controlled research methods, this adoption of scientific principles was by no means generally accepted. Making a case for research, Bastyr's first president wrote that it was important to concentrate on natural medicine, which could be scientifically verified, and less on anecdotal testimony. Leanna Standish, PhD, ND, the director of research at Bastyr College, argued that scholarly articles were important not only for the profession as a whole but for private practice as well. Wrote Tori Hudson, ND, the medical director of the NCNM, "A great motivating force for me in conducting research is wanting recognition for the profession. . . . We/I stand to gain [much] by doing research: treatments become more specific, more refined." In 1991 five naturopaths were awarded monetary prizes for their office-based research projects at the North American Convention of Naturopathic Physicians in Whistler, British Columbia; three of the five were women. The entire Whistler convention emphasized naturopathic research, advancing science, health, and activism for the cause. As part of the increased research effort, the Bastyr College Research Department set out to document a variety of treatment outcomes. It established protocol for collecting data and carefully chronicled interventions and postintervention patient functionality. Participants were urged to present their findings at the next AANP convention. By 1993, funding at Bastyr College provided research scholarships to students engaged in clinical training in the ND program. Awardees had demonstrated an ongoing commitment to research and showed promise of future scientific productivity.[58]

Entire issues of the *Journal of Naturopathic Medicine* reported on clinical trials or on individual physicians' original research. The data and analyses in these articles were drawn from allopathic research in the *New England Journal of Medicine*, the *Journal of the American Medical Association,* and the *American Journal of Obstetrics and Gynecology*. This did not signify outright agreement with biomedicine, but it signaled a willingness to compare the two systems and utilize both methods. (One can almost hear the verbal sparring between Benedict Lust and his nemesis, Morris Fishbein, editor of *JAMA* for twenty-six years, about the contradictions inherent in trying to meld the two systems.) From 2000 on, journals within and beyond the field of scientific naturopathy have continued research-based reporting. Two authors deemed the medical research agenda "the future and foundation of naturopathic medical science." At the same time, allopathic research and practice is catching up to naturopathic methods. When Rikli, the Lusts, and other naturopaths advocated morning nude sun baths for their urban, indoor-bound patients, they saw improved strength and vigor after a period of weeks—improvements MDs today see when they prescribe Vitamin D. Naturo-paths prescribed fasts one hundred years ago to improve healing, and scientists today are learning that fasts can boost the immune system. Throughout the twentieth century, naturopaths advised that births and other reproductive health issues should be handled at home with good preventive care because allopathic procedures and medications had negative effects. Today, MDs are beginning to acknowledge that there are too many cesareans in the United States (33% of births, compared with 26% in the United Kingdom), and at least one study has shown that healthy mothers "are better off staying out of the hospital to deliver their babies."[59]

Despite these developments and the increased rigor of ND research and of programs accredited by the Council on Naturopathic Medical Education and the US Department of Education, they compete with Internet offers. An endless array of professional degrees can be earned online in naturopathy, natural medi-cine, and natural healing, including an ND. They replicate the century-long struggle for naturopaths' professional standing—this time in cyberspace. Some of these online programs last only weeks or months and issue certificates and diplomas. Naturopaths are not alone in facing competition from online diploma mills. As of this writing, e-mail spam arrives daily offering easy online doctorates. The Obama administration is also attempting to curb for-profit schools from making prom-ises about the earning potential from online degrees. Many are prohibitively ex-pensive, overburden the poor, and have very meager graduation rates, and their graduates cannot compete with those from respected brick-and-mortar schools. The American Association of Naturopathic Medical Colleges states that one cannot become a naturopathic physician through online or correspondence courses. Education trains a physician to care for patients and thus requires years of academic preparation and hands-on clinical experience.[60]

Some naturopathic correspondence programs do offer valuable information, and at times scientifically trained naturopaths benefit from the cross-fertilization of ideas. But the correspondence programs' structure, requirements, length, rigor of coursework, faculty training, and lack of supervision constantly challenge the credibility of scientifically trained NDs. An afternoon's perusal of the Internet revealed that naturopathic correspondence courses are available throughout the United States and beyond with a few strokes of the keyboard. Administrators and faculty at online sites hold a variety of degrees, although their origin and legiti-macy are questionable. They identify themselves as NDs, PhDs, MHs (master herbalists), naturopaths, and Drs., among other titles. The Trinity School of Natu-ral Health (a nonprofit Christian-based professional program) offers a curriculum in "Classical Naturopathy" and an ND degree. The site states that the program does not provide a vocational curriculum in states where a license is required, but it does say that the program is accredited by an organization called the American Naturopathic Medical Accreditation Board in Las Vegas, Nevada. It is not recognized by the American Association of Naturopathic Medical Colleges. Other online programs require only a high-school diploma, twelve additional months of study, a good moral character, and English-language skills.[61]

The Herbal Academy typifies the credibility problems scientific naturopaths face. In 2002 the Arkansas Attorney General's Office sued the Herbal Academy for violating the state's Deceptive Trade Practices Act. The academy offered a two-week course, after completion of which, an ad promised, students could practice naturopathic medicine. However, Arkansas did not license naturopathic medicine. Today, Herbal Healer Academy Inc. still exists online, where one finds the proviso that "residents of Arkansas are not permitted by Arkansas State Law to order any of the HHA Correspondence courses."[62]

INTO THE FUTURE

There is an irony in the fact that naturopathy gained credibility and renewed notice amid the holistic health movement even though it was distinct from that movement. In 2001 Hans Baer, surveying extant scholarship, posed the following question: just how holistic could holistic health be without a critique of the existing American political economy, attention to the fragility of the earth, and acknowledgment of the interrelatedness of emotional, physical, and spiritual health? He observed that there was a lot of self-interest within holistic care and that as a social movement it lacked a cohesive approach or definition. Scientific naturopathy has a history of taking on these issues through distinct theories, principles, lifestyle, therapeutic methods, and activism. As Baer noted, the sociocultural solutions for improved health care may necessitate a blending of viable methods from all systems. Scientific naturopathy has achieved coherence and uniformity but is still professionally "buffeted about" by the underskilled and charlatans claiming the title "naturopath."[63]

In 2013 and 2014, seventeen US states (up from 14 when I began this project), Washington, DC, and two US territories (Puerto Rico and the US Virgin Islands) had statutes licensing and regulating naturopaths. They are overseen by sixteen different associations. In these states, naturopathic doctors are required to graduate from an accredited four-year naturopathic medical school and pass an extensive postdoctoral board examination overseen by the Naturopathic Physicians Licensing Examination Board and the North American Board of Naturopathic Examiners. Eleven additional US states plan to introduce licensure bills. Constant legal lobbying is a fundamental part of the AANP's national agenda through its State and Federal Advocacy Section. There is a "State Advocacy Toolbox" on the AANP website for anyone with questions about pursuing a campaign. Activism takes different forms, and consciousness-raising is one. The AANP has a "DC Federal Legislative Initiative," an annual three-day event that includes political leadership workshops and lobbying to increase the reach of the profession. In May 2012, naturopathic doctors and students from thirty-seven states traveled to Capitol Hill for the event. As one participant explained, they wanted support for a Naturopathic Medicine Awareness Week and future legislation that would allow NDs to work in the US Public Health Service and the Veterans Administration, as well as with other patient populations. The very first Naturopathic Medicine Week was celebrated in October 2013. In addition to inclusion in the Public Health Service and the VA, naturopaths' 2015 advocacy priorities are to support state licensure and full implementation of Section 2706 (the nondiscrimination provision of the Affordable Care Act) and to gain support for continued availability of safely compounded medications.[64]

The Patient Protection and Affordable Care Act, known widely as Obamacare, may well open up more opportunities for naturopathy. The program encourages wellness and preventive medicine and favorably mentions alternative medicine. In full effect in January 2014, the act forbids insurance companies from discriminating against any state-licensed health provider. This could lead to better coverage of naturopathic, chiropractic, and homeopathic care. The Bastyr University home page showed an image of President Barack Obama signing the bill into law and an article entitled "Health Care Reform Extends Reach of Naturopathic Medicine." The law's focus on preventive care, according to the article, allows NDs to demonstrate their medicine's efficacy and cost-efficiency.[65]

Alternative treatments could be more reimbursable and be used more often as a result of Obamacare. This inclusiveness was written in by Senator Tom Harkin, D-Iowa, who said, "Patients want good outcomes with good value, and complementary and alternative therapies can provide both." Jud Richland, the chief executive officer of the AANP, said, cautiously, that if the law were put into place as its authors intended, it could be very significant.[66]

The nondiscrimination clause does not apply to Medicaid and Medicare, and insurers can choose to pay providers at different rates. But potentially the act could facilitate naturopathic care's affordability for millions of new patients. As a preemptive move, there is concerted naturopathic lobbying of insurance carriers to recognize and include naturopathic doctors in their programs. There seem to

be clear and approved paths in Hawaii and Oregon, and Vermont is likely to follow; and in numerous licensed states, meetings are being held, with the AANP
providing state affiliates with useful resources and strategic counsel.[67]

Stumbling blocks still remain. If the language is found unclear or if implementation lacks sufficient oversight, insurance companies will balk. Senator
Harkin included language in the Department of Health and Human Services
appropriations bill to clarify the intent. This was necessary because guidance issued by Health and Human Services and the Labor Department had muddied
the waters.[68] Another threat looms as the 2015 Republican-led House and Senate
continue to threaten new challenges to Obamacare. Without the Affordable
Care Act's provisions, naturopathy will remain inaccessible to those who cannot
pay out of pocket. The Supreme Court in June 2015 upheld the Obama administration's intent of the act for insurance subsidies for the poor through "an exchange established by the state." The insurance exchange, or a marketplace of
insurance plans, can be instituted by a state, or the federal government can provide one for a state. The hope is that more insurers will cover naturopathy, as
modeled by the state of Washington, where naturopathy is licensed and legal. In
2012 the Bastyr University system treated thirty-five thousand patients with naturopathic medicine. Sixty percent of the patients billed their insurance companies for coverage.[69]

As of late 2014, the goals of the American Association of Naturopathic Physicians were licensure in every state; equitable reimbursement in the health care
marketplace; and integration into all aspects of the national health care system.
The AANP wants to increase awareness of naturopathy and of its safety, efficacy,
and cost-effectiveness. Additional goals are to serve diverse communities; to be
research-based but honor traditional naturopathic principles; to involve NDs in
federal, regional, and state health care initiatives; and to encourage countries
worldwide to include NDs in their health care systems to achieve "sustainable
human communities, and a healthy planetary ecosystem."[70]

Naturopaths today are continuing more than a century of activism to create a
cultural shift in the areas of health and wellness. The original goals of naturopaths to create a community giving individuals the freedom to choose the health
care best for them is still the national naturopathic agenda. Naturopathy still
provides significant opportunities for women. Cultural alliances provide a vibrant opportunity for naturopathy's future. The 2012–13 class at Bastyr University
had a striking demographic profile. Eighty-one percent of its 1,035 students were
female, with an average age of 30, and 80 of them represented thirty-one countries; 39 of the 63 faculty members were female. This uniquely high number of

female naturopaths can be a central strength in forging liaisons with gender-conscious Americans concerned with prevention-based care and with women's health. Forging additional alliances with networks of women's groups could produce new activist-allies with NDs on state licensing efforts. Liaisons with like-minded cultural organizations have always been productive.[71]

Public opinion and publicity continue to be key problem areas. When the San Diego campus of Bastyr University opened in September 2012, it was reported in the *San Diego Business Journal*, the *Natural Standard*, and *Mother Earth News* but it was not published in the newspaper with the largest circulation in the county, the *U-T San Diego*, until three months later. Nor was it covered in the *Los Angeles Times*. The *U-T*'s failure to cover the event robbed the school opening of its due attention. The December article was thorough and lengthy and no doubt took much political maneuvering to arrange. San Diego's higher-education institutions sometimes, with great effort, receive coverage of news, events, research, lectures, and so on. Bastyr is one of several universities in Southern California competing for the media spotlight.[72]

Naturopaths always knew dissemination of information was key and a top priority. The AANP has a fine tool in their official publication, the *Natural Medicine Journal*, and its distribution to college and university libraries and career counseling centers around the country could result in new liaisons. Securing corporate partners, already a successful campaign, is a prime method for obtaining additional funding to advance the profession. The *Naturopathic Doctor News & Review* provides case studies, discussions of ailments and therapeutics, and a regular column, Nature Cure Clinical Pearls, by the naturopath Sussanna Czeranko, rare books curator at the NCNM. This column connects today's naturopaths with their past through articles such as "The Yungborn" and "Homeopathy Meets Naturopathy," so that practitioners not only stay grounded in the fundamentals of nature cure but also are motivated to create change by the stories of those who paved the way. To underscore the importance of their history, Czeranko quotes Benedict Lust: "Naturopathy, with the fundamental basis of a healing art, Naturology, natural living, and . . . the great physical culture movement . . . constitute a great revolutionary movement for rational medicine, a true prevention of disease, and combined with medical freedom, which we must and will get, it will usher in the new day."[73]

Hope for increased equality for naturopaths to practice lies in two relatively new venues. One is the Patient-Centered Outcomes Research Institute, under the Secretary of Health and Human Services, started in 2010. Its mission is to improve health care and informed patient decision making based on evidence-based research. It is also a possible source of research funding for naturopaths, but only one of its twenty-one board members is a chiropractor, while the others hold MD, PhD, and JD degrees, suggesting that naturopathic practitioners would not benefit as much as others from these funds. The second is The *Integrator Blog*, published and edited by John Weeks. Weeks's life-long activism and writing on behalf of alternative medicine have gained him three honorary doctorates from naturopathic colleges and universities. The site offers news, reports, opinion, and networking for the business, education, policy, and practice of integrative medicine, CAM, and integrated health care. Both the Patient-Centered Outcomes Research Institute and *The Integrator Blog* are important and promising, the latter providing a wealth of information on the efficacy of naturopathy.[74]

HISTORICAL SIGNIFICANCE

Early naturopaths advocated for the power of natural healing methods and the citizens' right to choose a healer. They fought the medical monopoly and its allies in government. They worked to ameliorate urban health dangers and social class hegemony. They denounced the dominant scientific trends of vivisection and enforced vaccination as illogical, inhumane, and unnecessary if proper attention were paid to building up personal health and the environment. They saw through the panacea promises of germ theory and drugs. Women cocreated their movement, thanks to prominent female leaders, authors, and organizers and male professionals who supported women as colleagues. They had a collective consciousness (a willingness to downplay personal profit in favor of a larger vision) based on the belief that both nature and the spirit would promote healing. And they collaborated with an eclectic, often contradictory set of drugless doctors espousing varied therapeutics. At the same time, naïve, utopian ideas, insensitivity to the rigors, constraints, and oppression of immigrant and African American life, and the promotion by some of eugenics limited the applicability and appeal of naturopathy to many Americans during its development.

As naturopaths professionalized at midcentury and gained some legal recognition over the decades, they also struggled with their dual roots of practitioner inclusivity and widespread societal reform. To gain legal and social legitimacy during their fight against the American Medical Association, they shifted their emphasis to improving and securing the profession's licensure and distinguishing naturopathy from other natural healing systems. Prominent leaders began to accept some of the structures and methods of allopathy. While this brought valuable insights and therapeutics to naturopathy, it also distanced the movement from its radical roots. Aligning with social and professional norms also meant a decrease in female leadership and avowed agency for women in the mid-twentieth century.

In both the early and middle twentieth century, naturopathy, like biomedicine, struggled to define the causes of ill health and how best to address them. As one scholar recently suggested, both systems too often adhere to the "individualist philosophy which prevails in capitalism." In other words, the focus on licensure so that individuals could practice naturopathy meant moving away from placing equal energy on patients' environmental and socioeconomic factors. In addition, correcting an individual patient's inner physical, psychological, and spiritual health came to overshadow the critical sociopolitical factors that impacted everyone's health. Naturopaths cannot be faulted for this; their numbers were small, and their quest for legitimacy has been a steep uphill battle that is not over. However, the history—the heritage, if you will—of naturopaths is that they have been in the forefront of identifying environmental toxins, chemical pollutants, the questionable effects of pharmaceuticals and food processing, and bodily deterioration brought on by poor habits.[75]

In the present day, along with new issues, some old issues remain for naturopaths to negotiate: gender disparities in health care; the white male body as the central medical research model; poverty and accessibility to care; the deleterious impact of racism and homo- or transphobia within health care on a patient's ability to live well; and environmental racism, which threatens many communities of color. Naturopaths have their work cut out for them as they continue to deal with these issues along with their right to practice.[76]

This history of naturopathy demonstrates the movement's rocky path from inclusive eclecticism, or "therapeutic universalism," to articulate and more precise philosophies, therapeutics, and scientific expertise. The sources and internal debates in the field reveal that a "big tent" approach, over 120 years, was winnowed to an organized profession. But it was not an orderly progression, nor is it presently fully cohesive. Some of the same schisms persist, and for all those trained as scientific naturopaths, there are thousands more who claim the title "naturopath" without having met the same standards of training.

This history tells parallel tales. It illuminates the ongoing tension between allopathic medicine and traditions that emphasize the body's ability to heal itself, coupled with less invasive methods and good caretaking. It also demonstrates the vibrant scope—and flashpoints—of natural healing practitioners and their persistence in the face of traditional medicine's successful professionalization and dominance. In part, this history could be read as a history of progress, from a diffused mix of healing approaches to clarity and acceptance, with naturopathy now on the brink of gaining full legitimacy. But those who reject restrictive regulations can read this as capitulation and cooptation.

The beauty of naturopaths' history has been that their successes were built on bonds forged with other progressive thinkers and on their insistence on leading, not following, when it came to sociopolitical issues that intersected with medical care. The challenges naturopaths have faced will never go away; theirs will always be work in progress.

Notes

Frequently cited journals are identified by the following abbreviations:

AANPQN	*American Association of Naturopathic Physicians Quarterly Newsletter*
BHM	*Bulletin of the History of Medicine*
HHN	*Herald of Health and Naturopath*
JAMA	*Journal of the American Medical Association*
JANA	*Journal of the American Naturopathic Association*
JN	*Journal of Naturopathy*
JNM	*Journal of Naturopathic Medicine*
KWCM	*Kneipp Water Cure Monthly*
MAQ	*Medical Anthropology Quarterly*
NHH	*Naturopath and Herald of Health*
NP	*Nature's Path*
NPhys	*Naturopathic Physician*

Introduction

1. Susan E. Cayleff, *Wash and Be Healed: The Water-Cure Movement and Women's Health* (1987; repr., Philadelphia: Temple Univ. Press, 1992).

2. Michael Murray, ND, and Joseph Pizzorno, ND, *Encyclopedia of Natural Medicine*, 2nd ed. (New York: Three Rivers, 1998), 1.

3. See, e.g., Paul Boyer's classic *Urban Masses and Moral Order in America, 1820–1920* (Cambridge, MA: Harvard Univ. Press, 1992).

4. Harriet A. Washington, *Medical Apartheid: The Dark History of Medical Experimentation on Black Americans from Colonial Times to the Present* (New York: Random House, 2006); Erika Brady, ed., *Healing Logics: Culture and Medicine in Modern Belief Systems* (Logan: Utah State Univ. Press, 2001); Stephanie Mitchem, *African-American Folk Healing* (New York: New York Univ. Press, 2007).

5. Note the popularity of articles in which these topics overlap, e.g., from *Time* magazine, Alexandra Sifferin, "Aloha to the Healthiest State: Hawaii Ranks Highest in Well-Being" (28 Feb. 2013), Sifferin, "Sick before Their Time: More Kids Diagnosed with Adult Diseases" (16 July 2013), and Alice Park, "Why Electric Cars Are More Polluting Than Gas Guzzlers—at Least in China" (14 Feb. 2012); and Leslie Taylor, *The Healing Power of Rainforest Herbs: A Guide to Understanding and Using Herbal Medicinals* (Garden City Park, NY: Square One, 2004).

6. "In 2007, approximately 38 percent of US adults aged 18 years and over and approximately 12 percent of children used some form of CAM." National Center for Complementary and Alternative Medicine, National Institutes of Health, accessed 8 Feb. 2014, http://nccam.nih.gov/news/camstats/NHIS.htm. This represented an increase since 2002. P. M. Barnes, B. Bloom, and R. Nahin, "Complementary and Alternative Medicine Use among Adults and Children: United States, 2007," *CDC National Health Statistics Report* 12, 10 Dec. 2008, http://nccam.nih.gov/news/camstats /2007.

7. There has been a proliferation of articles, but as of this writing no comprehensive study has gathered data on the number of MDs using complementary and alternative medicine. See, e.g., Gerard E. Mullin, MD, "CAM Safety," *Nutrition in Clinical Practice* 27 (Dec. 2012): 832–33; John Kulig, Heike Rolle-Daya, and College Zacharyczuk, "CAM Practices in Practice," *Infectious Diseases in Children* 24 (Nov. 2011): 1; and "Considerations for CAM Use," *Current Psychiatry* 8 (Dec. 2009): 5–6. The increase in the budget of the National Center for Complementary and Alternative Medicine (a subdepartment of the National Institutes of Health) indicates the increase in therapeutics in the United States. The budget grew from $2 million at its creation in 1992 to $120.7 million in 2013, a marked increase but only a small part of the full NIH budget of $30.86 billion that year. The 2013 budget was down from its peak of $128 million in 2012. See National Center for Complementary and Alternative Medicine, National Institutes of Health, "NCCAM Funding: Appropriations History," accessed 6 Dec. 2013, www.nccam.nih.gov/about/budget/appropriations.htm.

8. American Association of Naturopathic Physicians website, accessed 19 Dec. 2007, www.alternativemedicinedirectory.org/naturopathic-organizations.html; "Definition of Naturopathic Medicine," House of Delegates position paper, accessed 1 Feb. 2014, www.naturopathic.org/files/Committees/HOD/Position%20Paper%20Docs /Definition%20Naturopathic%20Medicine.pdf.

9. Kathy Wong, ND, "12 Common Questions about Insurance and Complementary/Alternative Medicine," accessed 1 Feb. 2014, http://altmedicine.about.com/od /alternativemedicinebasics/a/Insurance.htm; "Health Care Reform Extends Reach of Naturopathic Medicine," accessed 12 June 2015, www.bastyr.edu/news/general-news -home-page/2013/01/health-care-reform-extends-reach-naturopathic-medicine.

10. "Licensed States & Licensing Authorities," accessed 1 Feb. 2014, www.naturo pathic.org/content.asp?contentid=57.

11. See www.naturopathic.org/viewbulletin.php?id=19, accessed 23 Jan. 2009; and "Setting Priority Policies," accessed 1 Feb. 2014, http://naturopathic.org/article_content .asp?edition=101&section=150&article=862.

12. "Licensed States & Licensing Authorities."

13. Basic science includes anatomy, physiology, biochemistry, pathology, microbiology, immunology, pharmacology, laboratory diagnosis, clinical and physical diagnosis, cardiology, neurology, radiology, minor surgery, obstetrics, gynecology, pediatrics, dermatology, and other clinical sciences. "Professional Education," accessed 1 Feb. 2014, http://naturopathic.org/content.asp?contentid=56. In states where licensure exists, NDs, like family practice MDs, are permitted to perform minor in-office surgeries such as repair of superficial wounds, removal of foreign bodies, cysts, and so on. "Academic Curriculum," accessed 9 Feb. 2014, www.aanmc.org/education/academic-curriculum

.php; "Comparing ND and MD Curricula: A Question of Degree," accessed 9 Feb. 2014, http://aanmc.org/schools/comparing_nd_md_curricula/.

14. "Doctor of Naturopathic Medicine: Dual-Degree Options," accessed 9 Feb. 2014, www.bastyr.edu/academics/areas-study/study-naturopathic-medicine/naturopathic-doctor-degree-program#Dual-Degree-Options.

15. "Birth Center (Natural Childbirth)," accessed 8 Feb. 2014, http://health.ucsd.edu/specialties/obgyn/maternity/facilities/birth-center/Pages/default.aspx; American Association of Birth Centers website, accessed 8 Feb. 2014, www.birthcenters.org/?page=NBCSII.

16. "Comparative Curricula: Naturopathic Medical Education," accessed 19 Dec. 2007, www.ncnm.edu/images/Factbook/Nat-Med-Ed-Comp-Curricula.pdf.

17. American Association of Naturopathic Physicians website, accessed 19 Dec. 2007, www.naturopathic.org.

18. "Comparing ND & MD Curricula: A Question of Degree ND Curriculum—Improvements and Quality Assurance," accessed 1 Feb. 2014, http://aanmc.org/schools/comparing_nd_md_curricula/.

19. Hans Baer, "Partially Professionalized and Lay Heterodox Medical Systems within the Context of the Holistic Health Movement," in *Biomedicine and Alternative Healing Systems in America: Issues of Class, Race, Ethnicity, and Gender* (Madison: Univ. of Wisconsin Press, 2001), 103–20. In 2010 there were more than 2,000 AANP members, but this number included students and supporting and corporate members as well as physicians. "About Us—Vision and Goals," accessed 9 June 2010, www.naturopathic.org/content.asp?contentid=19.

20. "Visions and Goals," accessed 8 Feb. 2014, www.naturopathic.org/content.asp?contentid=19.

CHAPTER 1: Following Nature's Path and Botanic Healing

1. Friedhelm Kirchfeld and Wade Boyle, "The Father of Naturopathy, Benedict Lust (1872–1945)," in *Nature Doctors: Pioneers in Naturopathic Medicine* (East Palestine, OH: Buckeye Naturopathic, 1993), 185–219, at 190. Kirchfeld and Boyle's explanation of the derivation of *naturopathy* differs from the earlier version put forth by Benedict Lust, who claimed *nature* and *illness* as the (contradictory) linguistic origin. Benedict Lust, ND, MD, "History of the Naturopathic Movement," *HHN* 26 (1921): 479, as cited in Kirchfeld and Boyle, *Nature Doctors*, 190–91. The changes in the name of Lust's journal reflect the difficulties of creating a trade publication for such disparate audiences. In 1896 he began with the *Kneipp Water Cure Monthly* (first in German, then English), adding the words *and Herald of Health* in 1901. Then it became the *Naturopath and Herald of Health* (1902–14), after which Lust's attempt at inclusiveness of other practitioners prompted him to reverse the order in the title to *Herald of Health and Naturopath* (1915–22). From 1923 to 1927, in a reversal of attitude toward professionalization, it was called *Naturopath*. For the following six years he only published *Nature's Path*, which had begun as a lay magazine in 1925 and continued until 1954. He brought back the *Naturopath and Herald of Health* from 1934 to 1944, for some reason jumping from volume 33 in 1934 to volume 40 the following year. Kirchfeld and Boyle, *Nature Doctors*, 313. For consistency in the text, I use only the title *Naturopath and Herald of Health*, but the notes reflect accurate title changes.

2. Yet another version is that only Lust and the fledgling American Naturopathic Association (ANA) could take credit for the term. Herbert M. Shelton, "Keeping Naturopathy's Record Straight," *Dr. Shelton's Hygienic Review* 11 (1949): 88; James Faulkner, "Speaking of Naturopaths," *NHH* 33 (1934): 102–3, as cited in Kirchfeld and Boyle, *Nature Doctors,* 189.

3. Lust, "History of the Naturopathic Movement," 479.

4. Shelton, "Keeping Naturopathy's Record Straight," 88; Harry Riley Spitler, ND, MD, PhD, ed., *Basic Naturopathy* (New York: American Naturopathic Association, 1948); Bernarr MacFadden, *Building of Vital Power: Deep Breathing and a Complete System for Strengthening the Heart, Lungs, Stomach, and All the Great Vital Organs* (Physical Culture City, NJ: Physical Culture, 1904); Henry Lindlahr, *Nature Cure: Philosophy and Practice Based on the Unity of Disease and Cure,* 2nd ed. (Chicago: Nature Cure, 1914); Friedrich Eduard Bilz, *The Natural Method of Healing,* vol. 1 (Leipzig, Germany: privately printed, 1898); Lindlahr, *The Philosophy of Natural Therapeutics,* vol. 2, *Practice* (1919; repr., ed. Jocelyn C. P. Proby, Saffron Walden, Essex, UK: C. W. Daniel, 1981); Louis Kuhne, *Neo Naturopathy: The New Science of Healing or The Doctrine of the Unity of Diseases* (Butler, NJ: Benedict Lust, 1917); Kirchfeld and Boyle, *Nature Doctors,* 257.

5. Lust, "History of the Naturopathic Movement," 480.

6. The timing of some of Lust's degrees is murky. He claims that he earned the DO degree in 1902, which could not be true, and that he began work on a degree in 1901. Yet Kirchfeld and Boyle say that he earned the DO in 1898. Cody says that Lust received his homeopathic degree from the New York Homeopathic Medical College in 1913 and a degree in eclectic medicine in 1914. According to Kirchfeld and Boyle, he earned this last degree in 1913 from the Eclectic Medical College of New York. Kirchfeld and Boyle, *Nature Doctors,* 200; Benedict Lust, *Yungborn: The Life and Times of Dr. Benedict Lust, Founder and Father of Naturopathy, The Cornucopia of Alternative Medicine: A Memoir,* with a foreword by Dr. Joseph E. Pizzorno Jr., ND, President of Bastyr University and author of *Total Wellness* (n.p.: privately printed by Anita Lust Boyd, 1997), 58, 72; George Cody, "History of Naturopathic Medicine," in *Textbook of Natural Medicine,* ed. Joseph Pizzorno and Michael Murray (New York: Churchill Livingstone, 1985), 7–37, at 16.

7. Kirchfeld and Boyle, *Nature Doctors,* 185–219; Cody, "History of Naturopathic Medicine"; Hans A. Baer, "The Potential Rejuvenation of American Naturopathy as a Consequence of the Holistic Health Movement," *MAQ* 13, no. 4 (1992): 369–83; James C. Whorton, "Therapeutic Universalism: Naturopathy," in *Nature Cures: The History of Alternative Medicine in America* (Oxford: Oxford Univ. Press, 2002); Robert McHenry, ed., "Woodhull, Victoria Claflin (1883–1927)," in *Famous American Women: A Biographical Dictionary from Colonial Times to the Present* (New York: Dover, 1980), 451–52; Victoria Claflin Woodhull and Tennessee Claflin Cook, *The Human Body the Temple of God; or The Philosophy of Sociology* (New York: Hyde Park Gate, 1890).

8. Kirchfeld and Boyle, *Nature Doctors,* 291; William A. Turska, "Botanics in Diseases of the Heart," *JNM* 5 (Apr. 1954): 16–18; Susan E. Cayleff, "Self-Help and the Patent Medicine Business," in *Women, Health, and Medicine in America: A Historical Handbook,* ed. Rima D. Apple (New York: Garland, 1990), 311–36.

9. E. Smith, *Compleat Housewife; or Accomplished Gentlewoman's Companion*, 4th ed. (London: J. and J. Peniberton, 1729); John Wesley, *Primitive Physic, or An Essay and Natural Method of Curing Most Diseases* (London, 1747); William Buchan, *Domestic Medicine; or, The Family Physician* (Edinburgh: Balfour, Auld & Smellie, 1769); John Gunn, *Domestic Medicine* (Knoxville, TN: Johnston & Edwards, 1830).

10. Francis Brinker, "The Role of Botanical Medicine in 100 Years of American Naturopathy," *HerbalGram* 42 (Spring 1998): 49–59, 49; Samuel Thomson, *New Guide to Health Prefixed by a Narrative of the Life and Medical Discoveries of the Author*, 2nd ed. (Boston: E. G. House, 1825); John B. Blake, "Health Reform," in *The Rise of Adventism: Religion and Society in Mid-Nineteenth-Century America*, ed. Edwin S. Gaustad (New York: Harper & Row, 1974), 31–33; Richard H. Shryock, "Empiricism vs. Rationalism in American Medicine, 1650–1950," *American Antiquarian Society Proceedings* 79 (Apr. 1969): 129–31; Frederick C. Waite, "American Sectarian Medical Colleges before the Civil War," *BHM* 19 (1946): 148–66; Cody, "History of Naturopathic Medicine," 8–9.

11. A. C. Twaddle and R. Hessler claim that American naturopathy was the heir to Thomsonianism, eclecticism, and homeopathy. Twaddle and Hessler, *A Sociology of Health Care* (New York: Macmillan, 1987).

12. Alex Berman, "The Thomsonian Movement and Its Relation to American Pharmacy and Medicine," *BHM* 25 (1951): 519, 524; Thomson, *New Guide to Health*; Waite, "American Sectarian Medical Colleges," 149.

13. Brinker, "Role of Botanical Medicine," 52–53; Cody, "History of Naturopathic Medicine," 9.

14. Wooster Beach, *The American Practice of Medicine: Being a Treatise of the Character, Causes, Symptoms, Morbid Appearance, and Treatment of the Diseases of Men, Women and Children, of All Climates, Vegetable or Botanical Preparations*, 2nd ed. (New York: Kelley & Tourrette, 1836); J. V. Lloyd, "A Review of the Principal Events in American Medicine," *Journal of American Pharmacists Association* 14 (1925): 621–31; John S. Haller Jr., *Medical Protestants: The Eclectics in American Medicine, 1825–1939* (Carbondale: Southern Illinois Univ. Press, 1994); Finley Ellingwood, "The Organic Remedies, The True Scientific and Rational Remedies," *Ellingwood's Therapeutist* 7 (1913): 215–17.

15. Laurie Aesoph, "The Roots Of Naturopathy," 8, "Naturopathy, History II" file, NCNM; Cody, "History of Naturopathic Medicine."

16. Brinker, "Role of Botanical Medicine," 52–53.

17. Ibid.

18. Baer, "Potential Rejuvenation," 372; Paul Wendel, *Standardized Naturopathy: The Science and Art of Natural Healing* (Brooklyn: privately printed, 1951). See also J. H. Tilden, MD, *Impaired Health: Its Cause and Cure; A Repudiation of the Conventional Treatment of Disease* (1921; repr., Whitefish, MT: Literary Licensing, 2014). Tilden was educated as an eclectic.

19. Brinker, "Role of Botanical Medicine," 50–51; "Herbs and Their Healing Power," *KWCM* 1 (1900): 11–21; "Kneipp's Popular Healing Remedies, and Their Application," *NHH* 3, no. 3, whole no. 27 (Mar. 1902): 348–49; Benedict Lust,

"Science of Herbo-Therapy; Its Unlimited Value in Connection with the Natural Method of Healing Disease," ibid. 9, no. 6, whole no. 101 (June 1908): 167–73.

20. M. G. Young, Phytotherapy Department: The American Herb Doctor, *HHN* 20, no. 12, whole no. 266 (Dec. 1915): 770–76.

21. William Joseph Simmonite, *Medical Botany: Or Herbal Guide to Health* (London: W. Foulsham, 1917), vii; "Selected Herbs," *HHN* 23, no. 12, whole no. 292 (Dec. 1918): 940.

22. C. E. Binck, "Drugs and Sickness," *NHH* 20, no. 1, whole no. 255 (Feb. 1915): 106; V. Greenwald, "Narcotic Poisons; Why Men Die in Hospitals," *HHN* 23, no. 1, whole no. 281 (Jan. 1918): 39–40.

23. See Lindlahr, *Nature Cure*, 7; Lindlahr, "How I Became Acquainted with Nature Cure," *HHN* 23, no. 2, whole no. 282 (Feb. 1918): 122; and Anna Lindlahr and Henry Lindlahr, *The Lindlahr Vegetarian Cook Book and ABC of Natural Dietetics* (Chicago: Lindlahr, 1915), which by 1918 was in its 15th edition.

24. Brinker, "Role of Botanical Medicine," 51n58; G. Bircham, "Some Poisonous Herbs and How to Identify Them," *NHH* 45, no. 1, current no. 542 (Jan. 1940): 27; E. Howard Tunison, "Experience as a 'Pill Peddler': What I Learned about the Public's Attitudes toward Drugs," ibid. 20, no. 8, whole no. 262 (Aug. 1915): 480–82.

25. "Florida Laws Prohibit Prescriptions of Narcotics by Naturopaths," ibid. 44, no. 12, current no. 541 (Dec. 1939): 382.

26. Brinker, "Role of Botanical Medicine," 52n58; J. C. Van Uchelen, "Heal with Herbs," *NP* 50, nos. 7–8, current no. 630 (July–Aug. 1947): 442–43; William A. Turska, "Catechism of Naturopathy," *JNM* 4 (Oct. 1953): 9–11, at 10.

27. A. W. Kuts-Cheraux, BS, MD, ND, ed., *Naturae Medicina and Naturopathic Dispensatory* (Des Moines: American Naturopathic Physicians and Surgeons Association, 1953), 3. Kuts-Cheraux delineates 14 categories of botanical agents: alkaloids, glucosides, carbohydrates, tannins, chlorophyll, volatile oils, fixed oils, saponins, resins, gums, balsams, vitamins, enzymes and hormones (10–14). Eugene Hans Lucas, "Folklore and Plant Drugs, Part I," *The Naturopath* 1 (Mar. 1963), the first in a four-part series on the topic that year.

28. R. L. Sanders, "Did Grandma Practice Medicine without a License?," *The Naturopath* 1 (Nov. 1963): 2–5; Joseph A. Cocannouer, "Pot Herbs My Nature-Loving Mother Taught Me to Enjoy," ibid. 3 (Mar. 1965): 8–9; Lucas, "Folklore and Plant Drugs," title page.

29. John W. Noble, "The Weed That Was," *The Naturopath* 3 (Mar. 1965): 2, 10.

30. Richard Lucas, *Nature's Medicines*, with a foreword by Harry Benjamin, MD (New York: Award Books, 1966).

31. Brinker, "Role of Botanical Medicine," 51. The linkages with American Indian herbology reemerge in other articles circa the 1930s. See D. O. Cauldwell, "More about Herbs," *NP* 32 (1933): 374, as cited in Brinker, "Role of Botanical Medicine," 51; and J.A.S., "Heap Big Medicine Man," *JN* 1 (Nov. 1957): 26.

32. Lauri Aesoph, "Dr. Jeanne Albin: First Native American Naturopathic Physician," *NPhys* 6 (Fall 1991): 23.

33. Ibid.

34. Paul Huson, *Mastering Herbalism: A Practical Guide* (New York: Stein & Day, 1974), 260–64. Dumas's presentation was part of David Whitehorse's course on

American Indian ethnobotany taught at San Diego State University's College of Extended Studies in January 1989.

35. Michael DiPalma, ND, "A Natural Way to Health," *New Frontier*, May 1984, 7, 30.

CHAPTER 2: Spokes of a Wheel

1. Harry B. Weiss and Howard Kemble, *The Great American Water-Cure Craze: A History of Hydropathy in the United States* (Trenton, NJ: Past Times, 1967); R. T. Trall, MD, *The Hydropathic Encyclopedia: A System of Hydropathy and Hygiene. In Eight Parts: Designed as a Guide to Families and Students, and a Text-Book for Physicians*, 2 vols. (New York: Fowler & Wells, 1851), 46–51; John B. Blake, "Health Reform," in *The Rise of Adventism: Religion and Society in Mid-Nineteenth-Century America*, ed. Edwin S. Gaustad (New York: Harper & Row, 1974), 44. Information on hydrotherapy is drawn largely from my book *Wash and Be Healed: The Water-Cure Movement and Women's Health* (1987; repr., Philadelphia: Temple Univ. Press, 1992); see esp. 185n5. See also T. K. Young, "Sweat Baths and the Indians," *Canadian Medical Association Journal* 119 (9 Sept. 1978): 406–8; and Leonard E. Barrett, "Healing in the Balmyard: The Practice of Folk Healing in Jamaica, W.I.," in *American Folk Medicine*, ed. W. Hand (Berkeley: Univ. of California Press, 1976), 285–330.

2. Cayleff, *Wash and Be Healed*, 21.

3. Joel Shew, *Hydropathy, Or, the Water-Cure: Its Principles, Modes of Treatment, etc.* (New York: Wiley & Putnam, 1845); Thomas L. Nichols, *An Introduction to the Water-Cure* (New York: Fowlers & Wells, 1850); Mrs. M. L. Shew, *Water-Cure for Ladies* (New York: Wiley & Putnam, 1844); Mary G. Nichols, *The Sick Cured without Medicine!* (Worcester, MA, n.d.); Trall, *Hydropathic Encyclopedia*. On water-cure philosophy, see Cayleff, *Wash and Be Healed*, 24–27, 29–31, 31–35. On the hydropathic leadership's encouragement of women self-doctoring and caring for their families' health, see ibid., 44–74.

4. Cayleff, *Wash and Be Healed*, 25; "The Pioneers of Health Reform," *Herald of Health* 5 (Jan. 1904): 17–18.

5. Cayleff, *Wash and Be Healed*, 30; James Caleb Jackson, in *Water-Cure Journal* 4 (Dec. 1847): 359–60.

6. Cayleff, *Wash and Be Healed*, 35–39; T. Nichols, *Introduction to the Water-Cure*, 34.

7. Cayleff, *Wash and Be Healed*, 39. Eric J. Cassell, in "The Nature of Suffering and the Goals of Medicine," *New England Journal of Medicine* 306 (Mar. 1982): 639–45, identified the dual obligation of the medical profession to relieve suffering and cure disease (639).

8. Benedict Lust, To Our Readers, *KWCM* 1 (Jan. 1900): 1. The *KWCM* was published by the Kneipp Magazines Publishing Company in New York City.

9. Benedict Lust, "A Brief History of Natural Healing," ibid., 4; Friedhelm Kirchfeld and Wade Boyle, *Nature Doctors: Pioneers in Naturopathic Medicine* (East Palestine, OH: Buckeye Naturopathic, 1993), detailing the lives of Priessnitz, 13–29, Johann Schroth, 31–39, and Sebastian Kneipp, 73–98; Lust, ND, MD, "History of the Naturopathic Movement," *HHN* 26 (1921): 479–80; Max Mader, "The Schroth Cure," *Naturopathy* 31, no. 11, whole no. 385 (Nov. 1926): 529–33; R. T. Trall, *Digestion and Dyspepsia* (New York: S. R. Wells, 1873); E. P. Miller, *Dyspepsia: Its Varieties, Causes,*

Symptoms, and Treatment by Hydropathy and Hygiene (New York: Miller, Haynes, 1870).

10. Lust, "Brief History of Natural Healing," 5.

11. Ibid.; "History of the Naturopathic Movement," *Naturopath*, n.d., 434, "Naturopathy, History II" file, NCNM; Kirchfeld and Boyle, *Nature Doctors*, 185.

12. Wade Boyle, ND, "Our Naturopathic Heritage: Sebastian Kneipp; His Water-Cure Became Our Naturopathic Medicine," *AANP Quarterly Magazine* 5 (1998): 38–39, at 38.

13. Ibid., 38; Kirchfeld and Boyle, *Nature Doctors*, 191.

14. Francis Brinker, "The Role of Botanical Medicine in 100 Years of American Naturopathy," *HerbalGram* 42 (Spring 1998): 40; Hans A. Baer, "The Potential Rejuvenation of American Naturopathy as a Consequence of the Holistic Health Movement," *MAQ* 13, no. 4 (1992): 369–83.

15. Earnest I. Cole, ND, DC, "Mark Twain, a Friend of Naturopathy," *HHN* 26 (Jan. 1921): 16.

16. Louisa Lust, ND, "The Curative Agency of Water and Herbs for Treatment of Fever," *Naturopath* 28, no. 9, whole no. 347 (Sept. 1923): 437; Robert Bieri, ND, "Hydrotherapy: Its Importance in the Field of All Rational Methods of Healing," *Herald of Health* 25, no. 10, whole no. 314 (Nov.–Dec. 1920): 484–85; Wade Boyle, ND, "Great Scott! A Lesson in Hydrotherapy," *AANPQN* 5 ([ca. 1991]): 30; Maryla DeChrapowicki, ND, "Dynamic Properties of Water," *NHH* 44, no. 9, current no. 538 (Sept. 1939): 231, 266–67. Specific water-cure applications were detailed repeatedly in the twentieth century, e.g., in "The Kneipp Full Douche," *HHN* 25, whole no. 306 (Jan. 1920): 34–36; and George Floden, "Naturopathic Therapeutics," *JNM*, Aug. 1960, 5–6.

17. James C. Whorton, *Nature Cures: The History of Alternative Medicine in America* (Oxford: Oxford Univ. Press, 2002), 85–86.

18. See Stephen Nissenbaum, *Sex, Diet, and Debility in Jacksonian America: Sylvester Graham and Health Reform* (Westport, CT: Greenwood, 1980). See also Richard H. Shryock, "Sylvester Graham and the Popular Health Movement, 1830–1870," *Mississippi Valley Historical Review* 18, no. 2 (1932); Blake, "Health Reform"; Sylvester Graham, *Lectures on the Science of Human Life*, 2 vols. (Boston: Marsh, Capen, Lyon, & Webb, 1839); Graham, "Fruits and Vegetables," *Water-Cure Journal* 2 (Feb. 1851): 37–39; and James C. Wharton, "'Tempest in a Flesh-Pot': The Formulation of a Physiological Rationale for Vegetarianism," in *Sickness and Health in America: Readings in the History of Medicine and Public Health*, ed. Judith Walzer Leavitt and Ronald L. Numbers (Madison: Univ. of Wisconsin Press, 1978), 315–30, 327, esp. 318–19.

19. William A. Alcott, *The House I Live In: or the Human Body; For the Use of Families and Schools*, 4th ed. (Boston: George W. Light, 1839); Blake, "Health Reform," 42; Alcott, *The Young Woman's Book of Health* (New York: Miller, Orton & Mulligan, 1855).

20. William A. Alcott, in *Health Journal and Independent Magazine* 1 (1843): 29, as cited in Shryock, "Sylvester Graham," 178.

21. Cayleff, *Wash and Be Healed*, 118–19.

22. "J. H. Kellogg Dies; Health Expert, 91," On this Day: 16 Dec. 1943, www .nytimes.com/learning/general/onthisday/bday/0226.html; J. H. Kellogg, MD, *Plain*

Facts for Old and Young: Embracing the Natural History of Hygiene of Organic Life (Burlington, IA: I. F. Segner, 1892); Cayleff, *Wash and Be Healed*, 94, 115, 177, 153, 165, 195n63, 204n96, 209n42; "Eugenics and the Race Betterment Movement," accessed 1 Jan. 2014, http://xroads.virginia.edu/~ma03/holmgren/ppie/eug.html.

23. Henry Lindlahr and Mrs. Anna Lindlahr, *The Nature Cure Cook Book and ABC of Natural Dietetics* (Chicago: Nature Cure, 1915); Henry Lindlahr, *The Philosophy of Natural Therapeutics*, vol. 2, *Practice* (1919; repr., ed. Jocelyn C. P. Proby, Saffron Walden, Essex, UK: C. W. Daniel, 1981).

24. William F. Havard, ND, "A Course in Basic Diagnosis: The Restoration of Impaired Function," *HHN* 25, no. 5, whole no. 309 (June 1920): 235; "How Do We Free Ourselves from the Care of Clothing?," *NHH* 19, no. 1, whole no. 243 (Jan. 1914): 58; Edward Earle Purinton, "The Moral Value of Exercise," ibid. 19, no. 2, whole no. 244 (Feb. 1914): 70; Benedict Lust, ND, "Advice Relative to Sun-Baths," ibid., n.p.

25. Martin Kaufman, *Homeopathy in America: The Rise and Fall of a Medical Heresy* (Baltimore: Johns Hopkins Univ. Press, 1971); Harris L. Coulter, "Homeopathic Medicine," in *Ways of Health*, ed. David Sobel (New York: Harcourt Brace Jovanovich, 1979), 289–310.

26. For a detailed explanation, see "The State of Nineteenth-Century Health Care," in Cayleff, *Wash and Be Healed*, 5–9.

27. Samuel Hahnemann, *Organon of the Art of Healing*, 5th American ed. (New York: Boericke & Tafel, 1875).

28. John B. Blake, "Women and Medicine in Ante-Bellum America," *BHM* 39 (Mar.–Apr. 1965): 117.

29. J. Laurie, MD, *Homeopathic Domestic Medicine* (New York: William Radde, 1848), vii; Hahnemann, *Organon*, 32; J. S. Douglass, *Practical Homeopathy for the People: Adapted to the Comprehension of the Non-Professional, and for Reference by the Young Practitioner including a Number of Most Valuable New Remedies*, 15th ed. (1992; repr., Milwaukee: Lewis Sherman, 1894).

30. Frederick C. Waite, "American Sectarian Medical Colleges before the Civil War," *BHM* 19 (1946): 148–66; Blake, "Health Reform," 35; Martin Kaufman, "Homeopathy in America: The Rise and Fall and Persistence of a Medical Heresy," in *Other Healers: Unorthodox Medicine in America*, ed. Norman Gevitz (Baltimore: Johns Hopkins Univ. Press, 1988), 104–6.

31. George Cody, "History of Naturopathic Medicine," in *Textbook of Natural Medicine*, ed. Joseph Pizzorno and Michael Murray (New York: Churchill Livingstone, 1985), 7–37, at 16; John T. Richter, DC, ND, "Biography," in *Nature—The Healer*, ed. Henry W. Splitter (Los Angeles: privately printed, 1946), [3].

32. Mary Baker Eddy, *Science and Health with Key to the Scriptures* (1875; repr., Boston: First Church of Christ, Scientist, 1875); Rennie B. Schoepflin, *Christian Science on Trial: Religious Healing in America* (Baltimore: Johns Hopkins Univ. Press, 2003); Stephen Gottschalk, *Rolling Away the Stone: Mary Baker Eddy's Challenge to Materialism* (Bloomington: Indiana Univ. Press, 2005).

33. Norman Gevitz, "Osteopathic Medicine: From Deviance to Difference," in Gevitz, *Other Healers*, 124–25. For the best and most complete study of osteopathy, see Gevitz, *The D.O.'s: Osteopathic Medicine in America* (Baltimore: Johns Hopkins Univ. Press, 1982).

34. Gevitz, "Osteopathic Medicine," 126–29, citing Andrew Taylor Still, *The Autobiography of Andrew Taylor Still* (Kirksville, MO: privately printed, 1897); Arthur Hildreth, *The Lengthening Shadow of Andrew Taylor Still* (Kirksville, MO: Journal Printing, 1942).

35. Paul Starr, *The Social Transformation of American Medicine: The Rise of a Sovereign Profession and the Making of a Vast Industry* (New York: Basic Books, 1982), 108; Still, *Autobiography*, 286–87; Still, "Obstetrics," in *Philosophy of Osteopathy* (Kirksville, MO: privately printed, 1899), 234–50.

36. Starr, *Social Transformation*, 108; Still, *Philosophy of Osteopathy*, 214.

37. Elmer D. Barber, *Osteopathy Complete* (Kansas City, MO: Hudson-Kimberly, 1898), 463, 506–37.

38. Cody, "History of Naturopathic Medicine," 15, summarizing Starr's argument in *Social Transformation*, 108–9. On osteopaths' legal battles, see Gevitz, "Osteopathic Medicine," 131–33. For insights into chiropractic legal battles, see Walter I. Wardwell, "Chiropractors: Evolution to Acceptance," in Gevitz, *Other Healers*, 178–82.

39. Wardwell, "Chiropractors," 157.

40. Ibid., 158; Brian Inglis, "Osteopathy and Chiropractic," in *The Case for Unorthodox Medicine* (New York: Putnam, 1965), 107–37, at 125.

41. Inglis, "Osteopathy and Chiropractic," 125–26.

42. B. J. Palmer, DC, PhC, *A Text Book on the Palmer Technique of Chiropractic* (Davenport, IA: privately printed, 1920).

43. See Walter I. Wardwell, *Chiropractic: History and Evolution of a New Profession* (Saint Louis: Mosley, 1992), 59–60, 66, 86, 89, 91, 110, 115, on Oakley Smith; and Andrew P. Davis, MD, OPHD, ND, CP, DO, *Neuropathy: The New Science of Drugless Healing Amply Illustrated and Explained* (Cincinnati: F. L. Rowe, 1909), preface and 94–97.

44. Thomas T. Lake, ND, DC, *Treatment by Neuropathy and the Encyclopedia of Physical and Manipulative Therapeutics* (Bottelo, WA: Bastyr Rare Books, 1946), xi.

45. Wardwell, "Chiropractors," 160–61.

46. Joseph C. Keating Jr., PhD, and William S. Rehm, DC, "The Origins and Early History of the National Chiropractic Association," *Journal of the California Chiropractic Association* 37 (Mar. 1993): 27–51, at 27.

47. Ibid.; Major Bertrand DeJarnette, DD, *Techniques and Practice of Bloodless Surgery* (Nebraska City, NE: privately printed, 1939).

48. Benedict Lust, "The Blood-Washing or Regenerating Shower Bath," *Naturopath* 28, no. 8, whole no. 346 (Aug. 1923): 370–71; Kirchfeld and Boyle, *Nature Doctors*, 214; Henry Lindlahr, "The New Blood Wash Treatment: A Great Discovery in Hydrotherapy—A Wonderful Rejuvenating and Regenerating Treatment," *Naturopath* 28, no. 10, whole no. 348 (Oct. 1923): 527–30; Arles G. Bahl, PhT, "Zone Therapy May Help You!," *NHH* 49, no. 2, current no. 583 (Feb. 1944): 54.

49. Carl E. Bohman, "Requisites and Instructions for Colon Irrigations," *Naturopath* 28, no. 10, whole no. 348 (Oct. 1923): 567; ibid. 31, no. 3, whole no. 377 (Mar. 1926), with three articles and six advertisements stressing bowel cleansing; Fred A. Mayberry, "Practice Building with Colonic Irrigations," address before 39th annual congress of the American Naturopathic Association, San Diego, CA, 13 June 1935, *NHH* 40, no. 9 (Sept. 1935): 266; advertisement, "YOUR PATIENTS NEED THE MAYBERRY COLON

IRRIGATOR: DR. FRED A. MAYBERRY, Wichita, Kansas," ibid. 42, no. 4, current no. 509 (Apr. 1937): 127. Among dozens of articles about cleansing the bowels, see Carroll James Mather, ND, DBM, PhG, "More about the Treatment of Constipation by Herbalism," ibid. 49, no. 3, current no. 584 (Mar. 1944): 79; and Maydene Williamson, BSc, DC, "Cause and Correction of Constipation," ibid. 49, no. 1, current no. 582 (Jan. 1944): 9.

50. [Benedict Lust?], "The Great Bloodless Operation," *Naturopath* 31, no. 7, whole no. 381 (July 1926): 318–21.

51. [Benedict Lust?], "Fasting a Curative Agent," *NHH* 20, no. 7, whole no. 261 (July 1915): 416–17. See also Edward Earle Purinton, "Twenty Rules for Sane Fasting," ibid. 20, no. 11, whole no. 265 (Nov. 1915): 674–82.

52. "To Fast or Not to Fast—That Is the Question," ibid. 40, no. 6 (June 1935): 169; "Bloodless Surgery Conference Announced for December 29, 30 and 31, 1948," *JANA* 1 (Dec. 1948–Jan. 1949): 17; Changhan Lee, Lizzia Raffaghello, Sebastian Brandhorst, et al., "Fasting Cycles Retard Growth of Tumors and Sensitize a Range of Cancer Cell Types to Chemotherapy," *Science Translational Medicine* 4, no. 124 (2012): 124–27.

53. Kirchfeld and Boyle, *Nature Doctors*, 129–41, at 135.

54. Iridiagnosis revealed eye color, signs of suppressed milk scurf and scabies, medicine colors, presence of allopathic toxins, signs of vaccination, defective organs, inflammations, catarrhal defects, and loss of nerve or cramps rings. H. E. Lane, MD, advertisement, "A Practical Course in Self-Diagnosis and Self-Healing by Means of THE DIAGNOSIS FROM THE EYE," *NHH* 8, no. 5, whole no. 89 (June 1907).

55. An advertisement for Lindlahr's College of Natural Therapeutics touts iridiagnosis as one of the subjects offered. "Lindlahr College of Natural Therapeutics," *HHN* 23, no. 11, whole no. 291 (Nov. 1918): 900. See also Henry Lindlahr, MD, *Iridiagnosis and Other Diagnostic Methods* (Chicago: Lindlahr, 1919), 12–13; and F. W. Collins, MD, *Disease Diagnosed by Observation of the Eye* (Newark, NJ: privately printed, 1919).

56. Rex Alto, "Iridiagnosis," *HHN* 23, no. 1, whole no. 281 (Jan. 1918): 63–65, at 64; J. Haskel Kritzer, MD, "Iridiagnosis," *Naturopath* 28, no. 2, whole no. 340 (Feb. 1923): 76.

57. State News: Illinois, *Naturopath* 31, no. 8, whole no. 382 (Aug. 1926): 408; H. McGolphin, DNT, "Iridiagnosis—Is It Reliable?," ibid. 28, no. 10, whole no. 348 (Oct. 1923): 551–56; F. W. Collins, "Cause, Removal of the Cause, and Prevention of Cancer by Natural Methods," ibid. 28, no. 12, whole no. 350 (Dec. 1923): 740–49, at 749. Recent scholarship has revisited the value of information that can be read from the eye. An experimental test to detect signs of Alzheimer's images the retina; in mice beta amyloid plaques can be detected in the retina before they impact the brain and mental function. Deborah Kotz, "What Your Eyes Say about Your Health," *Daily Dose Health News, Advice and Information*, 15 Oct. 2012, www.boston.com/dailydose/2012/10/15/what-your-eyes-say-about-your-health/B8L5lUblSKxxookxXbwXjL/story.html. Numerous practitioners, some certified iridologists, claim they can locate chemical and other toxic accumulations, inflammation, nerve depletion, mineral deficiencies, tissue degeneration, lymph congestions, stress, and inherently weak areas. See Amie Mosley, iridologist, *Iridology: Iris Photo Analysis, a 150 Year Old Science* (Anchorage: Mariposa Whole Life Center, n.d.), pamphlet distributed by the practitioner, collected ca. 2005; and

"Retinal Imaging Offers a Better View and Early Detection," accessed 7 Dec. 2014, www.vsp.com/retinal-exam.html.

58. Marko Petinak, "Ocular Diagnosis," *NHH* 40, no. 1 (Jan. 1935): 16, 30; R. M. McLain, DC, ND, "Iridiagnosis," ibid. 41, no. 2, current no. 507 (Feb. 1937): 46; Louis D. McKenney, "The Eyes Are the Windows," ibid. 44, no. 3, current no. 532 (Mar. 1939): 70, 88.

59. Lance C. Dalleck, MS, and Len Kravitz, PhD, "The History of Fitness," accessed 3 May 2005, www.unm.edu/~lkravitz/Article%20folder/history.html; Shelly McKenzie, *Getting Physical: The Rise of Fitness Culture in American* (Lawrence: Univ. Press of Kansas, 2013).

60. James C. Whorton, "Patient, Heal Thyself: Popular Health Reform Movements as Unorthodox Medicine," in Gevitz, *Other Healers*, 76–79. See also Bernarr MacFadden and Felix Oswald, *Fasting, Hydrotherapy and Exercise: Nature's Wonderful Remedies for the Cure of All Chronic and Acute Diseases* (London: Bernarr MacFadden, [ca. 1900]); MacFadden, *Preparing for Motherhood* (New York: MacFadden, 1936); and advertisement, "The Proposed Tour of Bernarr MacFadden and Family," [ca. 1905] (see, e.g., www.bernarrmacfadden.com/macfadden4.html).

61. Benedict Lust, *Yungborn: The Life and Times of Dr. Benedict Lust, Founder and Father of Naturopathy, The Cornucopia of Alternative Medicine: A Memoir*, with a foreword by Dr. Joseph E. Pizzorno Jr., ND, President of Bastyr University and author of *Total Wellness* (n.p.: privately printed by Anita Lust Boyd, 1997), 75; Whorton, *Nature Cures*, 205.

62. Benedict Lust, ND, MD, "Massage, Masseurs and Doctors," *Naturopath* 28, no. 10, whole no. 348 (Oct. 1923): 530; P. Puderbach, "Massage of the Prostata Gland," The Massage Operator, ibid. 31, no. 3, whole no. 377 (Mar. 1926): 136; Puderbach, "Swedish Gymnastics," The Massage Operator, ibid. 31, no. 10, whole no. 384 (Oct. 1926): 498.

63. James W. Barton, MD, "Effects of Massage," *Naturopath* 31, no. 7, whole no. 381 (July 1926): 328.

64. Per Nelson, "Radiant Light in Naturopathic Practice," *HHN* 23, no. 5, whole no. 285 (May 1918): 469–71, at 470; George Starr White, MD, FRSA, of London, speaking in Los Angeles, "High Lights of Intensive Study of Radiant Energy," *Naturopath* 31, no. 4, whole no. 378 (Apr. 1926): 173–75; P. Puderbach, of Brooklyn, The Massage Operator, ibid. 31, no. 10, whole no. 384 (Oct. 1926): 498–501.

65. F. W. Collins, "Lights, Colors and Nature's Finer Forces," *Naturopath* 28, no. 12, whole no. 350 (Dec. 1923): 749–50; E. A. Ernest, "Potency of Light and Color," *NHH* 44, no. 4, current no. 533 (Apr. 1939): 102, 118. In the same issue of *NHH* the Ernest Distributing Company's *Color Manual* lists 45 diseases alphabetically, with treatments and colors stipulated for each. *Color Energy Healing* (Milwaukee: Ernest Distributing, n.d.), 125. See also Carl Loeb, *A Course in Specific Light Therapy* (Chicago: Actino Laboratories, 1939).

66. Ernest J. Stevens, MSc, PhD, "Light and Chromo-Science Date," *NHH* 44, no. 7, current no. 536 (July 1939): 198, 214–25; "Our New 1950 List of the Greatest Health Book Collection," *NP* 54, no. 3 (Mar. 1950): 24.

67. Sidney H. Beard, "A Remarkable Scientific Discovery," *NHH* 9, no. 7, whole no. 102 (July 1908): 199–202, at 199; Samuel G. Tracy, BSC, MD, "High Frequency

Electricity to Retard Old Age: Influence of High Frequency Currents in Retarding Senility, with Remarks on a New Apparatus," ibid., 210–11, at 211. We do not know whether Tracy was the manufacturer or the salesman for this device, which does not bear his name.

68. J. P. Bean, "Electricity Is Life," *NHH* 20, no. 3, whole no. 257 (Mar. 1915): 185–86; Herbert A. Zettel, "Electropathy," ibid. 20, no. 10, whole no. 264 (Oct. 1915): 647–48; "Is Electro-Therapy in Harmony with the Principles of Nature?," *NP* 36, no. 2, current no. 436 (Feb. 1931): 40; "The Galvanic Current in Naturopathic Practice," *NHH* 42, no. 9, current no. 514 (Sept. 1937): 259; advertisement, "THE HW-100 POR-TABLE 6 METER Short Wave Apparatus Power Output 200 Watts, Ernest Distributing Co., Milwaukee, Wis. Exclusive World Distributors," ibid. 44, no. 3, current no. 532 (Mar. 1939): 87. Archivists at the online Electrotherapy Museum cited cases in which the devices were banned, such as *US v. August H. Riess* or FDC Case No. 30942 about the Violetta Kits. See "The Turn of the Century Electrotherapy Museum Violet Ray Misconceptions Article," accessed 31 Aug. 2014, www.electrotherapymuseum.com /Articles/VioletRayMisconceptions.htm.

69. Benedict Lust, ND, "Hypnotism and Naturopathy," *NHH* 9, no. 1, whole no. 96 (Jan. 1908): 7–9, at 9; Francis I. Regardie, FRSA, "Babies and Hypnotism," ibid. 44, no. 1, current no. 530 (Jan. 1939): 6, 18; J. Haskel Kritzer, MD, "Hypnotism—A Psychological Infection," ibid. 44, no. 5, current no. 534 (May 1939): 133, 150.

70. Kirchfeld and Boyle, *Nature Doctors*, 200; Adolph Just, *Return to Nature! The True Natural Method of Healing and Living and The True Salvation of the Soul: Paradise Regained*, 4th enlarged ed., trans. Benedict Lust (New York: Benedict Lust, 1903).

71. Friedrich Eduard Bilz, *The Natural Method of Healing* (Leipzig, Germany: privately printed, 1898).

72. Lust, To Our Readers, 1–5; Lust, "Editorial Drift," *NHH* 3, no. 1, whole no. 25 (Jan. 1902): 32–33.

73. J. H. Tilden, MD, *Diseases of Women and Easy Childbirth* (Denver: Smith-Brooks, 1912), 14; Kirchfeld and Boyle, *Nature Doctors*, 227–50, at 249; Henry Lindlahr, *Nature Cure: Philosophy and Practice Based on the Unity of Disease and Cure*, 2nd ed. (Chicago: Nature Cure, 1914), 13–16.

74. Benedict Lust, introduction to *Universal Naturopathic Encyclopedia, Directory and Buyers' Guide: Year Book of Drugless Therapy*, ed. Benedict Lust, vol. 1, for 1918–19 (New York: privately printed, 1918), 1, 9; Kirchfeld and Boyle, *Nature Doctors*, 209.

CHAPTER 3: "Nature Takes the Right Road"

1. Henry Lindlahr, *Nature Cure: Philosophy and Practice Based on the Unity of Disease and Cure*, 2nd ed. (Chicago: Nature Cure, 1914), 17; Horatio W. Dresser, *The Spirit of the New Thought* (New York: Thomas Y. Crowell, 1917), 1–2. Papers in Dresser's book came from the movement's middle period, when it was called Mental Science and taking shape as New Thought (v). There had been New Thought conventions in 1915 (San Francisco) and 1916 (Chicago). Lillian R. Carque, "Self-Control," *NHH* 45, no. 11, current no. 552 (Nov. 1940): 326, 338, at 326.

2. J. H. Tilden, MD, *Diseases of Women and Easy Childbirth* (Denver: Smith-Brooks, 1912), 15; Louis Kuhne, *Neo Naturopathy: The New Science of Healing or the*

Doctrine of the Unity of Diseases (Butler, NJ: Benedict Lust, 1917); Friedrich Eduard Bilz, *The Natural Method of Healing*, vol. 1 (Leipzig, Germany: privately printed, 1898), 641. Some of the foundational philosophical articles, from 1900–1923, are in Sussanna Czeranko, ND, BBE, ed., *Philosophy of Naturopathic Medicine: In Their Own Words* (Portland, OR: NCNM Press, 2013).

3. Benedict Lust, ed., *Universal Naturopathic Encyclopedia, Directory and Buyers' Guide: Year Book of Drugless Therapy*, vol. 1, for 1918–19 (New York: privately printed, 1918), 11, 13.

4. Lindlahr, *Nature Cure*, 17; William A. Turska, "Catechism of Naturopathy," *JNM* 4 (Oct. 1953): 23–24. Turska's article at times quotes verbatim from Lindlahr's *Nature Cure*.

5. Samuel Bloch, quoted in Czeranko, *Philosophy of Naturopathic Medicine*, 26; Lust, *Universal Naturopathic Encyclopedia*, 15; Louis Kuhne, "The New Science of Healing or the Doctrine of the Unity of Diseases," in Lust, *Universal Naturopathic Encyclopedia*, 244; Friedhelm Kirchfeld and Wade Boyle, *Nature Doctors: Pioneers in Naturopathic Medicine* (East Palestine, OH: Buckeye Naturopathic, 1993), 105; Jocelyn C. P. Proby, "Editor's Introduction," in *The Philosophy of Natural Therapeutics*, vol. 2, *Practice*, by Henry R. Lindlahr (1919; repr., ed. Jocelyn C. P. Proby, Saffron Walden, Essex, UK: C. W. Daniel, 1981), [3].

6. Benedict Lust, ND, MD, "History of the Naturopathic Movement," *HHN* 26 (1921): 581, "Naturopathy, History II" file, NCNM; "Stenographic Report of the First Meeting of the Naturopathic Society of America," *NHH* 4, nos. 1–2, whole nos. 37–38 (Jan.–Feb. 1903): 1. The American Naturopathic Association had begun as the Naturopathic Society of America in 1902. Benedict Lust, "Editorial Drift," ibid. 3, no. 1, whole no. 25 (Jan. 1902): 32–33, "Naturopathy, History II" file, NCNM; Robert P. Porter, *Compendium of the Eleventh Census: 1890: Part I: Population* (Washington, DC: GPO, 1892), 40.

7. Friedhelm Kirchfeld and Wade Boyle, "Discoverer of the Atmospheric Cure, Arnold Rikli, 1823–1906," in Kirchfeld and Boyle, *Nature Doctors*, 55–72; Adolph Just, *Return to Nature! The True Natural Method of Healing and Living and The True Salvation of the Soul: Paradise Regained*, 4th enlarged ed., trans. Benedict Lust (New York: Benedict Lust, 1903). See also "Return to Nature! Adolf Just (1859–1936)," in Kirchfeld and Boyle, *Nature Doctors*, 113–28; and John Hewins Kern, "The Naturopathic Credo," *NHH* 29 (1924): 102, "Naturopathy, History II" file, NCNM.

8. Henry Lindlahr, "How I Became Acquainted with Nature Cure," *HHN* 23, no. 2, whole no. 282 (Feb. 1918): 122–24, "Naturopathy, History II" file, NCNM. Lindlahr explained disease etiology in *Nature Cure*.

9. J. T. Work, "Cause and Cure of Disease from the Standpoint of Drugless Physicians," *HHN* 23, no. 12, whole no. 292 (Dec. 1918): 934; Louisa Lust, ND, "The Curative Agency of Water and Herbs for Treatment of Fever," *Naturopath* 28, no. 9, whole no. 347 (Sept. 1923): 437.

10. Kuhne, *Neo Naturopathy*, 206. On women as consumers at the turn of the century, see Alys Eve Weinbaum, Modern Girl around the World Research Group, Lynn M. Thomas, Priti Ramamurthy, Uta G. Poiger, and Madeleine Yue Dong, eds., *The Modern Girl around the World: Consumption, Modernity, and Globalization* (Durham, NC: Duke Univ. Press, 2008); and Elaine S. Abelson, *When Ladies Go*

A-Thieving: Middle-Class Shoplifters in the Victorian Department Store (New York: Oxford Univ. Press, 1989).

11. Harry Finkel, ND, DC, *Health via Nature: The Health Book for the Layman, Including a Rational System of Health Culture and the Prevention, Treatment and Cure of Disease by Natural Methods, Also the Study of Natural Diatetics, the Preparation and Combination of Foods for Health and Disease* (New York: Barnes, 1925), 17; Harry Benjamin, ND, *Everybody's Guide to Nature Cure* (1936; repr., New York: Benedict Lust, 1958), 9; Harry Riley Spitler, MD, ND, PhD, ed., *Basic Naturopathy* (New York: American Naturopathic Association, 1948), 34.

12. James Hewlett-Parsons, *Naturopathic Practice* (New York: Arco, 1969), 24; Roger Newman Turner, "Toxemia Theories," in *Naturopathic Medicine: Treating the Whole Person* (Wellingborough, Northamptonshire, UK: Thorsons, 1984), 34–36, at 34. Hewlett-Parsons was also secretary of the Register of Consultant Herbalist and Homeopathic Practitioners.

13. Bilz, *Natural Method of Healing*, 7; Lindlahr, *Nature Cure*, 19; Henry Lindlahr, "The Law of Crises," in ibid., 11–14, at 11; Roger Newman Turner, "Vis Medicatrix Nature," in *Naturopathic Medicine*, 21; A. R. Hedges, ND, editorial, *JANA* 3 (July 1950): 5. Mid- to late-nineteenth-century regular physicians also studied nerve force, a different theoretical concept. The theory was that everyone possessed a finite amount of nerve force and that various forms of overexertion depleted the body's nerve force, leading to neurasthenia or other diseases. There was debate about the extent to which nerve force was associated with electricity as doctors attempted to understand the relationships between electricity and the nervous system and between biochemistry and nerves. Dr. H. P. Bowditch, "What is Nerve-Force?," *Science* 8 (27 Aug. 1886): 196; Janet Oppenheim, "Nerve Force and Neurasthenia," in *"Shattered Nerves": Doctors, Patients, and Depression in Victorian England* (Oxford: Oxford Univ. Press, 1991).

14. See William A. Turska, "Naturopathic Medicine," *Oregon Naturopathic Review*, Oct. 1949, 9; and Lindlahr, "Law of Crises," 11–12.

15. "Louis Pasteur (1822–1895)," British Broadcasting Corporation, accessed 4 Nov. 2011, www.bbc.co.uk/history/historic_figures/pasteur_louis.shtml.

16. William Esser, ND, "The Morons of America," *NHH* 41, no. 2, current no. 507 (Feb. 1937): 39.

17. [Benedict Lust?], "The Germ Theory of Disease, a Fallacy," ibid. 16, no. 1, whole no. 207 (Jan. 1911): 29; George Starr White, MD, FSSc, "Health versus the Germ Theory," *HHN* 25, no. 3, whole no. 307 (Mar. 1920): 189–90; [Benedict Lust?], "20th Century Superstitions," ibid. 23, no. 11 (Nov. 1918): 883; H. E. Bliler, "Typhoid Fever Illusions," ibid. 25, no. 3, whole no. 307 (Mar. 1920): 124–25; D. O. Cauldwell, "Disease, Serum, Poison and Experience," *NHH* 44, no. 3, current no. 532 (Mar. 1939): 72, 79, at 72.

18. F. James Cook, "Cancer a National Menace as Knife and Germ Theory Fail; Prevention is the Only Hope," *NP* 3, no. 3 (Mar. 1927): 140; Frank E. Dorchester, "Responsibility of Naturopathy," ibid. 36, no. 2, current no. 436 (Feb. 1931): 41, 42, 51, at 41, 42. See also H. A. Swenson, ND, "Diagnosis and Prognosis," *NHH* 40, no. 2 (Feb. 1935): 40; and American Cancer Society, "Guidelines on Nutrition and Physical Activity for Cancer Prevention," accessed 8 Nov. 2014, www.cancer.org/healthy

/eathealthygetactive/acsguidelinesonnutritionphysicalactivityforcancerprevention
/nupa-guidelines-toc.

19. "Recantation of Virchow," *NP* 36, no. 10, current no. 444 (Oct. 1931): 297;
Myron Schultz, "Rudolf Virchow," accessed 31 Jan. 2014, www.ncbi.nlm.nih.gov/pmc
/articles/PMC2603088/. The quotation is found in books and on alternative medical
websites, e.g., Henry G. Bieler, MD, *Food Is Your Best Medicine* (1965; repr., New York:
Random House, 2010), 40.

20. Examples of germ critiques in the 1930s through the 1950s include Benedict
Lust, Dr. Lust Speaking, "Return to Nature? Microbe Theory Challenged," *NHH* 40,
no. 10 (Oct. 1935): 291; review of *Exploding the Germ Theory*, in *NP* 52, no. 6, current
no. 641 (July 1948): 380; and William A. Turska, "Some Observations of Naturopathic
Philosophy," *JN*, Oct. 1956, 12, rejecting the notion of germs invading a healthy body.
On the value of germs, see A. R. Williamson Jr., "A Tribute To Germs," *NHH* 42, no. 3,
current no. 508 (Mar. 1937): 79; Robert V. Carroll Jr., "The Germ Theory," ibid. 42,
no. 7, current no. 512 (July 1937): 201; Jesse Mercer Gehman, "Culture Media and the
Human Body as a Breeding Ground for Germs," ibid. 42, no. 11, current no. 516 (Nov.
1937): 371–72, at 371; and Gehman, "Germ Theory vs. Microsymian Theory," ibid. 42,
no. 3, current no. 508 (Mar. 1937) 79.

21. I. Schmid, ND, "Naturopathy—Ancient and Modern," *Journal of the American
Association of Naturopathic Physicians and Surgeons*, June 1951, 1.

22. Francis Brinker, "The Role of Botanical Medicine in 100 Years of American
Naturopathy," *HerbalGram* 42 (Spring 1998): 55; 58n33; 59nn55, 66, 67.

23. Friedhelm Kirchfeld and Wade Boyle, "Father of Modern Naturopathic
Medicine: John Bastyr (1912–)," in *Nature Doctors*, 303–12; "About," Bastyr University
website, accessed 22 Dec. 2011, www.bastyr.edu/about/default.asp?view=Dr-John
-Bastyr; John B. Bastyr, ND, "Antibiotics," *JANA* 3 (Dec. 1950), 7.

24. Brinker, "Role of Botanical Medicine," 51, quoting Florence Daniel, "Are
Herbs Drugs?—An Answer," *HHN* 26 (1921): 291; Brinker, "Role of Botanical Medi-
cine," 51, quoting from "Wild Spring Herbs as Food and Medicine," *NP* 3, no. 3
(Mar. 1927): 129.

25. Otto Carque, "Errors of Biochemistry," *NHH* 9, no. 11, whole number 106 (Nov.
1908): 331–34. See also S. T. Erieg, "Eating for Health," ibid. 9, no. 8, whole no. 103
(Aug. 1908): 243. Two examples of naturopathic advice are "The Simple Life in a
Nutshell: Mental Hygiene," ibid. 9, no. 10, whole no. 105 (Oct. 1908): 318–23; and
Walter R. Hadwen, "In Defense of Food Reform" (speech, London, 5 Mar. 1908), as
reprinted in ibid. 9, no. 11, whole no. 106 (Nov. 1908): 356–60, at 360.

26. Lindlahr, *Nature Cure*, 271–89, 301–211.

27. Louis Kuhne, "What Shall We Eat? What Shall We Drink? The Digestive
Process," in Lust, *Universal Naturopathic Encyclopedia*, 296–97; Kuhne, "The
Indigestibility of Denatured Foods," in ibid., 299.

28. N. S. Hanoka, "The Philosophy of Naturopathy," *JNM* 9 (Aug. 1959): 16;
Michael DiPalma, ND, citing Harry Riley Spitler in "A Natural Way to Health," *New
Frontier*, May 1984, 7, 30.

29. J. Cherrith Daniels, ND, DP, DC, SCT, "Unadulterated Nature Cure,"
Naturopath 29, no. 7 (July 1924): 648–51, at 648. Naturopaths saw in the power of the
mind something similar to what was observed by late-twentieth-century scientists like

Bernie Siegel in *Love, Medicine and Miracles: Lessons Learned about Self-Healing from a Surgeon's Experience with Exceptional Patients* (New York: HarperCollins, 1990) or what was utilized in traditional folk-healing belief systems that valued the synchronicity of mental and spiritual well-being. Lindlahr, *Nature Cure*, 7.

30. Adolph Just, "The Soul's Life," in *Return to Nature!*, 277–81; Benedict Lust, "Editorial Drift," *NHH* 3, no. 1, whole no. 25 (Jan. 1902): 32, "Naturopathy, History II" file, NCNM. An advertisement in a 1918 issue of *NHH* listed 30 lectures and articles on New Thought, 16 of them by Purinton. *NHH* 23, no. 2, whole no. 282 (Feb. 1918): 196.

31. Dresser, *Spirit of the New Thought*, v, vi, 5; Lorenzo Cohen, Patricia Parker, Qi Wei, et al., "PreSurgical Stress Management Boosts Immune Function, Lowers Mood Disturbance for Prostate Cancer Patients," accessed 7 Dec. 2014, www.mdanderson.org/newsroom/news-releases/2011/pre-surgical-stress-management-boosts-immune-function-lowers-mood-disturbance-for-prostate-cancer-patients.html; O. Eremin, M. B. Walker, E. Simpson, et al., "Immuno-modulatory Effects of Relaxation Training and Guided Imagery in Women with Locally Advanced Breast Cancer Undergoing Multimodality Therapy: A Randomised Controlled Trial," *Breast*, 11 Nov. 2008, www.ncbi.nlm.nih.gov/pubmed/19008099.

32. J. Waterloo Dinsdale, MD, "Modern Miracles of Healing by Following Christ's Teachings: Miracles of the Year 1911 A.D.," *NHH* 16, no. 10, whole no. 216 (Oct. 1911): 665; Edward Earle Purinton, "The Growth of the Soul," ibid. 9, no. 7, whole no. 102 (July 1908): 214–17.

33. Dresser, *Spirit of the New Thought*, vii; Walter Rauschenbusch, *Christianity and the Social Crisis* (New York: Macmillan, 1907); J. Augustus Weiner, "The Rediscovery of the Lost Fountain of Health and Happiness: For Nervous Affliction and Nerve Exhaustion, Including Mental Illness and Sexual Disease," *NHH* 16, no. 3, whole no. 209 (Mar. 1911).

34. Charles H. Gesser, "Would You Demonstrate Spirituality?," *NHH* 44, no. 1, current no. 530 (Jan. 1939): 15; William R. Bradshaw, "The Gospel of Physical Regeneration," *HHN* 23, no. 1, whole no. 281 (Jan. 1918): 34–37, at 34; Lindlahr, *Nature Cure*, 15; A. C. Biggs (head doctor, Biggs Health Resort, Asheville, NC), "Rational Treatment of Neurasthenia," *HHN* 23, no. 2, whole no. 282 (Feb. 1918): 136–38.

35. Henry Lindlahr, "Catechism of Nature Cure," in *Nature Cure*, 21–23; Lorne A. Summers, "Man's Greatest Asset," *HHN* 25, no. 7, whole no. 311 (Aug. 1920): 396; Richard Hofstadter, *Social Darwinism in American Thought, 1860–1915* (Philadelphia: Univ. of Pennsylvania Press, 1944); Gregory Caeys, "The 'Survival of the Fittest' and the Origins of Social Darwinism," *Journal of the History of Ideas* 61, no. 2 (Apr. 2000): 223–40.

36. "Helen Wilmans," New Thought Library, accessed 10 Jan. 2014, http://newthoughtlibrary.com/wilmansHelen/bio_wilmans.htm#TopOfBio; "Helen Wilmans—Home Course in Mental Science—1921," New Thought Library, accessed 10 Jan. 2014, http://newthoughtlibrary.com/wilmansHelen/homecourse/; "Short and Mild Criticism of Mental Science Lesson I by Helen Wilmans: Explanation," *HHN* 23, no. 6, whole no. 286 (June 1918): 571–73, at 571, 572.

37. Helen Wilmans, "Home Course in Mental Science: Omnipresent Life, Lesson I," *HHN* 23, no. 2, whole no. 282 (Feb. 1918): 168–71; "Affirmations," ibid. 23, no. 1, whole

no. 281 (Jan. 1918): 33; Helen Wilmans, "Home Course in Mental Science: Affirmation Lesson V," ibid. 23, no. 8, whole no. 288 (Aug. 1918): 733–36; John B. Lust, "Dr. Lust Speaking: Three Sources of Life Power," NP 51, no. 5, current no. 628 (May 1947): 260, 292. See also above, n. 13, about nerve force and neurasthenia.

38. Advertisement, "The New Psychology and Brain Building," NP 34, no. 3, current no. 413 (Mar. 1929): 15; Edward Earle Purinton, "Diagnosis of Your Temperament," ibid. 36, no. 5, current no. 439 (May 1931): 133.

39. Daniela Dantas Lima, Vera Lucia Pereira Alves, and Egberto Ribeiro Turato, "The Phenomenological-Existential Comprehension of Chronic Pain: Going beyond the Standing Healthcare Models," Philosophy, Ethics, and Humanities in Medicine 9, no. 2 (1914), www.peh-med.com/content/9/1/2.

40. Benedict Lust, Awaken to Complete Consciousness, Naturopathic Pamphlet Series, no. 25 (Butler, NJ: privately printed, 1915), reprinted from Naturopathic Magazine 20 (1915): 14; A. Adolphe Linke, ND, DC, "Mental Hygiene," NHH 40, no. 8 (Aug. 1935): 230–31, at 230.

41. Basil Fellrath, "Cheerfulness and Mental Therapeutics a Great Asset for the Drugless Healer," HHN 23, no. 10, whole no. 290 (Oct. 1918): 832–33, at 832; Gene Blossom, "The Optimist," NHH 18, no. 6 (June 1913): 585–86, at 586; Sam Rud. Cook, "The Art of Laughing," ibid. 20, no. 8, whole no. 262 (Aug. 1915): 471; J. Ogden Armour, "Enthusiasm," HHN 22, no. 4 (July–Aug. 1917): 216.

42. Edward Earle Purinton, "Play," NHH 20, no. 3, whole no. 257 (Mar. 1915): 172–74; Hobbies for Health, NP 54, no. 10 (Oct. 1949): 22; "Doll Painting," Hobbies for Health, ibid. 55, no. 2 (Feb. 1950): 17, 35.

43. Wilbur F. Prosser, "The Psychopathic Influences," Naturopath 28, no. 3, whole no. 341 (Mar. 1923): 123–24; Arthur W. Dennis, "Motherhood in Harmony with Nature," ibid., 130–31, at 130.

44. Lillian Russell, "Disease of Civilization," NHH 20, no. 11, whole no. 265 (Nov. 1915): 698; "The Man Who Sulks," ibid. 9, no. 9, whole no. 104 (Sept. 1908): 270; "Brain Fag," ibid. 19, no. 8, whole no. 250 (Aug. 1914): 544.

45. S. T. Erieg, "A Cure for Nervousness," ibid. 20, no. 11, whole no. 265 (Nov. 1915): 671–72; Robert Bieri, "Thoughts and Their Physical Effects," ibid. 9, no. 9, whole no. 104 (Sept. 1908): 276; Edwin J. Ross, "Anxiety," Naturopath 28, no. 3, whole no. 341 (Mar. 1923): 118–20; Lillian R. Carque, "The Human Mind in Health and Disease," NHH 44, no. 5, current no. 534 (May 1939): 134, 152; Benedict Lust, Dr. Lust Speaking, NP 51, no. 3, current no. 626 (Mar. 1947): 132, 163, at 132, posthumous reprint, original date unknown; J. W. Wigelsworth, ND, "Pills for Pessimists," HHN 23, no. 7, whole no. 287 (July 1918): 658–59, at 658. See also Jesse Mercer Gehman, "Fear," NP 55, no. 2 (Feb. 1950): 12, 38.

46. Benedict Lust, Editorial Comment, "Things for Consideration at the Convention," HHN 25, no. 8, whole no. 312 (Sept. 1920): 374.

47. Stephen Petrina, "Medical Liberty: Drugless Healers Confront Allopathic Doctors, 1910–1931," Journal of Medical Humanities 29 (Dec. 2008): 205–30; News Section, NHH 20, no. 10, whole no. 264 (Oct. 1915): 618; Edward Earle Purinton, "Cure by Suggestion," NP 3, no. 4 (Apr. 1927): 170; Milton Powell, FNCA, "Psychotherapy and Nature Cure," NHH 44, no. 11, current no. 540 (Nov. 1939): 329; E. Beran Wolff, "Synopsis of Hysteria," ibid., 350.

48. A. Adolphe Linke, ND, BS, "Mental Conflicts," *NHH* 49, no. 1, current no. 582 (Jan. 1944): 47, 53.

49. August Englehardt, ed., "For Sun, Tropics and Coconuts," in "Christ, Buddah [*sic*], Mohamed," special issue, printed by Benedict Lust, *NHH* 19 (May 1914): 337; "How Do We Free Ourselves from the Care of Clothing?," ibid. 19, no. 1, whole no. 243 (Jan. 1914): 58–59, at 58; "How Do We Free Ourselves from the Care of Shelter?," ibid., 60–61, at 60; "How the White Man's Diet Affects Natives of Africa," *Naturopath* 18 (Nov. 1913): 13; "More Healthy When Naked," *NHH* 20, no. 11, whole no. 265 (Nov. 1915): 713.

50. Claus Hansen, "The Common-Sense Law," *NHH* 20, no. 7, whole no. 261 (July 1915): 447–49, at 447.

51. Tell Berggren, "Natural Instincts the Best Guide to Health," ibid. 20, no. 4, whole no. 258 (Apr. 1915): 226–28, at 226, 227; Carl Strueh, "The Proper Way to Dress," ibid. 20, no. 3, whole no. 257 (Mar. 1915): 161–62, at 161.

52. John Summerfield, "The Value of Instinct," ibid. 20, no. 4, whole no. 258 (Apr. 1915): 234–37, at 236; "The Simple Life in a Nutshell: Mental Hygiene," ibid. 9, no. 10, whole no. 105 (Oct. 1908): 319–23, at 322.

53. Edythe Stoddard Seymour, "Reduce the High Cost of Living Yet Live Well, Even on a Small Salary," ibid. 20, no. 7, whole no. 261 (July 1915): 439–41.

54. Benedict Lust, "Are Your Nerves All Ragged? Nervousness and Neurasthenia," *NP* 36, no. 5, current no. 439 (May 1931): 134–35, 140. In 1935 W. R. Franklin hypothesized that the three ailments were essentially the same. Franklin, AB, OD, ND, "Neurasthenia," *NHH* 40, no. 10 (Oct. 1935): 294, 310.

55. Edward Earle Purinton, Efficiency in Drugless Healing, *NHH* 19, no. 9, whole no. 251 (Sept. 1914): 557–62, at 561; Jessie Allen Fowler, "Success, the Result of Harmony," *HHN* 23, no. 6, whole no. 28 (June 1918): 534–35.

56. Maurice Shefferman, "Exercise—For the Brain Worker," *NP* 52, no. 9, current no. 644 (Oct. 1948): 504, 530; Edward Earle Purinton, "Faith for Health and Wealth," ibid., 509, 525.

57. B. Coursin Black, "Cities Can Be Soothing," *NP* 55, no. 2 (Feb. 1959): 8, 40, at 40.

CHAPTER 4: Louisa Stroebele Lust, Benedict Lust, and Their Yungborn Sanatorium

1. Friedhelm Kirchfeld and Wade Boyle, "The Matriarch of New York: Louisa Lust (1868–1925)," in *Nature Doctors: Pioneers in Naturopathic Medicine* (East Palestine, OH: Buckeye Naturopathic, 1993), 221.

2. It is difficult to pin down the exact years of Stroebele's work with Claflin. However, by 1894 Louisa already owned and operated the Bellevue Sanitarium in Butler, New Jersey, as a successful and well-established business. Thus, it is likely Louisa's work at Bellevue began ca. 1892–93. Given her birth in 1868 and the likelihood that she was a young adult when she affiliated with Claflin, I estimate that she was with Claflin from roughly 1888 to 1891 or 1892. These three to four years would have allowed ample time for the three world tours, as well as for Stroebele to accumulate the cash she invested in Bellevue.

3. Bret E. Carrol, *Spiritualism in Antebellum America* (Bloomington: Indiana Univ. Press, 1997), 1; Emanie Sachs, *The Terrible Siren: Victoria Woodhull, 1838–1927* (New York: Harper, 1928), 19; Lois Beachy Underhill, *The Woman Who Ran for*

President: The Many Lives of Victoria Woodhull (New York: Penguin, 1996); Barbara Goldsmith, *Other Powers: The Age of Suffrage, Spiritualism, and the Scandalous Victoria Woodhull* (New York: Harper, 1999); Barbara Weisberg, *Talking to the Dead: Kate and Maggie Fox and the Rise of Spiritualism* (New York: Harper, 2004).

4. Joanna Johnston, *Mrs. Satan: The Incredible Saga of Victoria C. Woodhull* (New York: Putnam, 1967), 18, 22,31; Sachs, *Terrible Siren*, 16; Goldsmith, *Other Powers*, 32, 67.

5. Myra MacPherson, *The Scarlet Sisters: Sex, Suffrage and Scandal in the Gilded Age* (New York: Twelve, 2014).

6. Arlene Kisner, ed., *Woodhull and Claflin's Weekly: The Lives and Writings of the Notorious Victoria Woodhull and Her Sister Tennessee Claflin* (Washington, NJ: Times Change, 1972), 11.

7. Tennessee Claflin Cook, *The Ethics of Sexual Equality* (New York: Woodhull and Claflin, 1873), Microforms Center, F-704, reel 376, no. 2609, p. 4, Love Library, San Diego State University.

8. Ibid. See also Linda Gordon, "Voluntary Motherhood: The Beginnings of Feminist Birth Control Ideas in the United States," in *Women and Health in America*, ed. Judith Walzer Leavitt (Madison: Univ. of Wisconsin Press, 1999), 253–68.

9. Claflin, "Ethics of Sexual Equality," 13.

10. Nancy Cott coined the term *passionlessness* to refer to nineteenth-century middle-class Anglo ladies, who were presumed to be by nature asexual or without a sex drive. See Nancy Cott, "Passionlessness: An Interpretation of Victorian Sexual Ideology, 1790–1850," *Signs: Journal of Women in Culture and Society* 2 (1978): 219–36. See also Daniel S. Wright, *"The First Causes to Our Sex": The Female Moral Movement in the Antebellum Northeast, 1834–1848* (New York: Routledge, 2006).

11. Benedict Lust, "The Kneipp Water Cure Comes to New York—Marriage—Publishing," in "Yungborn: The Life and Times of Dr. Benedict Lust, Founder and Father of Naturopathy, The Cornucopia of Alternative Medicine: A Memoir," Jesse Gehman Papers, file 11, Bastyr University Special Collections, Kenmore, Washington, 1923. Lust's original manuscript was available in the Gehman Papers years before his niece Anita Lust Boyd (daughter of his brother Leo) privately published the edited version in 1997: Lust, *Yungborn: The Life and Times of Dr. Benedict Lust, Founder and Father of Naturopathy, The Cornucopia of Alternative Medicine: A Memoir*, with a foreword by Dr. Joseph E. Pizzorno Jr., ND, President of Bastyr University and author of *Total Wellness* (n.p.: privately printed by Anita Lust Boyd, 1997). I draw from both versions (distinguishing them by date). The 1997 version has also been reprinted as *Collected Works of Dr. Benedict Lust, Founder of Naturopathic Medicine*, ed. Anita Lust Boyd and Eric Yarnell, ND, RH (AHG) (Seattle: Healing Mountain, 2006).

12. Lust, *Yungborn* (1997), 50.

13. Ibid., 50, 50–51.

14. Ibid., 50–51; Anita Lust Boyd, telephone interview by author, 14 Sept. 2004.

15. Lust, *Yungborn* (1997), 52. In October 1895 Claflin had married a wealthy London dry goods merchant, Francis Cook, who was made a baron, whereupon she became Lady Cook. E. Danziger, "Appendix to 'The Cook collection, its founder and its inheritors,'" *Burlington Magazine* 146 (July 2004): 444–58.

16. Tennie C. Claflin, *Constitutional Equality: A Right of Woman; or A Consideration of the Various Relations which she sustains as a Necessary Part of the Body of*

Society and Humanity; With Her Duties to Herself—Together with a Review of the Constitution of the United States, Showing that the Right to Vote is Guaranteed to All Citizens. Also a Review of the Rights of Children (New York: Woodhull, Claflin, 1871), Microforms Center, F-704, reel 376, no. 2609, Love Library, San Diego State University. See also Madeline B. Stern, *The Victoria Woodhull Reader* (Weston, MA: MTS, 1974).

17. By the time Stroebele assisted Claflin on world tours, Claflin had stockpiled some of her money and financed her travel. Woodhull, meanwhile, lost most of her fortune after her run for president and the backlash she endured from the Beecher-Tilton scandal. When Vanderbilt died in 1877, he left the two sisters with a wealth of knowledge about successful business speculation, and it was widely believed that Vanderbilt's kin gave the sisters hush money to keep them from discussing their relationships with him. See Kisner, *Woodhull and Claflin's Weekly*, 61.

18. Goldsmith, *Other Powers*; Amanda Frisken, *Victoria Woodhull's Sexual Revolution: Political Theater and the Popular Press in Nineteenth-Century America* (Philadelphia: Univ. of Pennsylvania Press, 2011); Geoffrey Blodgett, "Woodhull, Victoria Claflin (Sept. 23, 1838–June 10, 1927), and her sister Tennessee Celeste Claflin (Oct. 26, 1845–Jan. 18, 1923)," in *Notable American Women, 1607–1950: A Biological Dictionary*, ed. Edward T. James, Janet Wilson James, and Paul S. Boyer, vol. 3 (Cambridge, MA: Belknap Press of Harvard Univ. Press, 1971), 654.

19. Goldsmith, *Other Powers*; Blodgett, "Woodhull, Victoria Claflin," 655.

20. Lust, "Kneipp Water Cure Comes to New York," in "Yungborn" (1923), 17. Benedict referred to her as both Louise and Louisa.

21. Blodgett, "Woodhull, Victoria Claflin," 643.

22. Sachs, *Terrible Siren*, 47; Goldsmith, *Other Powers*, 158–62; Johnston, *Mrs. Satan*, 45.

23. Kisner, *Woodhull and Claflin's Weekly*, 11.

24. Kirchfeld and Boyle, *Nature Doctors*, 221; F. W. Collins, "Report of Business Meeting of N.Y.S.S. of Naturopaths," *Naturopath* 29, no. 7 (July 1924): 659–65; Louisa Stroebele, "Mountain Air Resort 'Bellevue,' Butler, N.J." *Amerikanische Kneipp-Blätter* 4 (1899): 41.

25. Lust, "Kneipp Water Cure Comes to New York," in "Yungborn" (1923), 17–18.

26. Ibid., 18–20.

27. Ibid., 19.

28. Kirchfeld and Boyle, *Nature Doctors*, 222.

29. "Wedding Bells," *KWCM* 2 (1901): 163, as cited in Kirchfeld and Boyle, *Nature Doctors*, 222.

30. Lust, "Kneipp Water Cure Comes to New York," in "Yungborn" (1923), 2, 5.

31. Ibid., 19.

32. Ibid., 18; George Cody, "History of Naturopathic Medicine," in *Textbook of Natural Medicine*, ed. Joseph Pizzorno and Michael Murray (New York: Churchill Livingstone, 1985), 7–37, at 15.

33. Benedict Lust, "Intolerance of Official Medicine," *HHN* 24 (1919): 438–40; Lust, "Kneipp Water Cure Comes to New York," in "Yungborn" (1923), 10, 13.

34. Lust, "Kneipp Water Cure Comes to New York," in "Yungborn" (1923), 18–20. The "peasant food" included soups, vegetables, black bread, and clabbered milk.

35. Lust, "Kneipp Water Cure Comes to New York," in "Yungborn" (1923), 18; Kirchfeld and Boyle, *Nature Doctors*, 105.

36. Lust, "Kneipp Water Cure Comes to New York," in "Yungborn" (1923), 19.

37. Ibid., 10, 12, 13.

38. Friedhelm Kirchfeld and Wade Boyle, "The Father of Naturopathy, Benedict Lust (1872–1945)," in *Nature Doctors*, 187.

39. Ibid., 188; John Hewins Kern, "Dr. Louisa Lust: An Appreciation," *HHN* 30, no. 7, whole no. 369 (July 1925): 709.

40. Lust, *Collected Works*, 52.

41. Benedict Lust, "Marriage and Medical School," in "Yungborn" (1923), 6.

42. "Benedict Lust—The Man and His Work: By One to Whom He Gave a Lift in Life," *Naturopath* 30 (1925): 943–45, as cited in Kirchfeld and Boyle, "Father of Naturopathy," 189; Lust, "Marriage and Medical School," 3, 7, 13.

43. Lust, "Marriage and Medical School," 11, 13–14.

44. Benedict Lust, "I Go to Medical College—Persecution and Prosecution," in "Yungborn" (1923), 60.

45. Kirchfeld and Boyle, *Nature Doctors*, 200.

46. There is some dispute about the timing of Lust's degrees. Kirchfeld and Boyle state that Lust received his DO degree in 1898; Lust's memoir states that it was in 1902. According to his memoir, he began taking classes at the Eclectic Medical College of New York in 1901, but Kirchfeld and Boyle state that he did not receive the degree until 1913. Hans Baer asserts that Benedict earned a degree from the Eclectic Medical College in 1914. Kirchfeld and Boyle, *Nature Doctors*, 200; Lust, "I Go To Medical College," 60; Lust, *Collected Works*, 63; Hans A. Baer, "The Potential Rejuvenation of American Naturopathy as a Consequence of the Holistic Health Movement," *MAQ* 13, no. 4 (1992): 369–83.

47. Kirchfeld and Boyle, *Nature Doctors*, 221; Emil Posner, "In Memoriam to 'The Mother of Naturopathy,'" *NHH* 30 (1925): 707–8. The Lusts named Yungborn in homage to Adolph Just's (1859–1936) sanatorium in Germany. Louisa "was the leading financial partner in this enterprise." Collins, "Report of Business Meeting," as cited in Kirchfeld and Boyle, *Nature Doctors*, 221; Lauren Paschal Proctor (Benedict's niece), conversation with the author, 16 Jan. 2014; cover, *NHH* 8, whole no. 92 (Sept. 1907); Benedict Lust, "To My Friends and Readers," ibid. 30 (July 1925): 705; advertisement, "Dr. Benedict Lust's Recreation Resort 'Yungborn,' Butler, NJ," ibid. 23, no. 5, whole no. 285 (May 1918): 408.

48. Advertisement, "Sane at the American 'Yungborn' Park, Bellevue, Butler, NJ," *NHH* 8, whole no. 90 (July 1907); advertisement, "Sanitorium 'Yungborn,' Butler, NJ," ibid. 8, whole no. 89 (June 1907): 29; advertisement, "Yungborn," ibid. 22, no. 3 (Jan.–Feb. 1917).

49. J. Austin Shaw, "Austin Shaw Explains Yungborn Nature Cure," ibid. 16, whole no. 209 (Mar. 1911): 145–55.

50. Advertisement, "Sanitorium 'Yungborn,'" 29.

51. Shaw, "Austin Shaw Explains," 149; Kirchfeld and Boyle, *Nature Doctors*, 195–96; advertisement, "Dr. Benedict Lust's Recreation Resort 'Yungborn,' Butler, NJ," *NHH* 23, no. 5, whole no. 285 (May 1918): 408.

52. Advertisement for Syracuse Naturopathic Institute and Sanitarium, *NHH* 23, no. 6, whole no. 286 (June 1918): 511.

53. William R. Bradshaw, ND, "Yungborn, Tangerine, Fla., the Home of Health," ibid. 23, no. 11, whole no. 291 (Nov. 1918): 884–90; advertisement, "Dr. Benedict Lust's School of Diets; For the Art of Nourishment, the Art of Breathing, and the Training of Thought," in *Man, Learn to Think!*, Naturopathic Pamphlet Series, 18 (Butler, NJ: Kneipp-Naturopathic Health Resort, "Yungborn" and American School of Naturopathy, 1916), back cover; Kirchfeld and Boyle, *Nature Doctors*, 196.

54. "Planning a Vacation," *NHH* 8, whole no. 90 (July 1907): 32; Kirchfeld and Boyle, *Nature Doctors*, 195–96.

55. Samuel H. Williamson, "Seven Ways to Compute the Relative Value of a U.S. Dollar Amount, 1774 to present," MeasuringWorth, 2015, accessed 26 June 2015, www.measuringworth.com/uscompare/; "Median Income for 4-Person Families, by State," U.S. Census Bureau, Housing and Household Economic Statistics Division, 2005, accessed 8 Oct. 2008, www.census.gov/hhes/www/income/data/statistics/4person.html; Williamson, "Seven Ways to Compute"; advertisement, "Florida Nature Cure Resort—'Qui–Si–Sana' and Recreation Home Yungborn," left frontispiece of Benedict Lust, *Winds and Gases/Condition of Weakness/The Art of Nourishment, The Art of Breathing and the Training of Thought*, Naturopathic Pamphlet Series, 17 (Butler, NJ: Kneipp-Naturopathic Health Resort, 'Yungborn' and American School of Naturopathy, 1914).

56. Louisa Lust, *Practical Naturopathic-Vegetarian Cook Book: Cooked and Uncooked Foods* (New York: Benedict Lust, 1907); advertisement, "Have a Good Dinner?," *NHH* 8, whole no. 90 (July 1907): 3. The value of a dollar at the time may be measured by the following prices given in the *Portsmouth (NH) Herald*, 17 Apr. 1906: a quart of imported French brandy, $1.25; an evening at a moving-picture show, 15–50¢; a pound of coffee, 20¢; a pound of tea, 45¢; butter, 25¢; and an overcoat, $10. Louisa's cookbook seems moderately affordable in this context. See *Portsmouth (NH) Herald*, accessed 14 Sept. 2014, www.newspapers.com/newspage/57003964; and L. Lust, "Vegetarian Regime Important in Health and Disease," *Naturopath* 29 (1924): 812, as quoted in Kirchfeld and Boyle, *Nature Doctors*, 222. The *Good Dinner Cook Book* (Butler, NJ: Nature Cure, 1913) was a reprint of her 1907 book under a new name. See advertisement in *NHH* 18, no. 6 (June 1913); and advertisement for *The Practical Naturopathic-Vegetarian Cook Book*, ibid. 23, no. 4, whole no. 284 (Apr. 1918): 378.

57. L. Lust, *Practical Naturopathic-Vegetarian Cook Book*, 9; L. Lust, "Vegetarian Regime Important," 809–12.

58. Louisa Lust, ND, "Naturopath Health Kitchen," *NHH* 8, whole no. 90 (July 1907): 218–19.

59. Carole M. Counihan, *The Anthropology of Food and Body: Gender, Meaning and Power* (New York: Routledge, 1999).

60. Louisa Lust, "How to Avoid Pain and Sickness," *Naturopath* 16, no. 2, whole no. 208 (Feb. 1911): 99; L. Lust, *Practical Naturopathic-Vegetarian Cook Book*, 9.

61. L. Lust, *Practical Naturopathic-Vegetarian Cook Book*, 12.

62. Ibid., inside frontispiece. On costs, see Williamson, "Seven Ways to Compute." Pamphlet 17 was titled *Winds and Gases/Conditions of Weakness*, and pamphlet 18,

Man, Learn to Think! In order, other titles addressed fatigue, appropriate nourishment, the causes of cancer, training of thought, the conscious diet, and raw foods.

CHAPTER 5: Women, Naturopathy, and Power

1. Arthur W. Dennis, "Motherhood in Harmony with Nature," *Naturopath* 28, no. 3, whole no. 341 (Mar. 1923): 139; Edward Earle Purinton, "Silence," *HHN* 25, no. 4, whole no. 308 (May 1920): 167; Purinton, "Play," *NHH* 20, no. 3, whole no. 257 (Mar. 1915): 173.

2. Advertisement, "Have You These Symptoms of NERVE EXHAUSTION?," *NP* 34, no. 1, current no. 411 (Jan. 1929): 34; advertisement, "Are You Nervous as a Cat? There Is a Reason," ibid. 36, no. 8, current no. 442 (Aug. 1931): 252.

3. Benedict Lust to August Jensen, 19 Sept. 1944, *NHH* 49, no. 11, current no. 592 (Nov. 1944): 347; Roland Gustafson, "A Naturopathic Version of the Ten Commandments of Pregnancy," ibid. 49, no. 12, current no. 593 (Dec. 1944): 355.

4. The phenomenon of the New Woman was a product of the growing number of middle- and upper-class educated women after the 1870s. For changing notions of female sexuality and morality in this era, see John D'Emilio and Estelle B. Freedman, *Intimate Matters: A History of Sexuality in America* (Chicago: Univ. of Chicago Press, 2012); Carroll Smith-Rosenberg, "The New Woman as Androgyne: Social Disorder and Gender Crisis, 1870–1936," in *Disorderly Conduct: Visions of Gender in Victorian America*, ed. Smith-Rosenberg (New York: Knopf, 1985), 245–96; Paula S. Fass, *The Damned and the Beautiful: American Youth in the 1920s* (New York: Oxford Univ. Press, 1977); "The Influence of Women," *NHH* 3, no. 7, whole no. 31 (July 1902): 306; Edward Earle Purinton, "Anthony Comstock Angel Unawares," ibid. 8, whole no. 92 (Sept. 1907): 176, 179; and Purinton, "Duty vs. Desire," ibid. 9, no. 12, whole no. 107 (Dec. 1908): 382–88.

5. Catherine Gourley, *Flappers and the New American Woman: 1918 through 1920s* (Minneapolis: Twenty-First Century Books, 2007); Linda Gordon, *The Moral Property of Women: A History of Birth Control Politics in America* (Urbana: Univ. of Illinois Press, 2007); Kathy Peiss, *Cheap Amusements: Working Women and Leisure in Turn-of-the-Century New York* (Philadelphia: Temple Univ. Press, 1985); M. C. Goettler, ND, "Relation of Diet to Morals," *NHH* 16, no. 9, whole no. 215 (Sept. 1911): 575–76; J. H. Neff, "Social Purity," ibid. 16, no. 4, whole no. 210 (Apr. 1911): 240; J. W. Hodge, MD, "The Use of Tobacco: A Physical, Mental, Moral and Social Evil," ibid. 16, no. 11, whole no. 217 (Nov. 1911): 701–9; S. T. Erieg, "How Billions Are Wasted," ibid. 20, no. 9, whole no. 263 (Sept. 1915): 592.

6. "Treatment of the Infant during the First Months: Bringing-Up of Children," *HHN* 23, no. 7, whole no. 287 (July 1918): 649–51; Martha B. Opland, ND, For Mothers and Children, ibid. 25, no. 2, whole no. 306 (Feb. 1920): 70–71; and D. A. Thom, "Child Management: The 'Finicky' Appetite," *NP* 36, no. 5, current no. 439 (1931): 145, are examples of articles about children in the first half century. Advertisement, "THE NEW YORK PARENT HEALTH CENTER," ibid. 3, no. 5 (May 1927): 257.

7. Martha B. Opland, For Mothers and Children, *NHH* 25, no. 1, whole no. 305 (Jan. 1920): 70–71; Grace L. Dudley, ND, Motherhood, ibid. 49, no. 9, current no. 590 (Sept. 1944): 260, 281–82; John B. Lust, ND, Domestic Practice of Nature Cure, *NP* 54, no. 1 (Jan. 1950): 4.

8. G. W. Haas, "Mother, the Household Physician," *Naturopath* 30, no. 1 (Jan. 1925): 21–22.

9. "You Can't Understand—You're Not a Woman," *NHH* 3, no. 3, whole 35 (Nov. 1902): 107; "Keeping a Husband," ibid. 3, no. 1, whole no. 25 (Jan. 1902): 43; "Health Incarnate," ibid. 5, no. 4, whole no. 52 (Apr. 1904): 89; "Advice to Girls," ibid. 3, no. 3, whole no. 27 (Mar. 1902): 167—all from The Women's Column.

10. Louis Kuhne, "The New Science of Healing: Diseases of Women," ibid. 23, no. 7, whole no. 287 (July 1918): 635–41, at 635; Joseph A. Hoegen, ND, "Female Diseases," ibid. 21, no. 7 (Aug.–Sept. 1916): 469.

11. Bob Batchelor, "Food," in *The 1900s: American Popular Culture through History* (Westport, CT: Greenwood, 2002), 106, 109, 113–14; Otto Carque, "The Building Stones of the Body," *HHN* 23, no. 4, whole no. 284 (Apr. 1918): 332–33; William F. Kirik, "The Factory Girl," *NHH* 19, no. 8, whole no. 250 (Aug. 1914): 500.

12. See [Louisa Lust?], "Useful to Women," *NHH* 3, no. 2, whole no. 26 (Feb. 1902): 96; and "Naturopath Health Kitchen: A Series of Weekly Menus to Appear throughout the Year," ibid. 8, whole no. 92 (Sept. 1907): 185–88.

13. Goettler, "Relation of Diet to Morals," 575. See also, e.g., C. A. Lindenmeyer, "The Living Question," *HHN* 23, no. 6, whole no. 286 (June 1918): 535–37, on raw foods; "Mucusless-Diet Healing System," *Naturopath* 28, no. 12, whole no. 350 (Dec. 1923): 793; "Feeding Mothers and Babies," *NP* 3, no. 3 (Mar. 1927): 125–26; and State News: Alaska, *NHH* 40, no. 3 (Mar. 1935): 89, discussing a health food store in Juneau.

14. Helen Wilmans, author of the "Home Course in Mental Science," asserted her metaphysical "Life Principle" as the one law, the law of attraction. It was expressed in love or life in various ways and forms. Minna Beyer Madsen, ND, Occasional Intimate Little Chats to Women by a Woman, *NP* 3, no. 4 (Apr. 1927): 184, and no. 5 (May 1927): 231; Madsen, "Beauty Culture," ibid. 32, whole no. 394 (Aug. 1927): 384; Martha B. Opland, ND, "The Most Important Diagnosis—Urinalysis," *Naturopath* 28, no. 5, whole no. 343 (May 1923): 236–38; Louisa Lust, ND, "The Curative Agency of Water and Herbs for Treatment of Fever," ibid. 28, no. 9, whole no. 347 (Sept. 1923): 437; Alice M. Reinhold, ND, "Natural Treatment of Pneumonia," ibid. 29, no. 7 (July 1924): 628–29; Dorothy Dean Richardson, "Nerve Force," ibid., 647; Harriet Zoll, "How to Grow Younger as You Grow Older," *Naturopath* 30, no. 1 (Jan. 1925): 51–54; Lena Julia Sappia, "A Message," ibid. 31, no. 5, whole no. 379 (May 1926): 231–32.

15. Alice Chase, Your Health Problems, in *NP* 51, no. 3, current no. 626 (Mar. 1947): 147; 51, no. 5, current no. 628 (May 1947): 279, 290; and 52, no. 8, current no. 643 (Sept. 1948): 470. Barbara Ehrenreich and Deidre English, "The Rise of the Experts" and "The Reign of Experts," in in *For Her Own Good: 150 Years of the Experts' Advice to Women* (Garden City, NY: Doubleday/Anchor, 1978), 37–110 and 111–340, respectively.

16. Virginia S. Lust, "Health News for March," *NP* 51, no. 3 (Mar. 1947): 150; Elena Slade, "How to Attain Your Ideal Weight," ibid. 52, no. 6, current no. 641 (July 1948): 363, 384; Irma Goodrich Mazza, "How to Grow Herbs," ibid., 370, 391; Mina B. Hoffman, "I Make a Garden," ibid., 378; Edith Henne, "Is Life Prosaic?," ibid., 381.

17. Mrs. B. Lust, "Cheerfulness," *NHH* 3, no. 2, whole no. 26 (Feb. 1902): 91; Lust, "Provoking People," ibid. 3, no. 1, whole no. 25 (Jan. 1902): 42; Cynthia Westover Alden, "The Sunshine of Kind Deeds," *HHN* 23, no. 8, whole no. 288 (Aug. 1918): 73.

18. Mrs. Krueger, "Magnetism: Its Natural Curative Power," *NHH* 9, no. 10, whole no. 105 (Oct. 1908): 105–6; Mrs. A. S. Hunter, "With My Patients," ibid. 5, no. 8, whole no. 56 (Aug. 1904): 194–96; Mrs. Louisa A'Hmuty Nash, "My First Experience with Electricity," *HHN* 23, no. 6, whole no. 286 (June 1918): 560–61.

19. Julie La Salle Stevens, "How I Came Back," *Naturopath* 28, no. 8, whole no. 346 (Aug. 1923): 386–87; Mrs. Margaret A. Standish, letter to the editor, *HHN* 25, no. 4, whole no. 308 (May 1920): 196; Correspondence, *NP* 3, no. 4 (Apr. 1927): 171.

20. Ellen Schramm, "Obstetrics: Postpartum Care of Mother," *JNM* 4, no. 1 (Jan. 1953): 17–18; Schramm, "Obstetrics: Care of the Newborn," ibid. 4, no. 2 (Feb. 1953): 21; Nora E. Thrash, "Merchandising Health," *NP* 32, no. 12, whole no. 398 (Dec. 1927): 581; advertisement, "Be a Real Family Physician: Be Trained for This Service at the AMERICAN SCHOOL OF NATUROPATHY. . . . High Grade Men and Women Climb out of Commercialism," *Naturopath* 31, no. 1, whole no. 375 (Jan. 1926): 46; advertisement, "Health Is Power Book List," listing books by Helen Wilmans, Florence Daniel, Linda Hazzard, and Jenny Springer, MD, *NHH* 40, no. 12 (Dec. 1935): 382; "Books on Children," listing books by Florence Daniels, Mrs. Hester Pendleton, Hazel Corbin, and Carolyn C. VanBlarcom, ibid. 49, no. 1, current no. 582 (Jan. 1944): 25; *NP* 55, no. 2 (Feb. 1950): 3.

21. On eighteenth- and nineteenth-century ideology about female passionlessness, see Nancy F. Cott, "Passionlessness: An Interpretation of Victorian Sexual Ideology, 1790–1850," *Signs: Journal of Women in Culture and Society* 2 (1978): 219–36; and John D'Emilio and Estelle D. Freedman, "Seeds of Change," in D'Emilio and Freedman, *Intimate Matters*, 39–54. Two well-known proponents of ovarian determinism were Harvard's Edward Clarke (*Sex in Education: Or, A Fair Chance for the Girls* [Boston: James R. Osgood, 1873]) and S. Weir Mitchell ("Rest in the Treatment of Nervous Disease," in *A Series of American Clinical Lectures*, ed. E. C. Seguin, MD, vol. 1 [New York: G. P. Putnam's Sons, 1875]). See also Peiss, *Cheap Amusements*. The large influx of immigrants into working communities in the 1840s led to an increase in questions about the morality of working girls. See Thomas Dublin, *Women at Work: The Transformation of Work and Community in Lowell, Massachusetts, 1826–1860* (New York: Columbia Univ. Press, 1993); and Teresa Amott and Julie Matthaei, *Race, Gender and Work: A Multi-Cultural Economic History of Women in the United States* (Cambridge, MA: South End, 1999).

22. E. Seibert, "Prevention and Cure of Female Diseases by a Natural Method," *NHH* 9, no. 3, whole no. 98 (Mar. 1908): 79–81, at 79; "How Do We Free Ourselves from the Care of Clothing?," ibid. 19, no. 1, whole no. 243 (Jan. 1914): 58–60; Mitchell, "Rest in the Treatment of Nervous Disease"; Benedict Lust, "Hysteric Girls and Women," *HHN* 25, no. 6, whole no. 310 (July 1920): 284–85, attributing most cases to women; Richard F. Herzog, ND, DC, "Hysteria," *NHH* 44, no. 2, current no. 531 (Feb. 1939): 43, arguing that it occurred in both sexes and stemmed from wrong influences in sexual life and mental shocks; Carl E. Bittrich, ND, DN, "Hysteria," ibid. 49, no. 1, current no. 582 (Jan. 1944): 10, 21.

23. E. D. Titus, "The Crime of the Century," *HHN* 22, no. 3, whole no. 277 (May–June 1917): 129–33, at 133; "Feeding Mothers and Babies," 126; Edwin F. Bowers, "The Female Slacker," *NP* 34, no. 3, current no. 413 (Mar. 1929): 11; Benedict Lust, "Naturopathy: Nervous Disease," *NHH* 49, no. 12, current no. 593 (Dec. 1944): 359.

24. Bernarr MacFadden, *Preparing for Motherhood* (New York: MacFadden, 1936); David C. Long, "Naturopathy in Obstetrics," *NHH* 40, no. 9 (Sept. 1935): 266; "Noted Astrologer Reveals How to Predetermine Sex," ibid. 42, no. 7, current no. 512 (July 1937): 222.

25. "Feed Young Girls: Must Have Right Food While Growing," *NHH* 16, no. 6, whole no. 212 (June 1911): 385; Miss J. Rachel Walker, "The Physical Culture Girls Opportunity: Spiritually and Maternally," ibid. 20, no. 3, whole no. 257 (Mar. 1915): 196–98; untitled poem, ibid., 198. Edwin F. Bowers said childbirth troubles stemmed from "ignorance and unnatural living habits." Bowers, "The Greatest Boon to Women," *NP* 34, no. 5, current no. 415 (May 1929): 212, 228.

26. Announcement for a "Women's Column," *NHH* 3, no. 10, whole no. 35 (Nov. 1902): 480–81; Benedict Lust, ND, "Complaints during Pregnancy," ibid. 9, no. 4, whole no. 99 (Apr. 1908): 114; "Vomiting in Pregnancy," ibid. 44, no. 4, current no. 533 (Apr. 1939): 108; Gustafson, "Naturopathic Version of the Ten Commandments of Pregnancy," 355, 369; J. F. Blea, ND, "Pregnancy—Its Prenatal and Postnatal Care," *JANA* 4, no. 1 (Jan. 1951): 8, 13–14, 16; "How to Bring About Easy and Safe Parturition," *NHH* 23, no. 7, whole no. 287 (July 1918): 642–47; Otto Carque, "Infantile Paralysis," ibid. 22, no. 1, whole no. 275 (Jan.–Feb. 1917): 553–60, at 553. On naturopathic babies, see, e.g., "The First Naturopathic Babies of New Mexico," *NP* 26, no. 2, current no. 436 (Feb. 1931): 45. Examples of naturopaths advocating for high proficiency in obstetrics include David C. Long, ND, DC, "Naturopathy and Obstetrics," *NHH* 40, no. 5 (May 1935): 137, 159, and United Press, "Births Safest in Homes, Noted Obstetrician Says," ibid. 41, no. 1, current no. 506 (Jan. 1937): 23. John A. Stevenson, "News, Views, and Opinions: Artificial Insemination Is Adultery!," *JNM*, Dec. 1959, 23.

27. Victor H. Lindlahr, *An Outline of Hygiene for Women* (Chicago: Professional Writers, 1928); Alice Chase, here identifying as an osteopathic physician and nutritionist, "Perhaps You Have a Health Question That Is Bothering You," *NHH* 49, no. 2, current no. 583 (Feb. 1944): 50; Kuhne, "New Science of Healing: Diseases of Women."

28. Gilbert Patten Brown, ND, DO, DC, LLD, "Troubles of the Cervical Glands," *Naturopath* 28, no. 3, whole no. 341 (Mar. 1923): 134–35; Wilbur F. Prosser, "The Vaginal Douche," ibid. 28, no. 5, whole no. 343 (May 1923): 242. Others spoke in its favor in the case of leucorrhea.

29. Kuhne, "New Science of Healing: Diseases of Women," 635, 639–40; E. K. Stretch, "Gynecology—Minus the Knife," *NHH* 20, no. 12, whole no. 266 (Dec. 1915): 761–63.

30. John Xinos, ND, "Cancer of the Breast," *NHH* 49, no. 9, current no. 590 (Sept. 1944): 279; Alice Chase, "Breast Cancer," *NP* 55, no. 2 (Feb. 1950): 15, 37.

31. H. W. Peterson, "Cancer of the Womb," *NHH* 49, no. 7, current no. 588 (July 1944): 224; E. W. Cordingley, AM, ND, "Fibroid Tumors of the Uterus," *NP* 52, no. 4, current no. 639 (May 1948): 268, 298–99, at 298.

32. Edward Earle Purinton, "Happiness in Marriage," *NHH* 19, no. 4, whole no. 246 (Apr. 1914): 205–8; Purinton, "Fatherhood, the New Profession," ibid. 20, no. 5, whole no. 259 (May 1915): 297–98; [Louisa Lust?], "Keeping a Husband," ibid. 3, no. 1, whole no. 25 (Jan. 1902): 43.

33. Reverend A.S., "What Married People Should Know," ibid. 3, no. 3, whole no. 27 (Mar. 1902): 112. See also J. B. Caldwell, "The Gospel in Married Life," ibid. 9,

no. 12, whole no. 107 (Dec. 1908): 371–77. Caldwell argued that sex for newlyweds for any reason except reproduction damaged women's health and weakened the marital bond.

34. On allopathic ideas about masturbation, see John S. Haller and Robin M. Haller, *The Physician and Sexuality in Victorian America* (Urbana: Univ. of Illinois Press, 1974); and Benedict Lust, ND, "Masturbation, Its Causes, Its Consequences and Its Cure by Natural Methods," *NHH* 9, no. 12, whole no. 107 (Dec. 1908): 369. The history of allopaths' (and others') ideas about masturbation can be found in Thomas Laqueur, *Solitary Sex: A Cultural History of Masturbation* (Brooklyn: Zone Books, 2004).

35. Linda Gordon, "Voluntary Motherhood: The Beginnings of Feminist Birth Control Ideas in the United States," in *Women and Health in America*, ed. Judith Walzer Leavitt (Madison: Univ. of Wisconsin Press, 1999), 253–68; J. Waterloo Dinsdale, MD, "Why Is Marriage So Often a Failure?," *NHH* 16, no. 6, whole no. 212 (June 1911): 1; E. Rosch, "The Abuse of the Marital Relation: Explaining the Origin of Most Chronic Diseases, Especially the Chronic Diseases of Man and Woman," ibid. 22, no. 4 (July–Aug. 1917): 193–96; "Normal Sexual Impotency," ibid. 16, no. 9, whole no. 215 (Sept. 1911): 571.

36. W. B. Konkle, MD, "Sex Ethics: A Medical Orientation," *NHH* 20, no. 5, whole no. 259 (May 1915): 292–93. Female reformers had been arguing since the 1830s that men's inability to control their sex drive was a weakness, as noted in Carroll Smith-Rosenberg, "Beauty, the Beast and the Militant Woman: A Case Study in Sex Roles and Social Stress in Jacksonian America," *American Quarterly* 23 (1971): 562–84. See also Ruth Rosen, *The Lost Sisterhood: Prostitution in America, 1900–1918* (Baltimore: Johns Hopkins Univ. Press, 1983); David M. Kennedy, *Over Here: The First World War and American Society* (New York: Oxford Univ. Press, 1980); and Astro, "War Babies," *NHH* 20, no. 9, whole no. 263 (Sept. 1915): 588.

37. E. Rosch, "The Abuse of the Marriage Relation," *HHN* 23, no. 3, whole no. 283 (Mar. 1918): 290. The concept of rape has changed over time, socially and legally. See Estelle B. Freedman, *Redefining Rape: Sexual Violence in the Era of Suffrage and Segregation* (Cambridge, MA: Harvard Univ. Press, 2013); advertisement, "Sidney C. Tapp's Books on the Sex of the Bible," *HHN* 23, no. 3, whole no. 283 (Mar. 1918): 291; and Max Heindel, "The Sacrament of Marriage," ibid. 23, no. 10, whole no. 290 (Oct. 1918): 834–35.

38. L. E. Eubanks, "Health and Sexual Vigor," *Naturopath* 28, no. 4, whole no. 342 (Apr. 1923): 175; El Lernanto, "The Concealed Sins and Wastes of Health of Suffering Humanity, from a Medical and Biblical Standpoint," ibid. 28, no. 6, whole no. 344 (June 1923): 289–94; S. T. Erieg, "Chastity," ibid. 31, no. 5, whole no. 379 (May 1926): 225; "Marriage and Love," *NP* 34, no. 9, current no. 419 (Sept. 1929): 412–13, 442; Mayen Waldo, "The Man Who Is No Older Than His Children," ibid., 431.

39. B. Barker-Benfield, in "The Spermatic Economy: A Nineteenth-Century View of Sexuality," *Feminist Studies* 1, no. 1 (Summer 1972): 45–74, one of the first to note that middle-class men's sexual restraint was believed to lead to their economic success; Carl Loeb, "The Male Climacteric or Change of Life," *NHH* 44, no. 10, current no. 539 (Oct. 1939): 302; advertisement, "NATURE'S CURE: The Only Cure for Male Impotency; The Erectruss," ibid. 19, no. 2, whole no. 244 (Feb. 1914); advertisement, "MALE IMPOTENCE," ibid. 28, no. 1, whole no. 339 (Jan. 1923): 37; Eubanks, "Health and

Sexual Vigor," 175; "Drug Is Found to Offset Effect of Age on Men," Clippings, *NHH* 44, no. 7, current no. 536 (July 1939): 211.

40. Carl Strueh, "Sexual Diseases," *NHH* 9, no. 8, whole no. 103 (Aug. 1908): 237–41; Louis Kuhne, "The New Science of Healing: Sexual Diseases," *HHN* 23, no. 3, whole no. 283 (Mar. 1918): 235–37; State News: California, "Fletcher's Syphilitic Bills," *NHH* 45, no. 1, current no. 542 (Jan. 1940): 30, 63; Benedict Lust, Dr. Lust Speaking, ibid. 44, no. 5, current no. 534 (May 1939): 130, 154–55.

41. "We Observe and Comment: Birth Rate Drops to 16.6 in Nation," *NHH* 42, no. 10, current no. 515 (Oct. 1937): 32; "Rhythmic Birth Control Okayed by Church Guild," ibid. 40, no. 1 (Jan. 1935): 28; Lillian R. Carque, "Self-Control," ibid. 45, no. 11, current no. 552 (Nov. 1940): 338.

42. See Elena Slade, "Planned Parenthood," *NP* 52, no. 9, current no. 644 (Oct. 1948): 508, 535; "Sterilization Flayed," *NHH* 42, no. 6, current no. 511 (June 1937): 192; and Margaret Sanger, *Women and the New Race* (New York: Brentanos, 1920).

43. Wilh. Logan, "The Cry of the Haunted," *NHH* 9, no. 8, whole no. 103 (Aug. 1908): 235; B. J. Jones, "Democracy or Autocracy—Which?," *HHN* 25, no. 7, whole no. 311 (Aug. 1920): 351.

44. "The Little Mothers," *NHH* 16, no. 9, whole no. 215 (Sept. 1911): 559; "Outing Notes," Out-Door Life, ed. Max E. Peltzer, ibid. 21, no. 1 (Jan. 1916).

45. Max Eugene Peltzer, "The Out-Of-Door Girl," ibid. 21, no. 1 (Jan. 1916): 54–56; Dennis, "Motherhood in Harmony with Nature."

46. Martin Unterweger, "Cosmetics," *NHH* 5, no. 11, whole no. 59 (Nov. 1904): 283–84.

47. See W. N. Boldyreff, "Tobacco Smoke," *NP* 34, no. 5, current no. 415 (May 1929): 220–21, 244; advertisement, "A New Skin: The Blemishes All Vanish!" ibid., 233; "Can Girls Afford to Smoke?," ibid. 36, no. 12, current no. 446 (Dec. 1931): 367; and Roy S. Ashton, DC, PhC, "The Mechanical Factor: No. 7, Antero-Posterior Distortions," *NHH* 44, no. 9, current no. 538 (Sept. 1939): 270–71, regarding damages wrought by high heels.

48. Marguerite Agniel, advertisement, "BODY BEAUTY: A Woman's Guide to Charm, Poise and Personality," *NHH* 49, no. 8, current no. 589 (Aug. 1944): 249; "What Is the Best Hair Coloring for You? An Economic Necessity," *NP* 51, nos. 7–8, current no. 630 (July–Aug. 1947): 414, 429; advertisement, "New Hormone Cream Helps You Regain School Girl Complexion," ibid. 52, no. 6, current no. 641 (July 1948): 392; W. Robert Keashen, "Dubs and Debs," ibid. 52, no. 8, current no. 643 (Sept. 1948): 456, 480, 482.

49. Advertisement, "NEW Youth-Giving Belt Reduces Waistline—Quickly," *NP* 3, no. 5 (May 1927): 253; advertisement, "Getting a Little Too Fat? Mariba Obesity Tea," ibid. 34, no. 3, current no. 413 (Mar. 1929): 137; advertisement, "Eat Candy and get SLIM," ibid. 3, no. 1 (Jan. 1927): 33, suggesting a solution to sugar cravings; Edward Earle Purinton, "Mind Reading," *Naturopath* 31, no. 7, whole no. 381 (July 1926): 341–42, at 341; Berte Harris, "Reduce while Eating Plenty," *NP* 36, no. 12, current no. 446 (Dec. 1931): 364–66, esp. 364, 366.

50. E. W. Cordingley, ND, "Why Married Women Become Stout," *NP* 51, no. 5, current no. 628 (May 1947): 269, 300–301, at 300. Obesity in men was also addressed, but not with the contempt shown toward fat women. "Obesity and Its Treatment," *JNM* 2 (Sept.–Oct. 1951): 15.

51. "The World's Strongest Woman," *Naturopath* 20, no. 3, whole no. 257 (Mar. 1915): 270.

52. See photograph, "Delegates of Florida State Society Convention at Tampa, Florida," ibid. 28, no. 5, whole no. 343 (May 1923): 247; photograph, American School of Naturopathy graduates for 1927, *NP* 32, whole no. 394 (Aug. 1927): 394; and State News: New Mexico, *NHH* 40, no. 10 (Oct. 1935): 309.

53. [Louisa Lust?], "Woman's Freedom and Purity," *NHH* 8, no. 8, whole no. 92 (1907): 270; Purinton, "Play," 173; Purinton, "Fatherhood," 297.

54. M. E. Yergin, MD, "The Passing of the M.D.," *HHN* 25, no. 2, whole no. 306 (Feb. 1920): 85; Herbert M. Shelton, "Will Long Skirts Save Our Boys?," *Naturopath* 28, no. 5, whole no. 343 (May 1923): 230; "Finding the Together-Way," *NHH* 30, no. 1 (Jan. 1925): 22; Otis G. Carroll, in Correspondence, *Naturopath* 30, no. 3, whole no. 365 (Mar. 1925): 272–74.

55. Edward Earle Purinton, "The Third Sex," *Naturopath* 31, no. 11, whole no. 385 (Nov. 1926): 554; Minna Beyer Madsen, ND, "Beauty Culture," *NP* 32, whole no. 394 (Aug. 1927): 384; "Change Men's Clothes," ibid. 36, no. 6, current no. 40 (June 1931): 190.

56. Louise Nedvidek, "Address of Welcome by the National Hostess of the 39th Annual Congress of the A.N.A.," *NHH* 40, no. 9 (Sept. 1935): 265, with photograph of the officers at the ANA annual convention in 1935 (Nedvidek was the convention host, and Dr. Ruth B. Drown delivered a key address); Friedhelm Kirchfeld and Wade Boyle, *Nature Doctors: Pioneers in Naturopathic Medicine* (East Palestine, OH: Buckeye Naturopathic, 1993), 226; "National Naturopathic Activities," *JN* 1 (Sept. 1957): 29; Lorna Mae Murray (president, NANP), "Unity and Cooperation," ibid., May 1957, 7.

57. "15th Annual Convention: Report of the International Society of Naturopathic Physicians," *JNM* 4 (July 1953): 8–12; Julius Moses, MD, "The Death Dance of Lubeck," *NP* 36, no. 8, current no. 442 (Aug. 1931): 229.

58. Enes Campanella (secretary-treasurer, Ladies Auxiliary of the NANP), "We will be with you at the National and International Naturopathic Convention," *JN* 1 (June 1957): 25; Betty Hedges (president), "Greetings to the Ladies of the Auxiliary of the AANP," *JANA* 7 (Aug. 1954): 11.

59. Enes Campanella (secretary, Ladies Auxiliary of the NANP-ISNP), "Ladies Auxiliary of the NANP-ISNP," *JN*, Oct. 1956, 27.

CHAPTER 6: Culture Wars

1. Susan E. Cayleff, *Wash and Be Healed: The Water-Cure Movement and Women's Health* (1987; repr., Philadelphia: Temple Univ. Press, 1992); John B. Blake, "Health Reform," in *The Rise of Adventism: Religion and Society in Mid-Nineteenth-Century America*, ed. Edwin S. Gaustad (New York: Harper & Row, 1974); George Cody, "History of Naturopathic Medicine," in *Textbook of Natural Medicine*, ed. Joseph Pizzorno and Michael Murray (New York: Churchill Livingstone, 1985), 7–37; Hans A. Baer, "The Potential Rejuvenation of American Naturopathy as a Consequence of the Holistic Health Movement," *MAQ* 13, no. 4 (1992): 369–83; Paul Starr, *The Social Transformation of American Medicine: The Rise of a Sovereign Profession and the Making of a Vast Industry* (New York: Basic Books, 1982), 99; E. Richard Brown,

Rockefeller Medicine Men: Medicine and Capitalism in America (Berkeley: Univ. of California Press, 1979).

2. *Proceedings of the National Medical Conventions held in New York, May, 1846, and in Philadelphia, May, 1847* (Philadelphia: T. K. & P. G. Collins for the AMA, 1847), 17; Brown, *Rockefeller Medicine Men*, 75; Michael Sappol, *A Traffic of Dead Bodies: Anatomy and Embodied Social Identity in Nineteenth-Century America* (Princeton, NJ: Princeton Univ. Press, 2002). For more on conditions for allopaths, see Richard Shryock, "Public Confidence Lost," in *The Development of Modern Medicine: An Interpretation of the Social and Scientific Factors Involved* (Philadelphia: Univ. of Pennsylvania Press, 1936), 241–64; Todd L. Savitt, *Medicine and Slavery: The Disease and Health Care of Blacks in Antebellum Virginia* (Urbana: Univ. of Illinois Press, 1978); Morris J. Vogel and Charles Rosenberg, eds., *The Therapeutic Revolution: Essays in the Social History of American Medicine* (Philadelphia: Univ. of Pennsylvania Press, 1979).

3. Joseph F. Kett, *The Formation of the American Medical Profession: The Role of Institutions, 1780–1860* (New Haven, CT: Yale Univ. Press, 1968); William J. Rothstein, *American Physicians in the 19th Century: From Sects to Science* (Baltimore: Johns Hopkins Press, 1972). On women's status, see Martin Kaufman, "The Admission of Women to 19th Century Medical Societies," *BHM* 50 (Summer 1976): 251–60; and Mary Roth Walsh, *Doctors Wanted: No Women Need Apply; Sexual Barriers in the Medical Profession, 1835–1975* (New Haven, CT: Yale Univ. Press, 1977). On public health and AMA institutionalization, see Barbara Gutman Rosenkrantz, *Public Health and the State: Changing Views in Massachusetts, 1842–1936* (Cambridge, MA: Harvard Univ. Press, 1972); "AAMC History," accessed 31 Aug. 2012, www.aamc.org /about/history/; Ernie Gross, *Advances and Innovations in American Daily Life, 1600–1930s* (Jefferson, NC: McFarland, 2002), 106–7; G. Markowitz and D. K. Rosner, "Doctors in Crisis: A Study of the Use of Medical Education Reform to Establish Modern Professional Elitism in Medicine," *American Quarterly* 25 (1973): 83–107; and James G. Burrow, *Organized Medicine in the Progressive Era: The Move toward Monopoly* (Baltimore: Johns Hopkins Univ. Press, 1977).

4. Barbara Ehrenreich and Deidre English, "Science and the Ascent of the Experts," in *For Her Own Good: 150 Years of the Experts' Advice to Women* (Garden City, NY: Doubleday/Anchor, 1978), 69–70; Frederick C. Waite, "The First Medical Diploma Mill in the United States," *BHM* 20 (Nov. 1946): 495–504.

5. "Total and Urban Population at Each Census," in U.S. Bureau of the Census, *Twelfth Census of the U.S. Taken in the Year 1900, Statistical Atlas* (Washington, DC: U.S. Census Office, 1903), 17, plate 16; *Statistical History of the United States from Colonial Times to the Present* (Stamford, CT: Fairfield, 1965); Cayleff, *Wash and Be Healed*, 169–70.

6. Paul Starr, "The Consolidation of Professional Authority, 1850–1920," in Starr, *Social Transformation*, 79–144; Morris Fishbein, *A History of the American Medical Association, 1847 to 1947* (Philadelphia: W. B. Saunders, 1947); Brown, *Rockefeller Medicine Men*, 74–75; Ehrenreich and English, *For Her Own Good*, 69–70.

7. Markowitz and Rosner, "Doctors in Crisis"; Starr, *Social Transformation*, 112–14; "To What Cause Are We to Attribute the Diminished Respectability of the Medical

Profession in the Estimation of the American Public?," *Medical and Surgical Reporter* 1 (1858): 141–43.

8. Susan E. Cayleff, "Self-Help and the Patent Medicine Business," in *Women, Health, and Medicine in America: A Historical Handbook*, ed. Rima D. Apple (New York: Garland, 1990), 311–36.

9. Abraham Flexner, *Medical Education in the United States and Canada: Bulletin Number Four* (1910; repr., New York: Carnegie Foundation, 1972); Carleton Chapman, "The Flexner Report," *Daedalus* 103 (1974): 105–17.

10. Cody, "History of Naturopathic Medicine," 14; Starr, *Social Transformation*, 127–41; F. Campion, *AMA and U.S. Health Policy since 1940* (Chicago: AMA, 1984), 467–73; Cayleff, "Self-Help."

11. Cayleff, "Self-Help," 322–24.

12. Louis S. Reed, *The Healing Cults: A Study of Sectarian Medical Practice; Its Extent, Cause, and Control* (Chicago: Univ. of Chicago Press, 1932); James Harvey Young, *The Toadstool Millionaires* (Princeton, NJ: Princeton Univ. Press, 1961); Sarah Stage, *Female Complaints: Lydia Pinkham and the Business of Women's Medicine* (New York: Norton, 1979).

13. Samuel Hopkins Adams, *The Great American Fraud: Articles on the Nostrum Evil and Quacks, in Two Series, Reprinted from Collier's Weekly* (New York: P. F. Collier & Son, 1906); Starr, *Social Transformation*, 131; Cayleff, "Self-Help," 323.

14. Austin Smith, "The Council on Pharmacy and Chemistry," in Fishbein, *History of the American Medical Association*, 876, as cited in Starr, *Social Transformation*, 131; American Medical Association, *New and Nonofficial Remedies, 1912: Containing Descriptions of the Articles Which Have Been Accepted by the Council on Pharmacy and Chemistry of the American Medical Association Prior to January 1, 1912* (Chicago: AMA, 1912).

15. Arrell Gibson, "Medicine Show," *American West* 4 (Feb. 1967): 39; Cayleff, "Self-Help," 324; David L. Dykstra, "The Medical Profession and Patent Proprietary Medicine during the Nineteenth Century," *BHM* 29 (Sept.–Oct. 1955): 413.

16. Samuel J. Thomas, "Nostrum Advertising and the Image of Woman as Invalid in Late Victorian America," *Journal of American Culture* 5, no. 3 (Fall 1982): 104–12, at 104; Susan E. Cayleff, "Prisoners of Their Own Feebleness: Women, Nerves and Western Medicine—A Historical Overview," *Social Science and Medicine* 26, no. 12 (1988): 1199–1208; Diane Price Herndl, *Invalid Women: Figuring Feminine Illness in American Fiction and Culture, 1840–1890* (Durham, NC: Univ. of North Carolina Press, 1993).

17. Cayleff, "Self-Help," 325; Stage, *Female Complaints*, 27; Sears, Roebuck & Company, "Drug Department," in *1897 Sears Roebuck & Co. Catalogue* (1897; repr., with an introduction by Nick Lyons, New York: Skyhorse, 2007), 26–52; David T. Courtwright, *Dark Paradise: Opiate Addiction in America before 1940* (Cambridge, MA: Harvard Univ. Press, 1982).

18. Starr, *Social Transformation*, 132–33.

19. Ibid., 133; Stewart H. Holbrook, *The Golden Age of Quackery* (New York: Macmillan, 1959), 12; Morris Fishbein, MD, *The New Medical Follies: An Encyclopedia of Cultism and Quackery in the United States with Essays on the Cult of Beauty, the Craze for Reduction, Rejuvenation, Eclecticism, Bread and Dietary Fads, Physical*

Therapy, and a Forecast as to the Physician of the Future (New York: Boni & Liveright, 1927); Arthur Kallet and F. J. Schlink, *100,000,000 Guinea Pigs: Dangers in Everyday Foods, Drugs and Cosmetics* (1933; repr., New York: Grosset & Dunlap, 1937); James Cook, *Remedies and Rockets: The Truth about Patent Medicines Today* (New York: Norton, 1958); Holbrook, *Golden Age of Quackery*.

20. A. A. Erz, "Official Medicine—As It Is—And What It Is Not," *NHH* 19, no. 7, whole no. 249 (July 1914): 431–35; "If Jesus Were on Earth Now What Would the Doctors Do with Him?," ibid. 20, no. 10, whole no. 264 (Oct. 1915): 610.

21. The naturopathic journals are filled with references to these groups, and their coalition with naturopaths is evidenced throughout this book.

22. Benedict Lust, "Opinions of Famous Physicians on Medical Science," *NHH* 9, no. 3, whole no. 98 (Mar. 1908): 77–78; Robert Bieri, ND, "Maxims for the Allopaths," ibid. 9, no. 10, whole no. 105 (Oct. 1908): 301.

23. Otto Carque, "The Rockefeller Institute of Medical Research: How the Oil King's Money Is Spent in a Useless Manner," ibid. 9, no. 5, whole no. 100 (May 1908): 144–52; Abraham Flexner, *Medical Education in the United States and Canada: Bulletin Number Four* (1910; repr., New York: Carnegie Foundation, 1972); Starr, *Social Transformation*, 119; Brown, *Rockefeller Medicine Men*, 145.

24. Starr, *Social Transformation*, 120; Benedict Lust, *Yungborn: The Life and Times of Dr. Benedict Lust, Founder and Father of Naturopathy, the Cornucopia of Alternative Medicine: A Memoir*, with a foreword by Dr. Joseph E. Pizzorno Jr., ND, President of Bastyr University and author of *Total Wellness* (n.p.: privately printed by Anita Lust Boyd, 1997), 58.

25. Chapman, "Flexner Report"; "Medical Education in the United States," *JAMA* 57 (9 Aug. 1911): 654–57; Starr, *Social Transformation*, 119, 120.

26. Chapman, "Flexner Report." On the distribution of funds, see Starr, *Social Transformation*, 120–21. Rockefeller provided $91 million, and the remainder came from other philanthropies. Brown, *Rockefeller Medicine Men*, 155. Cody, "History of Naturopathic Medicine," 14, discusses the closures. The experiences of women are detailed in Walsh, *Doctors Wanted*; "Women Medical Students," *Women's Medical Journal* 19 (Dec. 1909): 258; and "Medical Education in the United States," *JAMA* 59 (24 Aug. 1912): 650.

27. Gilbert Bowman, "Vaccination and Serum Therapy," Department of Medical Freedom, *HHN* 22, no. 6 (Nov.–Dec. 1917): 372–76, at 372.

28. Pieces by Lust calling for less government control include an editorial in ibid. 25, no. 2, whole no. 306 (Feb. 1920): front page; and "Back to the Wood-Shed," ibid., 69. Countless others joined him. See Gilbert Patten Brown, PhD, DO, ND, DC, DPOS, "Why Many Allopaths Are Criminals," ibid., 76; Per Nelson, ND, "Naturopathy versus Medicine," ibid., 78; and M. E. Yergin, MD, "The Passing of the M.D.," ibid., 85.

29. William F. Havard, "The New Party," *HHN* 25, no. 5, whole no. 309 (June 1920): 16. Pieces calling for less governmental control include Benedict Lust, editorial in ibid., 217; F. W. Collins, "Nomination of Dr. F. W. Collins for President of the U.S.," ibid., 221; "Platform Adopted by Constitutional Liberty League for Dr. F. W. Collins, Presidential Candidate on the American Drugless Platform," ibid., 221–22; and B. J. Jones, "Democracy or Autocracy—Which?," *HHN* 25, no. 7, whole no. 311 (Aug. 1920): 351.

30. Edward Earle Purinton, "Modern Sale of Health," *NP* 36, no. 4, current no. 438 (Apr. 1931): 100, 106; Benedict Lust, Dr. Lust Speaking, "Let the Poor Die," *NHH* 40, no. 3 (Mar. 1935): 67; Clippings, ibid. 44, no. 1, current no. 530 (Jan. 1939): 30, on the Tampa hospital; "Costs of Proposed Health Insurance," ibid., 32; I. S. Weinger, "The Basic Science Act: How Innocent It Looks—How Vicious It Is!," *NHH* 44, no. 5, current no. 534 (May 1939): 136; "Welfare Doctors Put on Fixed Fees," ibid. 44, no. 8, current no. 537 (Aug. 1939): 254; Chas A. Toll, PhT, State News: Illinois, ibid. 45, no. 10, current no. 551 (Oct. 1940): 317.

31. Benedict Lust, Dr. Lust Speaking, *NHH* 45, no. 3, current no. 544 (Mar. 1940): 66; William R. Lucas, DNPh, Mail Bag, ibid., 79; Clippings, *NHH* 42, no. 4, current no. 509 (Apr. 1937): 120.

32. B. H. Jones, "Letting Well Enough Alone" and "Believing We Have It All," Correspondence, *NHH* 20, no. 2, whole no. 256 (Feb. 1915): 130; Edward Earle Purinton, Efficiency in Drugless Healing, "The Importance of Good Letters," ibid. 20, no. 4, whole no. 258 (Apr. 1915): 205; Purinton, Efficiency in Drugless Healing, "Improving Health Correspondence," ibid. 20, no. 5, whole no. 259 (May 1915): 269; "Large Business," Correspondence, ibid. 20, no. 9, whole no. 263 (Sept. 1915): 553.

33. M. J. Rodermund, "Dr. Rodermund vs. The Columbus 'Medical Journal,'" ibid. 16, no. 6, whole no. 212 (June 1911): 353; Erz, "Official Medicine," 367.

34. Harriet A. Washington, "The Black Stork: The Eugenic Control of African American Reproduction," in *Medical Apartheid: The Dark History of Medical Experimentation on Black Americans from Colonial Times to the Present* (New York: Random House, 2006), 189–215, at 192; Friedhelm Kirchfeld and Wade Boyle, "Founder of Scientific Naturopathy: Henry Lindlahr (1862–1924)," in *Nature Doctors: Pioneers in Naturopathic Medicine* (East Palestine, OH: Buckeye Naturopathic, 1993), 227–50. Lindlahr received his drugless practitioner's license in 1902 and became a licensed MD in 1904. He owned the Nature Cure Publishing Company in Chicago. See review of *The Black Stork*, by Henry Lindlahr, Naturopathic Book Shelf, *HHN* 23, no. 2, whole no. 282 (Feb. 1918): 189. Naturopaths discredited allopaths by rejecting their clandestine practice of abortion. One reprinted article indicting allopaths was based on a 1912 piece in the Practical Medical Series. "Medical Ethics; a Pretense and a Sham," ibid. 23, no. 10, whole no. 290 (Oct. 1918): 836.

35. Benedict Lust, "Red Cross and Vivisection," *HHN* 23, no. 3, whole no. 283 (Mar. 1918): 219; Lust, "The Point of View," ibid. 23, no. 5, whole no. 285 (May 1918): 419; Harrison H. Lynn, "The Crime of the Century" (extracted), ibid. 441.

36. Harry W. Nelson, DN (president of the Illinois Naprapathic Association), "A Modern(?)Crusade," *NHH* 42, no. 10, current no. 515 (Oct. 1937): 293, 304.

37. B. Stanford Claunch, ND, "Was President Harding's Death Due to Medical Ignorance?," *Naturopath* 28, no. 9, whole no. 347 (Sept. 1923): 444; Benedict Lust, "The Drugger's Panic," *NP* 3, no. 3 (Mar. 1927): 109; Lust, "Millions of Tax Money Wasted on Medical Examinations," *Naturopath* 30, no. 5, whole no. 379 (May 1926): 213; Lust, "Medical Octopus," ibid. 29, no. 7 (July 1924): 615.

38. "A Health Library," *New York Times*, 19 June 1904.

39. "Held as Bogus Doctor: Starken Told Sleuth He Was Ill and Prescribed Oil and Whisky," ibid., 9 Oct. 1909.

40. "Find Dynamite Man in Ship Bomb Plot," ibid., 2 Nov. 1915.

41. Associated Press, "Connecticut Moves for Wide Exposure of Diploma Frauds," ibid., 19 Nov. 1923, 1.

42. "Move to Restrain Mclaughlin and Harris: Chiropractors and Naturopaths Want Commissioners to Let Them Practice in Peace," ibid., 3 July 1926; "Protests Licensing of Massage School: County Medical Society Says Political Pressure Won the Health Board's Approval; Wants a General Inquiry; Harris Condemns Charge of Outside Influence—Says He Will Back His Aide's Findings," ibid., 28 Dec. 1926.

43. "Those Who Heal without Drugs," ibid., 26 June 1932.

44. "Naturopath Sentenced: Practitioner without License Sent to Workhouse," ibid., 19 Oct. 1938.

45. Foster Hailey, "Defense Attacks Case of Bridges: Naturopath Contends That a Government Affidavit Came from a Paranoiac; Two Witnesses Assailed," ibid., 9 May 1941, 31; "Bridges Witness Sticks to Story: Naturopathic Physician Gives Only Series of Denials on Cross-Examination," ibid., 13 May 1941, 19.

46. Charles William Thompson, "Misguided Pride of Philadelphia: A City Blameworthy, Not for Excess but for Astigmatism," ibid., 16 Sept. 1923; "Naturopaths Re-Elect Dr. Lust," ibid., 17 June 1935; "News, Missouri," *NHH* 40, no. 6 (May 1935): 155; advertisement, "REPRESENTATIVES," ibid. 40, no. 3 (Mar. 1935): 93.

47. Poem, *NHH* 5, no. 5, whole no. 53 (May 1904): 110; "Medicine amongst the Ancient Hebrews," *Naturopath* 31, no. 7, whole no. 381 (July 1926): 350.

48. "Our Trip to the Fatherland," *Naturopath* 31, no. 12, whole no. 386 (Dec. 1926): 580–85; Martin H. Glynn, "The Part That German and German-Americans Have Played in the Making of the United States" (speech, Academy Hall, Albany, NY), ibid. 28, no. 1, whole no. 339 (Jan. 1923): 16–17.

49. *Oxford English Dictionary Online*, s.v. "eugenics," accessed 17 Nov. 2013, www .oxforddictionaries.com/us/definition/american_english/eugenics.

50. See Stefan Kuhl, *The Nazi Connection: Eugenics, American Racism and German National Socialism* (New York: Oxford Univ. Press, 1994); and "History of Genetics: Eugenics," Dartmouth University, Biological Sciences Department, accessed 28 June 2015, www.dartmouth.edu/~bio70/eugenics.html.

51. Advertisement, "PRACTICAL EUGENICS," *NHH* 20, no. 1, whole no. 255 (Jan. 1915); William Lienhard, "A Higher Civilization Movement," ibid. 20, no. 6, whole no. 255 (June 1915): 374–76, at 375.

52. M. E. Yergin, "Health Necessary to Universal Government and Peace," *Naturopath* 28, no. 8, whole no. 346 (Aug. 1923): 372–74. Yergin also advocated naturism, the promotion of nudity and strength building.

53. "European Youth Movement Aims at Regeneration of Society," *NP* 36, no. 3, current no. 437 (Mar. 1930): 71–72, 83, at 71; Ellis Chadbourne, MA, LLB, "In Defense of Modern Youth; The Revolt of Youth; The Return to Nature," ibid. 36, no. 7, current no. 441 (July 1931): 208–10; Chadbourne, "Youth Culture a Cultural Renaissance," ibid. 36, no. 10, current no. 444 (Oct. 1931): 300–303. Historians claim that the German Youth Movement emphasized habits that were not in line with the extremely conservative traditional customs and habits of the German people. Hitler Youth took on many aspects of the Bundische Jugend after 1933, when the National Socialist German Workers' Party (the Nazi Party) was established. The German Youth Movement had a complex relationship with Nazi politics. Some of its followers were

committed to destroying the Nazi regime; the anti-Nazi author Hans Scholl and Claus von Stauffenber were part of the failed plot to assassinate Hitler. While the German Youth Movement produced anti-Nazi activists, it also produced Adolph Eichmann, a member between 1930 and 1931, who became a major organizer of the Holocaust. Peter D. Stachura, *The German Youth Movement, 1900–1945: An Interpretive and Documentary History* (London: Macmillan, 1981).

54. Arne L. Suominen, DN, "Observing the German Methods," *NHH* 31, no. 2, current no. 507 (Feb. 1937): 40. See also, for example, President Franklin Roosevelt to Herbert Lehman (governor of New York), 13 Nov. 1935, www.fdrlibrary.marist.edu /archives/pdfs/holocaust.pdf. Many scholars have criticized Roosevelt for not taking more decisive action on behalf of European Jews.

55. Anonymous wireless to *New York Times*, 1937; "Hitler Called 'History's Greatest Physician' for Lighting Rays of Hope in the Despairing," Clippings, *NHH* 42, no. 9, current no. 514 (Sept. 1937): 262.

56. Lillian R. Carque, "Are We Multiplying Morons?," *NHH* 44, no. 6, current no. 535 (June 1939): 165, 184.

57. "Nazified Medicine: German Doctors Are Falling Back on Hitler's Eugenic Ideas," *New York Times*, 6 Dec. 1942.

58. Benedict Lust, "Attention—Veterans of World Wars I and II," *JANA* 2 (Sept. 1949): 20; "The True Solution of the Social Question," *NHH* 19, no. 8, whole no. 250 (Aug. 1914): 554; American Humanity League, "To the President of the United State: A Petition," ibid. 20, no. 8, whole no. 262 (Aug. 1915): 520–21.

59. The greatest number of articles, many of them by women NDs, recommended treatments for specific diseases, including, among countless others, nerve force, coughs, colds and chest complaints, heart disease, vitamin deficiency, diabetes, cancer, thyroid, blood conditions, sexual diseases, pneumonia, bladder issues, depression, obesity, eye disease, and the flu. See, e.g., Carl Strueh, "Sexual Diseases," *NHH* 9, no. 8, whole no. 103 (Aug. 1908): 237–41; Dorothy Dean Richardson, "Nerve Force," *Naturopath* 29, no. 7 (July 1924): 647; Alice M. Reinhold, ND, "Natural Treatment of Pneumonia," ibid., 628; Berte Harris, "Reduce while Eating Plenty," *NP* 36, no. 12, current no. 446 (Dec. 1931): 364–66; and Jesse Mercer Gehman, "Leucopenia—Its Nature and Correction," *NHH* 44, no. 9, current no. 538 (Sept. 1939): 262. International snippets began appearing ca. 1915, and by the 1920s there was an international news section.

60. J. W. Bush, DC, "The Sheep and the Goat," *NHH* 20, no. 5, whole no. 259 (May 1915): 279; Frances Benzecry, "What I Have Found Out as a Medical Detective?" [*sic*], *Ladies Home Journal* 32, no. 4 (1915): 14; Benedict Lust, "Letter to Ladies Home Journal by the American Naturopathic Association," and "Answer from the Ladies Home Journal," *NHH* 20, no. 6, whole no. 260 (June 1915): 385; Kirchfeld and Boyle, *Nature Doctors*, 189, 257, quoting Andre Saine, "Constitutional Food Intolerance" (lecture, NCNM, 8 Dec. 1983, audiotape).

61. Edward Earle Purinton, Efficiency in Drugless Healing, "The Ethics of Advertising," *NHH* 20, no. 7, whole no. 261 (July 1915): 401.

62. Naturopathic News, "Missionaries," *HHN* 23, no. 11, whole no. 291 (Nov. 1918): 900; Harriet Schwartz, Correspondence, ibid., 901.

63. Kirchfeld and Boyle, *Nature Doctors*, 202; Benedict Lust, Editorial Comment, "Things for Consideration at the Convention," *HHN* 25, no. 8, whole no. 312 (Sept. 1920): 374; Lust, "Gone! Never to Be Forgotten: A Report of the Great Convention," *NHH* 28, no. 9, whole no. 347 (Sept. 1923): 485; Naturopathic News, *Naturopath* 21, no. 6, whole no. 380 (June 1926): 286; Jean Guthrie, "A Letter to Dr. Lust," *NHH* 49, no. 5, current no. 586 (May 1944): 155.

64. A brief sampling of advertisements includes "DR. MASON'S LAXATIVE MALT FOOD," *NHH* 8, no. 5, whole no. 89 (June 1907): 10; "NATURE'S CURE: The Only Cure for Male Impotency," ibid. 19, no. 2, whole no. 244 (Feb. 1914); "WHY YOU SHOULD COME TO BURKE SANITARIUM (Sonoma County, CA)," *HHN* 25, no. 7, whole no. 311 (Aug. 1920): 357; Lust, *Universal Naturopathic Encyclopedia*, with vendors ranging from Benedict Lust to others in Philadelphia, Texas, Cuba, New Jersey, Wisconsin, Virginia, Los Angeles, New York, and Washington State, *Naturopath* 28, no. 2, whole no. 340 (Feb. 1923): 102; "BOOKS, MAGAZINES AND PAMPHLETS: World Reform, Naturology, Natural Life, Naturopathy, Nature Cure, Physiotherapy, etc." (all for sale by Benedict Lust), *HHN* 25, no. 7, whole no. 311 (Aug. 1920): 204; and "ANABOLIC FOOD PRODUCTS" (a Chicago company that advertised consistently over the decades), *NHH* 40, no. 6 (June 1935): 155.

65. "To the Members of the A.N.A. and Progressive Doctors of Every Kind," *NHH* 30, no. 9 (Sept. 1925): 866. The price of a yearly subscription was $3 within the United States and $4 internationally, that of a single copy, 25¢. Naturopathic News, ibid. 30, no. 1 (Jan. 1925): 68.

66. Masthead, *NP* 30, no. 1 (Feb. 1927); "Miss Clare Meyer and Dr. Joel H. Jackson," Correspondence, ibid. 3, no. 2 (Feb. 1927): 85; Correspondence, ibid. 30, no. 4 (Apr. 1927): 171.

67. Masthead, ibid. 31, no. 4, current no. 438 (Apr. 1931).

68. A. L. Allen, PhD, "Scientific Health Restoration," ibid. 51, nos. 7–8, current no. 630 (July–Aug. 1947): 429; Patrick Lackey, "How to Treat Chronic Illness," ibid. 52, no. 6, current no. 641 (1948): 362, 392; Irma Goodrich Mazza, "How to Grow Herbs," ibid., 370; Alice Chase, Your Health Problems, *NP* 51, nos. 7–8 (July–Aug. 1947): 407, calling saccharine a coal tar product; Samuel Goldenstein, MD, "Atomic Energy and Cure," in ibid. 51, no. 3, current no. 626 (Mar. 1947): 146 ("All of it points only to the destructive nature of atomic energy on life"); Elizabeth Hassell, "Next to the Angels," ibid. 52, no. 4, current no. 639 (May 1948): 304; "Cigarette Smoking," ibid. 52, no. 8, current no. 643 (Sept. 1948): 484; E. W. Cordingley, AM, MD, "Fibroid Tumors of the Uterus," ibid. 52, no. 4, current no. 639 (May 1948): 268, 298; Benard C. Gindes, DC, ND, "Alcoholism Can Be Cured," ibid. 52, no. 8, current no. 643 (Sept. 1948): 464, 474.

69. K. Weitzel, "So You're Going On Fifty!," ibid. 51, no. 5, current no. 628 (May 1947): 314–15; Paul C. Bragg (author, lecturer, and diet adviser to Hollywood movie stars), "The New Science of Healthier, Longer Life for Those Past 40," ibid. 51, nos. 7–8, current no. 630 (July–Aug. 1947): 390; "'Prostatitis?' Electronic Stimulator with Soothing Heat and Dilation," ibid. 52, no. 6, current no. 641 (July 1948): 392; "Oster-Stim-U-Lax-Junior" (which attached to the hand for use on the scalp, face, and muscles), ibid., 354. For a fascinating investigation of the use of vibrators supposedly for medical

purposes and their impact on women's sexuality, see Rachel P. Maines, *The Technology of Orgasm: "Hysteria," the Vibrator, and Women's Sexual Satisfaction* (Baltimore: Johns Hopkins Univ. Press, 1999). Resort advertisements include "First Naturist Resort and Colony, Inc. Santa Fe, NM and Isles of Pines, Cuba, Bergholz Health Institutes, Inc. Milwaukee, WI," *NP* 52, no. 4, current no. 639 (May 1948): 303; and "Hygiology Country Club in Rhinebeck NY," ibid. 52, no. 8, current no. 643 (Sept. 1948): 489.

70. Advertisement, "What Is the Best Hair Color for You? An Economic Necessity," *NP* 52, no. 6, current no. 641 (July 1948): 392; "New Hormone Cream Helps You Regain School Girl Complexion," ibid.; Alice Chase, Your Health Problems, *NP* 51, nos. 7–8, current no. 630 (July–Aug. 1947): 407, advising against drinking salt water or taking salt pills to replace lost body water; advertisement for salt tablets, "KOOLEEZ: Don't Risk Heat Cramps and Exhaustion," ibid., 429; advertisement, "Health Schools: Learn at Home," *NP* 51, no. 3, current no. 626 (Mar. 1947): 319; advertisement, "Learn at Home: Institute of Drugless Therapy [Tama, IA]," ibid. 52, no. 4, current 639 (Apr. 1948): 303; advertisement, "Home Correspondence Course by the Late Dr. Joe Shelby Riley," ibid. 52, no. 8, current no. 643 (Sept. 1948): 489.

71. The *NHH* cost $5 yearly and 50¢ per issue; *NP* (1948) cost half that annually, down from $5 in 1947. Charles Silk, "Results," *NHH* 49, no. 5, current no. 586 (May 1944): 139; Benedict Lust, Dr. Lust Speaking, ibid. 49, no. 1, current no. 582 (Jan. 1944): 22. On possible changes to the *NHH*, see Lust, Dr. Lust Speaking, ibid. 48, no. 4, current no. 581 (Oct. 1943): 99.

CHAPTER 7: Medical Monsters

1. R. E. Brandman, "A Message to the People: Practical Ideals Embodied in Naturopathy," *NHH* 19, no. 3, whole no. 245 (Mar. 1914): 149.

2. René Descartes, *Discourse on the Method of Rightly Conducting One's Reason and of Seeking Truth in the Sciences* (1637; repr., Provo, UT: Renaissance Classics, 2012); "Perils of Vaccination—'A Disease,'" *Naturopath* 28, no. 12, whole no. 350 (Dec. 1923): 745; Maud S. Weeks, "Why the Church Should Oppose Vivisection," ibid. 49, no. 12, current no. 593 (Dec. 1944): 375.

3. As early as 1663 Robert Hooke, a fellow of the Royal Society in Britain, took a conscious dog, "cut away its ribs and diaphragm, severed its trachea so the nose of a bellows could be inserted, and blew air into the lungs to keep the animal live." This procedure demonstrated that the motion of lungs was not essential to respiration. Hooke's performance gained his colleagues' appreciation, and numerous colleagues would perform the procedure over the next thirty years. Macdonald Daly, "All Heaven in a Rage," *History Today* 37 (May 1987): 7–9, at 7. The ascendance of vivisection as an experimental school was driven by the publication of Burdon Sanderson's *Handbook for the Physiological Laboratory* in 1873. Stewart Richards, "Drawing the Life-Blood of Physiology: Vivisection and the Physiologists' Dilemma, 1870–1900," *Annals of Science* 43 (1986): 27–56, at 27.

4. Daly, "All Heaven in a Rage," 9.

5. Mark Twain, "A Dog's Tale," *Harper's Monthly*, Dec. 1903.

6. Daniel Engber, "Pepper Goes to Washington: The Most Important Animal Welfare Law in America Began with a Stolen Dog," 3 June 2009, www.slate.com /articles/health_and_science/pepper/2009/06/wheres_pepper.html.

7. Benedict Lust, "Vivisection and What It Really Is," *NHH* 3, no. 6, whole no. 30 (June 1902): 245; Lust, "Why We Oppose Vivisection," ibid. 5, no. 6, whole no. 54 (June 1904): 121–25; [Lust?], "A Plea for Our Second Best Friends, the Dumb Animals," ibid. 9, no. 10, whole no. 105 (Oct. 1908): 301.

8. Benedict Lust, Dr. Lust Speaking, "Return to Nature; Vivisection," ibid. 40, no. 10 (Oct. 1935): 291; James H. Doyle, "Vivisection," ibid. 45, no. 4, current no. 545 (Apr. 1940): 103; "The Meaning of Vivisection," ibid. 44, no. 6, current no. 535 (June 1939): 171.

9. George Allen White, "Majorities and Vivisection," *NHH* 19, no. 4, whole no. 246 (Apr. 1914): 209–13.

10. "On Animal Food," *NHH* 16, no. 11, whole no. 27 (Nov. 1911): 731; Herman Schildkraut, ND, "Carried to Slaughter," ibid. 25, no. 9 (Sept. 1920): 392; Alice Chase, "Correct Feeding for Healthful Living," *NP* 52, no. 9, current no. 644 (Oct. 1948): 503, 523, at 503; Jesse Mercer Gehman, "The Fetish of Meat," ibid. 51, no. 3, current no. 626 (Mar. 1947): 192.

11. Wilh. Logan, "The Cry of the Haunted," *NHH* 9, no. 8, current no. 103 (Aug. 1908): 235.

12. Craig Buettinger, "Antivivisection and the Charge of Zoophil-psychosis," *Historian* 55 (Winter 1993): 281.

13. Craig Buettinger, "Women and Antivivisection in Late Nineteenth-Century America," *Journal of Social History* 30 (Summer 1997): 421, 437, 857–72, at 862. Anna Sewell's book also detailed the overwork, physical abuse, underfeeding, grooms' neglect, and debilitating harnesses that destroyed horses' health and temperaments. Sewell, *Black Beauty: His Grooms and Companions; The Autobiography of a Horse* (London: Jarrold & Sons, 1877). See also Coral Lansbury, "Gynecology, Pornography and the Antivivisection Movement," *Victorian Studies* 28 (Spring 1985): 413–37.

14. Buettinger, "Antivivisection and the Charge of Zoophil-psychosis," 278.

15. Buettinger, "Women and Antivivisection," 857–69.

16. Ibid., 864; "Vivisection Valueless," *NHH* 45, no. 9, current no. 550 (Sept. 1940): 288.

17. Cartoon, "The Dream of the Vivisectionist," *NHH* 19, no. 6, whole no. 248 (June 1914): 402.

18. Weeks, "Why the Church Should Oppose Vivisection," 375; "Vivisection Valueless," 288; Caryl B. Abbott, "Save Your Pet from Torture," *NHH* 49, no. 7, current no. 588 (July 1944): 201, 206; "American Medical Liberty League Notes," ibid. 44, no. 8, current no. 537 (Aug. 1939): 243; "How About 'Driving' That Horse 'to Death,'" ibid. 25, no. 10, whole no. 314 (Nov.–Dec. 1920): 499.

19. Caroline Earle White, "Protecting Animals: A Nineteenth-Century Woman's Take," National Museum of Animals and Society, accessed 4 Nov. 2013, www .museumofanimals.org/#/caroline-earle-white/3788770; Susan Pearson, *The Rights of the Defenseless: Protecting Animals and Children in the Gilded Age* (Chicago: Univ. of Chicago Press, 2011).

20. Buettinger, "Antivivisection and the Charge of Zoophil-psychosis," 277, 285, quoting from *New York Times*, 11 July 1909, 6.

21. Susan Orlean, *Rin Tin Tin: The Life and the Legend* (New York: Simon & Schuster, 2011), 65–79; "About the Iditarod: The Spirit of Alaska! More Than a Race . . . ," accessed 19 Dec. 2011, http://iditarod.com/about/.

22. S. S. Rather, "A Dog That Died of a Broken Heart," *Naturopath* 30 (Mar. 1925): 227.

23. Edward Earle Purinton, "Animals Our Teacher," *NP* 3, no. 3 (Mar. 1927): 118–19; Naturopathic Veterinarian, *NHH* 3, no. 5, whole no. 29 (May 1902): 236–37; "If Horses Ate White Bread," *NP* 52, no. 4, current no. 639 (May 1948): 295; Jesse Mercer Gehman, "Building Good Animals vs. Good Humans," ibid. 51, no. 5, current no. 628 (May 1947): 266.

24. Doyle, "Vivisection," 103; "Anti-Vivisection Movement," *NHH* 9, no. 9, whole no. 104 (Sept. 1908): 272; "The Effects of Fright on Paralyzed Monkeys Studied," ibid. 44, no. 5, current no. 534 (May 1939): 155.

25. "Boston Paper Exposes Dog-Thieving Ring," *NHH* 44, no. 1, current no. 530 (Jan. 1939): 28; "Dogs Wanted at Once," ibid. 45, no. 12, current no. 553 (Dec. 1940): 367.

26. Abbott, "Save Your Pet from Torture," 201, 206.

27. Adelaide Hecker, "Facts for Anti-Vivisectionists to Remember," *NHH* 49, no. 4, current no. 585 (Apr. 1944): 111; *New York Times*, 2 Oct. 1943; Abbott, "Save Your Pet from Torture," 201.

28. Mrs. Stephanie Chez, "Will the Bronx Zoo Become a Hell?," *NHH* 49, no. 10, current no. 591 (Oct. 1944): 293, 316, at 293; Steve G. A. Cushing, "Must 'Science' Torture Helpless Animals?," ibid. 49, no. 11, current no. 593 (Nov. 1944): 334, 352, at 342; Old Doc Cherokee, "How to Treat a Cow," *Naturopath* 1 (Nov. 1963): 3.

29. Hecker, "Facts for Anti-Vivisectionists to Remember," 73.

30. H. B. Anderson (secretary, Citizens Medical Reference Bureau), "Wagner Health Bill," *NHH* 44, no. 8, current no. 537 (Aug. 1939): 228, 250, at 228. For the history of the Wagner-Murray-Dingell Bill, see Samuel W. Bloom, *The Word as Scalpel: A History of Medical Sociology* (New York: Oxford Univ. Press, 2002), 61, 95, 117; and Karen S. Palmer, MPH, MS, "A Brief History: Universal Health Care Efforts in the US" (paper, meeting of the Physicians for a National Health Program, 1999), accessed 2 July 2012, www.pnhp.org/facts/a-brief-history-universal-health-care-efforts-in-the-us. See also "The Wagner-Murray-Dingell Bill," *NHH* 48, no. 4, current no. 581 (Oct. 1943): 102–3.

31. H. P. Garden, "Why I Thank God I Am an Antivivisectionist," *NHH* 49, no. 5, current no. 586 (May 1944): 134, 150, 158, at 134 and 150.

32. Andrew N. Rowan and Franklin M. Loew, "Animal Research: A Review of Developments, 1950–2000," in *The State of Animals*, ed. D. J. Salem and A. N. Rowan (Washington, DC: Humane Society, 2001), 111–20.

33. Hearst was sued for libel in 1950 by a veterinarian who claimed that five articles had slandered him and others who favored vivisection. The case was dismissed because the veterinarian had not been named or pictured in the articles. *Brewer v. Hearst Pub. Co. et al.*, no. 10174, US Court of Appeals, 7th Cir., 1 Dec. 1950, http://openjurist.org/185/f2d/846/brewer-v-hearst-pub-co; Rowan and Loew, "Animal Research," 115.

34. Rowan and Loew, "Animal Research," 113.

35. Engber, "Pepper Goes to Washington." Pepper's law was passed on the same day that the Voting Rights Act was passed and that the Senate agreed to add Medicare as Title XVII of the Social Security Act, yet Pepper's story and the resultant law created a media sensation and gained more public recognition in the months that followed than either of those other two historical acts.

36. Rowan and Loew, "Animal Research," 113.

37. Ibid. A series of studies generated by the National Science Board found that public support for animal testing had declined from 63% in 1985 to 50% in 1996.

38. Ibid., 113–16. I worked in medical humanities at the University of Texas Medical Branch in Galveston from 1983 to 1987. I was a nonscientific member of the Institutional Animal Care and Use Committee, which ensured cage size that allowed for standing and turning, regular feces removal, limits to the number of times an animal could be used in experiments, and humane killing techniques for all mammals. We monitored noise levels and animal distress, as observed from escape behavior, call sounding, and other fear-based behavior. At that time the medical center was physically connected to the city animal pound, and the transfer of animals from one to the other was rumored to be routine. On the USDA crackdown, see National Research Council of the National Academies, "Use of Dogs and Cats in Research: Public Perception and Evolution of Laws and Guidelines," in *Scientific and Humane Issues in the Use of Random Source Dogs and Cats in Research* (Washington, DC: National Academies Press, 2009) accessed 2 July 2012, www.nap.edu /openbook.php?record_id=12641&page=31. Since 1987 a quarter of the states have had bills introduced to end animals' use in education. In this charged cultural context, research scientists responded with efforts to protect research facilities against vandalism and break-ins; thousands of stray dogs are used as research animals each year, down from half a million in the late 1960s. Currently advocates work under the AWA to protect rats, mice, and birds, to improve primates' psychological well-being, and to guarantee adequate exercise for dogs. The USDA is being forced to comply through court rulings and under amendments to the Farm Bill, S. 714 and H.R. 1280.

39. Institute for Laboratory Animal Research, "Use of Dogs and Cats in Research"; Rowan and Loew, "Animal Research," 115–18.

40. American Anti-Vivisection Society website, accessed 6 Jan. 2014, www.aavs.org.

41. Arthur Allen, *Vaccine: The Controversial Story of Medicine's Greatest Lifesaver* (New York: Norton, 2007), 14.

42. Ibid., 46–47, 49–50; Donald R. Hopkins, *The Greatest Killer: Smallpox in History* (Chicago: Univ. of Chicago Press, 2002).

43. Allen, *Vaccine*, 27–44; Kevin M. Malone and Alan R. Hinman, "Vaccination Mandates: The Public Health Imperative and Individual Rights," in *Law in Public Health Practice*, ed. Richard A. Goodman, 2nd ed. (Oxford: Oxford Univ. Press, 2007), www.cdc.gov/vaccines/imz-managers/guides-pubs/downloads/vacc_mandates_chptr13 .pdf.

44. The first laws requiring compulsory vaccination were passed in England in 1853. In response, the Anti-Compulsory Vaccination League was formed in 1866, followed by the Anti-Vaccination League in 1896. See "Smallpox and Anti-vaccination Leagues in England," in "History of Anti-Vaccination Movements," The History of Vaccines: A Project of the College of Physicians of Philadelphia, accessed 4 July 2012, www.history ofvaccines.org/content/articles/history-anti-vaccination-movements; "The Public Health Movement," doc. 7 of the online exhibit "From Quackery to Bacteriology: The Emergence of Modern Medicine in 19th Century America," University of Toledo Libraries, accessed 4 Dec. 2013, www.utoledo.edu/library/canaday/exhibits/quackery /quack7.html.

45. Allen, *Vaccine*, 51, 60; Kathleen S. Swendiman, "Mandatory Vaccinations: Precedent and Current Laws," Congressional Research Service report, 24 Feb. 2011, www.fas.org/sgp/crs/misc/RS21414.pdf, pg. 1.

46. R. M. Wolfe and L. K. Sharpe, "Anti-Vaccinationists Past and Present," *British Medical Journal* 325 (Aug. 2002): 430–32.

47. Herbert Harvey, DO, DC, ND, "Is Fever a Destructive or Constructive Process?," *NHH* 20, no. 7, whole no. 261 (July 1915): 434–36, at 435.

48. [Benedict Lust?], "Influenza in Children," *Naturopath* 30, no. 9 (Sept. 1925): 843; Alice M. Reinhold, ND, "Natural Treatment of Pneumonia," ibid. 29, no. 7 (July 1924): 628.

49. A. A. Erz, "Medical Laws vs. Human Rights Constitution," *NHH* 19, no. 1, whole no. 243 (Jan. 1914): 17–25, at 17, 22, 25; H. E. Bliler, "Vaccination Is a Gigantic Delusion," *HHN* 25, no. 6, whole no. 310 (July 1920): 285; "Who Are the Quacks?," *NHH* 44, no. 12, current no. 541 (Dec. 1939): 362, 376, at 376.

50. On these two movements, see Lawrence Goodwyn, *The Populist Moment: A Short History of the Agrarian Revolt in America* (New York: Oxford Univ. Press, 1978); and Michael McGerr, *A Fierce Discontent: The Rise and Fall of the Progressive Movement in America, 1870–1920* (New York: Oxford Univ. Press, 2005).

51. George Rosen, "The Evolution of Social Medicine," in *Handbook of Medical Sociology*, ed. Howard Freeman, Sol Levine, and Leo G. Reeder, 3rd ed. (Englewood Cliffs, NJ: Prentice Hall, 1979), 166.

52. Robert Weeks DeForest and Lawrence Veiller, "The Tenement House Problem," in *The Tenement House Problem: Including the Report of the New York State Tenement House Commission of 1900*, ed. DeForest and Veiller (New York: Macmillan, 1903), 3–6, 10.

53. J. T. Robinson, MD, "Preventive and Cure for Epidemic Disease," *NHH* 19, no. 11, whole no. 253 (Nov. 1914): 748–51, at 749–50; "Man's Natural Resistance or Immunity," ibid. 42, no. 8, current no. 513 (Aug. 1937): 244.

54. John Duffy, *The Sanitarians: A History of Public Health* (Urbana-Champaign: Univ. of Illinois Press, 1990), 93–109. In 1811 the first national meeting to establish quarantine laws was held. See Albert H. Buck, ed., *A Treatise on Hygiene and Public Health* (New York: William Wood, 1879); and Bloom, *Word as Scalpel.*

55. Mark Bostridge, *Florence Nightingale: The Making of an Icon* (New York: Farrar, Straus & Giroux, 2008); Penny Colman, *Breaking the Chains: The Crusade of Dorothea Lynde Dix* (1992; repr., New York: American Society of Journalists and Authors, 2007); Stephen B. Oates, *Woman of Valor: Clara Barton and the Civil War* (New York: Free Press, 1994). Creation of the US Sanitary Commission was aided by the efforts of Dr. Elizabeth Blackwell, the first woman to receive an MD degree in the United States. John Duffy, "Public Health and the Civil War" and "The Institutionalization of Public Health," in *Sanitarians*, 110–25 and 126–37; *The Medical and Surgical History of the War of the Rebellion*, Prepared under the Direction of Surgeon General United States Army, Joseph K. Barnes (Washington, DC: GPO, 1870–88).

56. Ernie Gross, *Advances and Innovations in American Daily Life, 1600–1930s* (Jefferson, NC: McFarland, 2002), 108, 113; Allen, *Vaccine*, 72.

57. Allen, *Vaccine*, 97; Michael Willrich, *Pox: An American History* (New York: Penguin, 2011); Willrich, "How the 'Pox' Epidemic Changed Vaccination Rules,"

interview by Terry Gross, National Public Radio, 5 Apr. 2011, www.kqed.org/news/story/2011/04/05/49461/how_the_pox_epidemic_changed_vaccination_r.

58. "An Innocent Victim of That Medical Superstition, Vaccination," *NHH* 16, no. 5, whole no. 211 (May 1911): 312; "State Bureau for the Protection of Immigrants," ibid., 314.

59. Porter F. Cope (secretary, Anti-Vaccination League of America, Philadelphia), "Vaccination Disasters," *NHH* 19, no. 11, whole no. 253 (Nov. 1914): 272; A. A. Erz, "Official Medicine—As It Is—And What It Is Not: An Experience," ibid. 19, no. 12, whole no. 254 (Dec. 1914): 820; "Death after Vaccination," *Naturopath* 28, no. 4, whole no. 342 (Apr. 1923): 198; "Here Is What G. B. Shaw Thinks of Vaccination," *NP* 36, no. 9, current no. 443 (Sept. 1931): 270–71, at 270.

60. Willrich, "How the 'Pox' Epidemic Changed Vaccination Rules"; R. Swinburne Clymer, PhD, MD, "The Vaccination Curse," *NHH* 3, no. 4 (Apr. 1902): 161.

61. *Jacobson v. Com. of Massachusetts*, 197 U.S. 11 (1905), 197 U.S. 11, *Henning Jacobson, Plff. in Err., v. Commonwealth of Massachusetts*, no. 70, accessed 2 May 2012, http://caselaw.lp.findlaw.com/scripts/getcase.pl?court=us&vol=197&invol=11; Samuel H. Williamson, "Seven Ways to Compute the Relative Value of a U.S. Dollar Amount, 1774 to present," MeasuringWorth, 2015, accessed 26 June 2015, www.measuringworth.com/uscompare/; Allen, *Vaccine*, 98; Wendy K. Mariner, George J. Annas, and Leonard H. Glantz, "*Jacobson v Massachusetts*: It's Not Your Great-Great-Grandfather's Public Health Law," *American Journal of Public Health* 95 (2005), 583–84; Wendy E. Parmet, Richard A. Goodman, and Amy Farber, "Individual Rights versus the Public's Health—100 Years after *Jacobson v. Massachusetts*," *New England Journal of Medicine* 352 (Feb. 2005): 652; Allen, *Vaccine*, 60; Judith Walzer Leavitt, "Typhoid Mary: Villain or Victim?," "NOVA" website, created 12 Oct. 2004, accessed 4 Aug. 2014, www.pbs.org/wgbh/nova/body/typhoid-mary-villain-or-victim.html.

62. David J. Chalmers, *Hooded Americanism: The History of the Ku Klux Klan* (Durham, NC: Duke Univ. Press, 1987), for the conditions of the 1920s; Allen, *Vaccine*, 82, 99.

63. "Lockjaw from Vaccination," *NHH* 9, no. 4, whole no. 99 (Apr. 1908): 115; Frances B. Livesay, "Insomuch as Vaccination Cannot Be Abolished, Public Schools Must Be," ibid. 16, no. 10, whole no. 216 (Oct. 1911).

64. "Syphylis and Vaccination," ibid. 9, no. 5, whole no. 100 (May 1908): 143; Alfred Bolton, "An Anti-Vaccination Law," ibid. 9, no. 9, whole no. 104 (Sept. 1908): 272; "Vaccination a Curse," ibid. 9, no. 4, whole no. 99 (Apr. 1908): 108; Walter E. Elfrink, "Anti-Compulsory Vaccination," ibid. 9, no. 3, whole no. 98 (Mar. 1908): 95.

65. [Benedict Lust?], "State Medical Imposition," ibid. 9, no. 8, whole no. 103 (Aug. 1908): 246–47; "How Vaccination Kills," ibid. 9, no. 9, whole no. 104 (Sept. 1908): 289; Mary E. Calbour, "Another Victim of That Legalized Modern Superstition Vaccination," ibid. 16, no. 1, whole no. 207 (Jan. 1911): 44; H. Molenar, "The Vaccination Pest," ibid. 16, no. 2, whole no. 208 (Feb. 1911): 91.

66. "Why Medical Serumtherapy Is a Failure and a Detriment to Humanity," ibid. 19, no. 5, whole no. 247 (May 1914): 321–22; C. Oscar Beasley, "Vaccination," *HHN* 23, no. 1, whole no. 281 (Jan. 1918): 69.

67. Gilbert Patten Brown, PhD, DO, ND, DC, DPOS, "Why Many Allopaths Are Criminals," *HHN* 25, no. 2, whole no. 302 (Feb. 1920): 76–77, at 76; J. W. Hodge, "How a Physician Lost His Faith in Vaccination," *NHH* 19, no. 7, whole no. 249 (July 1914): 418.

68. Harry Weinberger, "Vaccination and the Law," *NHH* 19, no. 6, whole no. 248 (June 1914): 393–95; "Emma Goldman and A. Berkman behind the Bars," *New York Times*, 16 June 1917, http://ucblibrary3.berkeley.edu/goldman/Writings/Accounts /NYT61617.html; James Keith Colgrove, *State of Immunity: The Politics of Vaccination in Twentieth-Century America* (Berkeley: Univ. of California Press, 2006), 30; New York Supreme Court Appellate Division, *Reports of Cases Heard and Determined in the Appellate Division of the Supreme Court of the State of New York, Vol.* 11 (n.p.: Banks, 1897), 10.

69. F. Antonius, "Anti-Vaccination History in California," *NHH* 16, no. 7, whole no. 213 (July 1911): 429; Hecker, "Facts for Anti-Vivisectionists to Remember," 73, 95, at 95.

70. H. B. Anderson, Anti-Vaccination Reports, *NHH* 48, no. 1, current no. 579 (Jan. 1943): 25, quoting John A. Toomey, MD, "Active Immunity—Preventions against Smallpox, Diphtheria, Whooping Cough, Tetanus and Typhoid," *JAMA* 119 (2 May 1942). See also Anderson, *The Facts against Compulsory Vaccination* (New York: Citizens Medical Reference Bureau, 1929), at 2, 225; and H. A. Swenson, ND, "Diagnosis and Prognosis," *NHH* 40, no. 2 (Feb. 1935): 40.

71. J. W. Hodge, MD, "Cupidity, Invincible Ignorance and Credulity, the Bases of Professional Faith in Vaccination," *NHH* 16, no. 1, whole no. 207 (Jan. 1911): 40; Benedict Lust, Dr. Lust Speaking, "Medical Greed," ibid. 40, no. 7 (July 1935): 194; Jesse Mercer Gehman, "The Synthetic Immunization Craze," ibid. 42, no. 8, current no. 513 (Aug. 1937): 243–44, at 243. The journal had changed names twice since the 1920s, becoming the *Naturopath*, then *Nature's Path*, which was directed at a more popular audience. The return to the title *Naturopath and Herald of Health* is evidenced in this 1937 piece.

72. Morris A. Bealle, "Compulsory Vaccination of Boy Scouts," ibid. 42, no. 4, current no. 509 (Apr. 1937): 124, reprinted from *Plain Talk* magazine, 12 Jan. 1937; E. W. Cordingley, "Compulsory Vaccination of Boy Scouts," correspondence with *Plain Talk* magazine, reprinted in *NHH* 42, no. 4, current no. 509 (Apr. 1937): 124; "Boy Scouts Reply," ibid. 42, no. 5, current no. 510 (May 1937): 140.

73. H. D. Scanlan, State News, "The Compulsory Vaccination Bill," *NHH* 48, no. 1, current no. 578 (Jan. 1943): 30; [H. B. Anderson?], Anti-Vaccination News, ibid. 48, no. 2, current no. 579 (Apr. 1943): 55, 57, at 55.

74. "Hydrophobia of Dogs," ibid. 9, no. 10, whole no. 105 (Oct. 1908): 299; "Perils of Vaccination—'A Disease,'" *Naturopath* 28, no. 12, whole no. 350 (Dec. 1923): 745.

75. Peter Schrag, *Not Fit for Our Society: Immigration and Nativism in America* (Berkeley: Univ. of California Press, 2010).

76. Jessica Wang, "'He growled and snapped like a cur': Rabies and Animal Spirit Possession in 19th-Century America" (paper, annual meeting of American Association for the History of Medicine, Philadelphia, 1 May 2011); Michael Worboys, "'As if they had seen spooks': Canine Hysteria in the 1920s and 1930s," ibid.; John S. Haller and Robin M. Haller, *The Physician and Sexuality in Victorian America* (Urbana: Univ. of Illinois Press, 1974); S. Weir Mitchell, MD, "Rest in the Treatment of Nervous Disease," in *A Series of American Clinical Lectures*, ed. E. C. Seguin, MD, vol. 1 (New York: G. P. Putnam & Sons, 1875); Ellen L. Bassuk, "Rest Cure: Repetition or Resolution of Victorian Women's Conflicts?," in *The Female Body in Western Culture: Contempo-*

rary Perspectives, ed. Susan Rubin Suleiman (Cambridge, MA: Harvard Univ. Press, 1986).

77. J. W. Hodge, "Not a Single Case of Rabies Found among 56,000 Stray Dogs," *NHH* 16, no. 9, whole no. 215 (Sept. 1911): 554; R. H. Reid, "The Rabies Myth," ibid. 49, no. 8, current no. 589 (Aug. 1944): 233, 254, at 233.

78. Samuel Rud Cook, Lit B, ND, in *Naturopath* 22, no. 3, whole no. 277 (May–June 1917): 193–94; John D. Grabenstein et al., "Immunization to Protect the US Armed Forces: Heritage, Current Practice and Prospects," *Epidemiologic Review* 28, no. 1 (Aug. 2006): 3–26, at 8; "Bill to Make Vaccination Optional," *NP* 3, no. 3 (Mar. 1927): 143, table 2, "Immunizations used widely during major conflicts"; H. B. Anderson, "The New Vaccination Drive," *NHH* 44, no. 7, current no. 536 (July 1939): 224.

79. Gilbert Bowman, "Vaccination and Serum Therapy," Department of Medical Freedom, *HHN* 22, no. 6 (Nov.–Dec. 1917): 372–76, at 372; "The Compulsory Vaccination Law," ibid. 23, no. 8, whole no. 288 (Aug. 1918): 711; Herbert M. Shelton, *Vaccines and Serum Evils, and Natural Hygiene*, facsimile ed. (Hastings, East Sussex, UK: Society of MetaPhysicians, Jan. 1998).

80. "Vaccinations Dangerous," *NP* 36, no. 9, current no. 443 (Sept. 1931): 271. The countries reported on were Australia, Japan, England, and Wales. See also "Financier Balks at Vaccination," *NHH* 45, no. 8, current no. 549 (Aug. 1940): 253.

81. Wilbur A. Sawyer, "Controlling Disease during WWII, 1939–1944," Wilbur A. Sawyer Papers, Profiles in Science, National Library of Medicine, accessed 4 July 2012, www.profiles.nlm.nih.gov/ps/retrieve/Narrative/LW/P-nid139.

82. Routine requirements included inoculations for influenza, smallpox, tetanus, typhoid, and paratyphoid A and B. Situational vaccinations, administered according to location and likelihood of exposure, were given against cholera, diphtheria, plague, diphtheria antitoxin, immune globulin (hepatitis A and B), and measles prophylaxis. See Grabenstein et al., "Immunization to Protect the US Armed Forces," 8. The same source reports that as of 2006, 31 vaccines were required for US military personnel. Sawyer, "Controlling Disease during WWII"; Allen, *Vaccine*, 16; Anderson, Anti-Vaccination Reports, 25, quoting Toomey, "Active Immunity"; "Is Vaccination Detrimental?," *NHH* 49, no. 4, current no. 585 (Apr. 1944): 116.

83. Cash Asher, "A Medical Absurdity," *NHH* 48, no. 2, current no. 579 (Apr. 1943): 64; Asher, "Compensation for Serum Damage?," ibid. 48, no. 3, current no. 580 (July 1943): 95; "Vaccination and Inoculation Fatal to British Airwoman," *NHH* 48, no. 3, current no. 580 (July 1943): 91.

84. H. B. Anderson, "Protest against Compulsory Medicine," *NHH* 45, no. 11, current no. 552 (Nov. 1940): 345; National News, "Report War Secretary Misinformed on U.S. Army Health Situation," ibid. 48, no. 1, current no. 578 (Jan. 1943): 28; Albert Whiting, ND, "New Ration Board Ruling Disqualifies Non-Medical Physicians," ibid. 48, no. 3, current no. 580 (July 1943): 91.

85. Morris Fishbein, MD, "Army Health," ibid. 48, no. 4, current no. 581 (Oct. 1943): 119. In fact, Albert Deutsch was without formal training but had educated himself through archival research, Veterans Administration practices, and treatment of the masses of mentally challenged in the twentieth century. See Albert Deutsch, *The Mentally Ill in America* (New York: Columbia Univ. Press, 1937); and Kenneth

J. Weiss, MD, "Albert Deutsch, 1905–1961," *American Journal of Psychiatry* 168 (Mar. 2011): 252.

86. "Ways to Save Rubber," *NHH* 48, no. 1, current no. 578 (Jan. 1943): 27; "Honor the Dead—Help the Living," *NP* 52, no. 4, current no. 639 (May 1948): 289.

87. Michael B. Oldstone, "Poliomyelitis," in *Viruses, Plagues, and History: Past, Present and Future* (Oxford: Oxford Univ. Press, 2009), 159–98, crediting virologists; Suzanne Humphries, MD, and Roman Bystrianyk, *Dissolving Illusion: Disease, Vaccines, and the Forgotten History* (Create Space Independent Publishing Platform, 2013), www.drsuzanne.net/dr-suzanne-humphries-vaccines-vaccination/, deemphasizing the role of biomedicine in disease transmission; Allen, *Vaccine*, 163.

88. Allen, *Vaccine*, 167, 169.

89. Garden, "Why I Thank God I Am an Antivivisectionist," 134, 158, at 158; "Vaccination and Infantile Paralysis," *NHH* 49, no. 3, current no. 584 (Mar. 1944): 86.

90. Benedict Lust, MD, DC, ND, DO, "Are the American People Being Taken for a 'Buggy-Ride' by the 'March of Dimes'?," *NHH* 49, no. 1, current no. 582 (Jan. 1944): 7, 23, at 7; Allen, *Vaccine*, 166, 180, 182, 191.

91. Allen, *Vaccine*, 186.

92. Ibid., 169, 186, 192, 198–99, 209, 211, 213.

93. Joanne Silberner, "Jonas Salk's Polio Vaccine Makes a Comeback," NPR, accessed 1 July 2015, www.npr.org/sections/goatsandsoda/2015/04/12/398806324/jonas-salks-polio-vaccine-makes-a-comeback?sc=17&f=&utm_source=iosnewsapp&utm_medium=Email&utm_campaign=app.

94. Edward H. Kass, "Infectious Disease and Social Change," *Journal of Infectious Diseases* 123, no. 1 (Jan. 1971): 110–14.

95. Willrich, "How the 'Pox' Epidemic Changed Vaccination Rules"; Allen, *Vaccine*, 11, 16, 73.

96. Stephen Barrett, MD, "Comments on the AANP Position on Childhood Vaccinations," 29 Dec. 2001, www.quackwatch.org/01QuackeryRelatedTopics/Naturopathy/aanpimmu.html. The 1991 position paper can be found on various websites, including that of Chrystal Hannan, ND, www.ndaccess.com/Crystal HannanND/Forms/Vaccine%20Resources.pdf. The AANP website has the 2011 version, *House of Delegates Position Paper: Immunizations.* www.naturopathic.org/files/Committees/HOD/Position%20Paper%20Docs/Immunizations.pdf.

97. Kimball C. Atwood IV, MD, and Stephen Barrett, MD, "Naturopathic Opposition to Immunization," Quackwatch, accessed 2 Nov. 2011, www.quackwatch.com/01QuackeryRelatedTopics/Naturopathy/immu.html, citing B. Silbert, "Naturopathy: Recommended Licensure in Massachusetts" (paper, Massachusetts Special Commission on Complementary and Alternative Medical Practitioners, Boston, 17 Apr. 2001), 117.

98. Willrich, "How The 'Pox' Epidemic Changed Vaccination Rules."

99. Samuel Katz, MD, "Poliovaccine Policy—Time for a Change," Commentaries, *Pediatrics* 98 (July 1996): 116–17; Philip J. Smith, PhD, MS, Susan Y. Chu, PhD, MSPH, and Lawrence E. Barker, PhD, "Children Who Have Received No Vaccines: Who Are They and Where Do They Live?," ibid. 114 (July 2004): 187–95.

100. Helen Gao, "More Parents Reject Vaccines," *San Diego Union Tribune*, 24 Aug. 2010, 1.

101. Grabenstein et al., "Immunization to Protect the US Armed Forces," 8; Allen, *Vaccine*, 15.

102. Allen, *Vaccine*, 373; Scott Hensley, "Lancet Renounces Study Linking Autism and Vaccines," NPR, accessed 30 June 2015, www.npr.org/sections/health-shots/2010/02 /lancet_wakefield_autism_mmr_au.html; Lisa Aliferis, "Whooping Cough Vaccine's Protection Fades Quickly," NPR, accessed 1 July 2015, www.npr.org/sections/health -shots/2015/05/05/404407258/whooping-cough-vaccines-protection-fades-quickly?sc=17 &f=1001&utm_source=iosnewsapp&utm_medium=Email&utm_campaign=app; Scott Neuman, "California Lawmakers Vote to Remove Vaccine Exemptions for Schoolchildren," NPR, accessed 1 July 2015, www.npr.org/sections/thetwo-way/2015 /06/25/417492698/california-lawmakers-vote-to-remove-vaccine-exemptions-for -schoolchildren?sc=17&f=1001&utm_source=iosnewsapp&utm_medium=Email &utm_campaign=app.

CHAPTER 8: Legal Battles

1. See, e.g., James Burrow, *Organized Medicine in the Progressive Era: The Move toward Monopoly* (Baltimore: Johns Hopkins Univ. Press, 1977); and George Rosen, *The Structure of American Medical Practice, 1875–1941*, ed. Charles E. Rosenberg (Philadelphia: Univ. of Pennsylvania Press, 1983).

2. Arthur S. Link and Richard L. McCormick, *Progressivism* (Wheeling, IL: Harlan Davidson, 1983), 2.

3. See, e.g., George Starr White, MD, FSSc, "Health versus the Germ Theory," *HHN* 25, no. 4, whole no. 308 (May 1920): 189.

4. Friedhelm Kirchfeld and Wade Boyle, *Nature Doctors: Pioneers in Naturopathic Medicine* (East Palestine, OH: Buckeye Naturopathic, 1993), 210.

5. B. J. Jones, "Democracy or Autocracy—Which?," *HHN* 25, no. 7, whole no. 311 (Aug. 1920): 351–52; "A New Plan of Attack," ibid. 23, no. 12 (Dec. 1918): 949; "Without Trial by Jury," ibid. 25, no. 1, whole no. 305 (Jan. 1920): 1.

6. "Better Babies," ibid. 25, no. 2, whole no. 306 (Feb. 1920): 247; Jones, "Democracy or Autocracy," 351.

7. Gilbert Patten Brown, PhD, DO, ND, DC, DPOS, "Why Many Allopaths Are Criminals," *HHN* 25, no. 2, whole no. 306 (Feb. 1920): 76.

8. A. Matiaca, "Insurance Companies and the Drugless Profession," ibid. 22, no. 1, whole no. 275 (Jan.–Feb. 1917): 589–90.

9. Benedict Lust, Naturopathic News, *Naturopath* 30 (1925): 366, reprinted in Kirchfeld and Boyle, *Nature Doctors*, 198; A. A. Erz, "Official Medicine—As It Is— And What It Is Not," *NHH* 19, no. 6, whole no. 248 (June 1914): 362.

10. Edward Earle Purinton, Efficiency in Drugless Healing, *NHH* 19, no. 9, whole no. 251 (Sept. 1914): 557; S. T. Etreig, "Law and the Art of Healing," *HHN* 25, no. 4, whole no. 308 (May 1920): 194; Bernarr MacFadden, "Shall We Have Medical Freedom?," ibid. 25, no. 7, whole no. 311 (Aug. 1920): 336; Kirchfeld and Boyle, *Nature Doctors*, 215; George Cody, in "History of Naturopathic Medicine," in *Textbook of Natural Medicine*, ed. Joseph Pizzorno and Michael Murray (New York: Churchill Livingstone, 1985), 17–40, at 19, noting that after MacFadden became wealthy, he added to his publishing empire the popular magazines *True Confessions* and *True Detective* and campaigned for the 1936 Republican presidential nomination.

11. G. R. Clements, ND, "What Is Disease or Where Naturopathy and Allopathy Part," *NHH* 30, no. 9 (Sept. 1925): 836.

12. "Doctors and Their Exorbitant Fees," ibid. 9, no. 1, whole no. 96 (Jan. 1908): 14–16; "Eugene Christian, Food Scientist, Vindicated," ibid. 9, no. 3, whole no. 98 (Mar. 1908): 95; Otto Carque, "The Rockefeller Institute of Medical Research: How the Oil King's Money Is Spent in a Useless Manner," ibid. 9, no. 5, whole no. 100 (May 1908): 144–45; "Another Medical Persecution," ibid. 9, no. 6, whole no. 101 (June 1908): 190; Naturopathic News, "President Roosevelt to Lend Support to Dr. Bland," ibid. 8, no. 5, whole no. 89 (June 1907): 189.

13. Chas. I. White, "An Open Letter to the Drugless Physician," ibid. 8, no. 8, whole no. 92 (Sept. 1907): 265.

14. Benedict Lust, "The American Naturopathic Association," folder 8, Gehman Papers Jesse Gehman Papers, Bastyr University Special Collections, Kenmore, WA.

15. H. N. D. Parker, ND, "Legal Standing of Drugless Practitioners," *HHN* 22, no. 3 (May–June 1917): 159; William F. Havard, ND, "What Does the A.N.A. Mean to You?," ibid. 23, no. 5, whole no. 284 (Apr. 1918): 369; M. T. Council, "Drugless Persecution," *Naturopath* 28, no. 4, whole 342 (Apr. 1923): 194.

16. "Platform Adopted by Constitutional Liberty League," *HHN* 25, no. 2, whole no. 306 (Feb. 1920): 221.

17. "Eugene Christian, Food Scientist, Vindicated," 95; Eugene Christian, letter to the editor, *Naturopath and Herald of Health*, 4 Jan. 1908, in *NHH* 9, no. 3, whole no. 98 (Mar. 1908): 96–97.

18. Christian, letter to the editor, 97.

19. "Another Naturopath Vindicated in New York," *NHH* 9, no. 6, whole no. 101 (June 1908): 173; "A Naturopath Vindicated," ibid. 9, no. 11, whole no. 106 (Nov. 1908): 352, as quoted from *Los Angeles Times*, 25 Oct. 1908.

20. H. B. Anderson, "Fallacy of Plea for Medical Control through Medical Practice Acts," *HHN* 25, no. 2, whole no. 306 (Feb. 1920): 86–87; "Contributions to the Fund of A. Boerner's Family," *NHH* 19, no. 11, whole no. 253 (Nov. 1914): 754; Samuel H. Williamson, "Seven Ways to Compute the Relative Value of a U.S. Dollar Amount, 1774 to present," MeasuringWorth, 2015, accessed 26 July 2015, www.measuringworth .com/uscompare/.

21. Stephen Petrina, "Medical Liberty: Drugless Healers Confront Allopathic Doctors, 1910–1931," *Journal of Medical Humanities* 29 (Dec. 2008): 209; Benedict Lust, ND, MD, "History of the Naturopathic Movement," *Naturopath* 28, no. 9, whole no. 347 (Sept. 1923): 436; Williamson, "Seven Ways to Compute," using 1910 and 2014 as the comparative dates; Naturopathic News, *NHH* 20, no. 12, whole no. 266 (Dec. 1915): 737; Naturopathic News, ibid. 20, no. 2, whole no. 256 (Feb. 1915): 126; "Medical Ethics: A Pretense and a Sham," *HHN* 23, no. 9, whole no. 289 (Sept. 1918): 777–80; W.S.M., in *Life* magazine, reprinted as "The Dream of the Doctors' Trust," *NHH* 20, no. 2, whole no. 256 (Feb. 1915): 129; Naturopathic News, ibid. 20, no. 6, whole no. 260 (June 1915): 390.

22. Naturopathic News, "Florida," *HHN* 25, no. 2, whole no. 306 (Feb. 1920): 58–59.

23. Naturopathic News, *NHH* 19, no. 3, whole no. 245 (Mar. 1914).

24. Kirchfeld and Boyle, *Nature Doctors*, 188. The form of Frances Benzacry's name varies in publications. Benedict Lust, "Letter to Ladies Home Journal by the

American Naturopathic Association," and "Answer from the Ladies Home Journal," ibid. 20, no. 6, whole no. 260 (June 1915): 385; Kirchfeld and Boyle, *Nature Doctors*, 189.

25. Kirchfeld and Boyle, *Nature Doctors*, 188; Lust, "History of the Naturopathic Movement," 436.

26. Kirchfeld and Boyle, *Nature Doctors*, 191.

27. James C. Whorton, *Nature Cures: The History of Alternative Medicine in America* (Oxford: Oxford Univ. Press, 2002), 231; Benedict Lust, editorial, *HHN* 23, no. 6, whole no. 286 (June 1918): 520.

28. "A Report on Conditions in California," *NHH* 28, no. 5, whole no. 343 (May 1923): 250–51. The 1909 California Medical Act allowed just naturopaths to be licensed; it was amended in 1911 and 1913 as the Medical Practice Act to more broadly provide "for the issuance of . . . Drugless Practitioner's certificate." California Board of Medical Examiners, *Directory of Physicians and Surgeons, Osteopaths, Drugless Practitioners, Chiropodists Holding Certificates Issued Under the Medical Practice Acts of the State of California* (Sacramento: California State Printing Office 1922), 9–10.

29. Benedict Lust, "Proclamation to the Board of Directors, Trustees, Officers, State Representatives, Delegates and Members Assembled for the 31st Annual Convention of the American Naturopathic Association . . . September 14–18, 1927 Minneapolis, MN," *NP* 32, whole no. 396 (Oct. 1927): 473–77. Hans A. Baer concurs with Lust on the (semi)recognition of naturopathy in more than twenty states, but he questions the claim of "good schools with excellent standards." Baer, "The Sociopolitical Status of US Naturopathy at the Dawn of the 21st Century," *MAQ* 15, no. 3 (2001): 329–46.

30. Lust, "Proclamation," 474; William F. Havard, "Eclecticism in Drugless Healing," *NHH* 20, no. 4, whole no. 258 (Apr. 1915): 253; Naturopathic News, ibid. 20, no. 8, whole no. 262 (Aug. 1915): 519, 522, 519; Harry Riley Spitler, ND, DC (president, Ohio Association of Naturopaths), Correspondence, ibid. 20, no. 9, whole no. 263 (Sept. 1915): 552–53; J. Shelby Riley, "Chiropractic as Distinguished from Other Drugless Methods," ibid., 561–62; Wallace Fritz, MD, DDS, ND, "Neuropathy Department," ibid., 575–77.

31. E. W. Cordingley, ND, DNSc, "Let Us Standardize the Practice of Naturopathy," *Naturopath* 28, no. 11, whole no. 349 (Nov. 1923): 684.

32. W. A. Budden, "Basic Science Legislation: An Examination into Its Origin, Purposes and Effects!," *NHH* 41, no. 1, current no. 506 (Jan. 1937): 26.

33. [Benedict Lust?], "Naturopathy's Progress," ibid. 5, no. 5, whole no. 54 (May 1904): 110; Edward Earle Purinton, Efficiency in Drugless Healing, "Who Should Heal?," ibid. 20, no. 6, whole no. 260 (June 1915): 333–34; Purinton, "An Open Letter from Mr. Purinton to the Editor," 1 June 1915, ibid. 20, no. 7, whole no. 261 (July 1915): 449–50; E. D. Snyder, Correspondence "Letter Addressed to Benedict Lust," Correspondence, ibid. 20, no. 8, whole no. 262 (Aug. 1915): 522; Alexander D. Bateman, MTD, "Naturopathic Physician and His Ethics" (lecture, Naturopathic Convention, Pittsburgh, 1915), reprinted in ibid., 517.

34. "Pennsylvania Association of Naturopaths," *NHH* 16, no. 3, whole no. 209 (1911): 173; Laura C. Little, "The California Medical Law," ibid. 19, no. 1, whole no. 243 (Jan. 1914): 12; Oscar Evertz, "Curing the Nature Cure," Efficiency Department, ibid. 20, no. 10, whole no. 264 (Oct. 1915): 655; Evertz, "The Master Key," ibid. 20, no. 11, whole no. 265 (Nov. 1915): 721.

35. Benedict Lust, Naturopathic News, ibid. 20, no. 12, whole no. 266 (Dec. 1915): 735; Spitler, Correspondence, 552–53.

36. Cordingley, "Let Us Standardize the Practice of Naturopathy," 684; Herbert M. Shelton, DP, DNT, "Unity or Contabescence . . . Which Shall It Be?," *NHH* 40, no. 3 (Mar. 1935): 72.

37. "Minnesota's A.N.A. Convention: President's Annual Report, June 6, 1929," *NP* 34, no. 9, current no. 419 (Sept. 1929): 421; US Congress, House, Committee on the District of Columbia. *An Act to Amend Meaning And Intent of Healing Art Practice Act of February 27, 1929: Hearing Before the Committee On the District of Columbia, House of Representatives, Seventy-First Congress, Third Session, On H.R. 12169, a Bill to Amend the Meaning And Intention of an Act of Congress Entitled "An Act to Regulate the Practice of the Healing Art to Protect the Public Health of the District of Columbia," Approved February 27, 1929. January 28, 1931* (Washington, DC: GPO, 1931), clarifying amendments H.R. 12169.

38. "South Carolina Naturopathic Examining Board Meets," *NHH* 42, no. 8, current no. 513 (Aug. 1937): 244; South Carolina Naturopathic Association, "Laws of the South Carolina Naturopathic Association and the South Carolina Board of Naturopathic Examiners," ibid. 42, no. 9, current no. 514 (Sept. 1937): 264–65.

39. Lust defined naturopathy as including hydrotherapy and mineral baths, external applications, physiotherapy, mechanotherapy, massage, Swedish movements, corrective and orthopedic gymnastics, manipulations, spondylotherapy, zone therapy, electrotherapy, short wave therapy, thermotherapy, heliotherapy, phototherapy, chromotherapy, vibrotherapy, hygiene and sanitation, first aid, minor surgery, drugless gynecology, obstetrics, dietetics, phytotherapy (herbalism), biochemistry, psychotherapy, and other natural healing methods, excepting, however, materia medica and major surgery. Benedict Lust, ND, MD, "The Definition of Naturopathy in Accordance with the Tenets of the School Is as Follows," ibid. 42, no. 8, current no. 513 (Aug. 1937): 241–42.

40. J. Thompson, "State Board Questions," ibid. 42, no. 10, current no. 515 (Oct. 1937): 318; Thompson, "State Board Questions," ibid. 42, no. 11, current no. 516 (Nov. 1937): 350.

41. Israel Shmid, ND, "State Board Examinations," ibid. 44, no. 1, current no. 530 (Jan. 1939): 4; T. M. Schippell (secretary-treasurer, ANA), "The New Plan Adopted by the 42nd Annual Congress of the ANA in Seattle, WA. August 31–September 4, 1938," ibid. 44, no. 3, current no. 532 (Mar. 1939): 80.

42. Walter Seth Kipnis, State News: Pennsylvania, ibid. 44, no. 4, current no. 533 (Apr. 1939): 112; Kipnis, State News: Utah, ibid., 113.

43. Benedict Lust, Dr. Lust Speaking, ibid. 44, no. 8, current no. 537 (Aug. 1939): 248; Dr. I. S. Weinger, quoting Benedict Lust, in "Naturopathic Commemoration Day," ibid. 41, no. 1, current no. 506 (Jan. 1937): 19, 24–25.

44. Whorton, *Nature Cures*, 215–16, quoting Lust.

45. Walter Seth Kipnis, State News: Hawaii, *NHH* 45, no. 3, current no. 544 (Mar. 1940): 93; Committee on Education of the ANA, "Recommendations of the Committee on Education of the American Naturopathic Association," ibid. 48, no. 3, current no. 580 (July 1943): 81.

46. Association of Progressive Naturopathic Physicians of California, "What Is Naturopathy? And What Is the Status of the Naturopathic Physician?," ibid. 25, no. 5,

whole no. 309 (June 1920): 239; Kirchfeld and Boyle, *Nature Doctors*, 212, giving the complimentary appellations for Schultz; New York State Society of Naturopaths, "Naturopathic Legislation Series Part One—Brief I, II and III," *NHH* 19, no. 3, whole no. 245 (Mar. 1914): 143–50, 183–87, with the complete text of the proposed naturopathic law, to amend the public health law, on 188–90.

47. "Our Model Bill: Be It Enacted by the Legislature of the State of Florida," *Naturopath* 28, no. 3, whole no. 341 (Mar. 1923): 138–40, at 140; "Indiana Naturopathic Bill," ibid. 30, no. 3, whole no. 365 (Mar. 1925): 233–37, at 234; Bernarr MacFadden, "Drugless Methods Prove Their Worth," *NP* 3, no. 2 (Feb. 1927): 64; Benedict Lust, "Convention Greetings," ibid. 32, whole no. 395 (Sept. 1927): 421–22.

48. H. M. Shelton and B. Stanford Claunch, "Declaration of Medical Liberty Rights," *Naturopath* 30, no. 3, whole no. 365 (Mar. 1925): 246–51; Whorton, *Nature Cures*, 136–37; Alexander Wilder, *History of Medicine* (New Sharon, ME: New England Eclectic, 1901), 456, as quoted in Whorton, *Nature Cures*, 137.

49. "Move to Restrain M'laughlin and Harris: Chiropractors and Naturopaths Want Commissioners to Let Them Practice in Peace," *New York Times*, 3 July 1926.

50. US Congress, House, Committee on the District of Columbia, *Act to Amend Meaning And Intent of Healing Art Practice Act*; Benedict Lust, "President's Proclamation," *NP* 36, no. 11, current no. 445 (Nov. 1931): 322. The act was passed on 7 Feb. 1931.

51. J. Stuart Moore, *Chiropractic in America: The History of a Medical Alternative* (Baltimore: Johns Hopkins Univ. Press, 1993), 88, quoting the health commissioner, in Whorton, *Nature Cures*, 138; Benedict Lust, "41st Annual Convention of the American Naturopathic Ass'n," *NHH* 42, no. 6, current no. 511 (June 1937): 163.

52. Section 2 stated in part: "Diagnosis and practice of physiological, mechanical, and material sciences of healing as follows: The physiological and mechanical sciences such as mechanotherapy, articular manipulation, corrective orthopedic gymnastics, neurotherapy, psychotherapy, hydrotherapy, and mineral baths, electrotherapy, thermotherapy, phototherapy. Chromotherapy, vibrotherapy, thalmotherapy, and dietetics, which shall include the use of foods of such biochemical tissue-building products and cell salts as are found in the normal body: and the use of vegetable oils and dehydrated and pulverized fruits, flowers, seeds, barks, herbs, roots and vegetables uncompounded and in their natural state." US Congress, House, Committee on the District of Columbia, *Act to Amend Meaning And Intent of Healing Art Practice Act*.

53. R. L. Duffus, "Shall Medicine Be Socialized? A Big Issue Is Joined," *New York Times*, 4 Dec. 1932.

54. Bills 1057 (state senate) and 1444 (state assembly) were described in Benedict Lust, Dr. Lust Speaking, *NHH* 40, no. 3 (Mar. 1935): 66. See also Jacob H. Miller, ND, "North Dakota Basic Science Proposal," *NP* 36, no. 2, current no. 436 (Feb. 1931): 44; Naturopathic News, "Maine," ibid. 36, no. 6, current no. 440 (June 1931): 173; Walter Seth Kipnis, State News: Rhode Island, *NHH* 40, no. 7 (July 1935): 223; and Shelton, "Unity or Contabescence," 72.

55. Benedict Lust, "Basic Science Bill Passed: A Blow Aimed at Naturopathic Healing Profession," *NHH* 44, no. 7, current no. 536 (July 1939): 210; Dr. A. R. Hedges, "Oregon Takes It On the Chin!" ibid. 40, no. 1 (Jan. 1935): 6, 28; Dr. Linke, "The California Situation," 30 Nov. 1934, ibid., 15. In Oregon the AMA-backed bill gave the

governor power to appoint an examining board in the "basic sciences"; no drugless healers were included on the AMA board. Robert V. Carroll (chairman, Executive Committee of the ANA, Seattle), "To the Naturopaths of America," *NHH* 44, no. 2, current no. 531 (Feb. 1939): 62, on naturopaths and the Social Security Act; "New Federal Health Bill Proposed," ibid. 44, no. 11, current no. 540 (Nov. 1939): 347.

56. Kirchfeld and Boyle, *Nature Doctors*, 257–58; George Cody, "History of Naturopathic Medicine," 22.

57. Walter Seth Kipnis, State News: Hawaii, Honolulu, *NHH* 42, no. 3, current no. 508 (Mar. 1937): 80; Kipnis, State News: Utah, ibid. 44, no. 6, current no. 535 (June 1939): 180; Kipnis, State News: Illinois, ibid. 45, no. 4, current no. 545 (Apr. 1940): 125; Kipnis, State News: Idaho and Kansas, ibid. 42, no. 3, current no. 508 (Mar. 1937): 80; Kipnis, State News: Alaska, ibid. 44, no. 5, current no. 534 (May 1939): 141; "Proposed Naturopathic Act for the State of Texas," ibid. 40, no. 12 (Dec. 1935): 360–61, 374.

58. Shelton, "Unity or Contabescence," 72.

59. Benedict Lust, Dr. Lust Speaking, "Congratulations Arizona," *NHH* 40, no. 4 (Apr. 1935): 98–99; Walter Seth Kipnis, State News: Washington D.C., ibid. 45, no. 4, current no. 545 (Apr. 1940): 124.

60. The history of naturopathic licensing efforts would best be told in a separate publication for each state. The scope of this book only allows me to touch upon the most documented state issues in the national sources; I leave more in-depth studies to future scholars. According to the attorney Fred H. Hartwell, "Different jurisdictions [had] different defining clauses." This made for wide variation according to locale. See Hartwell, "What Is the Practice of Medicine?," *HHN* 23, no. 8, whole no. 288 (Aug. 1918): 723–26, detailing the variety of legal theories and language for drugless healers in Alabama, Alaska, Arizona, and Colorado.

61. Walter Seth Kipnis, Naturopathic News, "New York," *NP* 36, no. 7, current no. 441 (July 1931): 207; Naturopathic News, *NHH* 19, no. 5, whole no. 247 (May 1914): 329.

62. Kipnis, Naturopathic News, "New York," 207; Naturopathic News, *NHH* 19, no. 5, whole no. 247 (May 1914): 329; "Legal Status of Chiropractic in 1937," ibid. 42, no. 3, current no. 508 (Mar. 1937): 80; Fred H. Hartwell, "Laws Regulating Osteopathy," *HHN* 25, no. 6, whole no. 310 (July 1920): 346–53; [Benedict Lust?], Naturopathic News, ibid. 23, no. 1, whole no. 281 (Jan. 1918): 76.

63. Hartwell, "What Is the Practice of Medicine?," 723–26.

64. Walter Seth Kipnis, State News: New York, *NHH* 40, no. 1 (Jan. 1935): 23–24; Kipnis, State News: New York, *Naturopath* 41, no. 5, whole no. 379 (May 1926): 246.

65. Walter Seth Kipnis, State News: New York, *NHH* 40, no. 6 (June 1935): 182. Five hundred dollars in 1935 was the equivalent of $8,620 in 2014. Williamson, "Seven Ways to Compute."

66. William John Barrison, ND, "Medical Persecution," *NHH* 40, no. 9 (Sept. 1935): 262; Walter Seth Kipnis, State News: New York, ibid. 40, no. 11 (Nov. 1935): 34; Kipnis, State News: New York, "Court Curbs Beauty Shops on Hair Removal; Doctors Only to Use Electrolysis Method," ibid. 41, no. 2, current no. 507 (Feb. 1937): 47.

67. Israel Schmid, ND, "Medical and Naturopathic Association: (Jail Thoughts)," ibid. 42, no. 8, current no. 513 (Aug. 1937): 252–53.

68. Kirchfeld and Boyle, *Nature Doctors*, 212; Association of Progressive Naturopathic Physicians of California, "What Is Naturopathy?," 239–40; "Bits of News,

Gathered from Here and There, in the Naturopathic Field: California," *NHH* 49, no. 8, current no. 589 (Aug. 1944): 248–49; Carl Schultz, "The Association of Naturopaths of California," *HHN* 5, no. 7, whole no. 55 (July 1904): 203; "Naturopaths and Osteopaths in California," *NHH* 9, no. 1, whole no. 96 (Jan. 1908): 29; Association of Progressive Naturopathic Physicians of California, "What Is Naturopathy?," 240; Hartwell, "What Is the Practice of Medicine?," 723–26; Association of Progressive Naturopathic Physicians of California, "What Is Naturopathy?," 239–41.

69. The sixteen therapeutics in California naturopathy were physiotherapy, electrotherapy, mechanotherapy, Swedish exercises, medical massage, spinal adjustments, osteopathy, vibratory and other manipulative treatments, hydrotherapy, thermotherapy, radiant light and heat, phototherapy, refraction, dietetics, mental culture, and more. [Benedict Lust?], "'Drugless' Doctors Victorious in Ruling by Webb in Sacramento, Cal.," *Naturopath* 30, no. 1 (Jan. 1925): 39. Other news from the 1930s reported the formation of a research council for the California University of Liberal Physicians; the convening of the Naturopathic Congress at the Pacific International Exposition; the National Chiropractic Convention to be held in Hollywood (Fall 1935); chiropractic announcements; and the meeting of the United States Naturopathic Association. Walter Seth Kipnis, State News: California, *NHH* 40, no. 4 (Apr. 1935): 118, on the research council; Kipnis, State News: California, "Special Bulletin from the California Pacific International Exposition," ibid. 40, no. 6 (June 1935): 181; Kipnis, State News: California, ibid. 40, no. 8 (Aug. 1935): 245, on the twelfth annual meeting of various groups in Los Angeles.

70. Linda A. McCready and Billie Harris, *From Quackery to Quality Assurance: The First Twelve Decades of the Medical Board of California* (Sacramento: Medical Board of California, 1995), 20; Lust, "Report on Conditions in California."

71. The *Naturopathic Weekly* is credited with helping put naturopathy in California in the vanguard of naturopathic education. Walter Seth Kipnis, Naturopathic News, "Benedict Lust, Here Today; Plans College for Los Angeles," *Naturopath* 31, whole no. 382 (Aug. 1926): 408.

72. Linke, "California Situation," 15; Tonya Blood (chief, Bureau of Naturopathic Medicine), "Findings and Recommendations Regarding Minor Office Procedures," report presented to the California State Legislature, 1 Jan. 2007, www.naturopathic.ca .gov/formspubs/mop_report.pdf. California's Senate Bill 907 created the Naturopathic Doctors Act, which established criteria for licensure and scope of practice. The bill also established the Bureau of Naturopathic Medicine within the California Department of Consumer Affairs.

73. Walter Seth Kipnis, Naturopathic News, "Illinois," *Naturopath* 21, no. 2, whole no. 376 (Feb. 1926): 88; Henry Lindlahr, *Nature Cure: Philosophy and Practice Based on the Unity of Disease and Cure*, 2nd ed. (Chicago: Nature Cure, 1914); Kipnis, State News: Illinois, *NHH* 44, no. 10, current no. 539 (Oct. 1939): 305. On medical freedom activism, see Petrina, "Medical Liberty," 205–30. It is not surprising that the Medical Freedom League attracted members. At its 1918 annual meeting in North Dakota, the league called for medical freedom to be written into states' constitutions. Naturopaths in St. Petersburg, Florida, formed a state chapter of the Medical Freedom League on 1 April 1920 and hoped for medical freedom to be written into Florida law. "Resolutions Adopted by the Medical Freedom League of North Dakota, at Annual Meeting,

Fargo, North Dakota, June 11, 1918," *HHN* 23, no. 10, whole no. 290 (Oct. 1918): 844; Paul Dowd, ND, "Are Floridians Languid?," ibid. 25, no. 4, whole no. 308 (May 1920): 201; "Chicago Leads the World: Has Real Health Board," *Naturopath* 21, no. 4, whole no. 378 (Apr. 1926): 172.

74. Benedict Lust, "A New Use for the Injunction," *HHN* 25, no. 7, whole no. 311 (Aug. 1920): 321–22.

75. Walter Seth Kipnis, State News: Illinois, *NHH* 45, no. 5, current no. 546 (May 1940): 154; Kipnis, State News: Illinois, ibid. 44, no. 5, current no. 534 (May 1939): 141; Kipnis, State News: Illinois, "Doctors We Are Going to Present a Naturopathic Bill Next January? Are You Interested?," ibid. 45, no. 5, current no. 546 (May 1940): 154.

76. Friedhelm Kirchfeld and Wade Boyle, "The Father of Naturopathy, Benedict Lust (1872–1945)," in *Nature Doctors*, 185–219.

77. House Bill 58 was described in "Legislation Reports," *NHH* 28, no. 6, whole no. 344 (June 1923): 305. At that time the bill still awaited approval by the Florida state senate. See also "Law of the State of Washington," *Naturopath* 30, no. 3, whole no. 365 (Mar. 1925): 216–21, 272; Walter Seth Kipnis, State News: Maine, ibid. 31, no. 3, whole no. 377 (Mar. 1926): 149; Kipnis, State News: Virginia, ibid., 150; Kipnis, State News: Oregon, *Naturopath* 31, no. 8, whole no. 382 (Aug. 1926): 408; Kipnis, State News: Indiana, ibid. 31, no. 9, whole no. 383 (Sept. 1926): 460; and Kipnis, State News: Pennsylvania, ibid. 31, no. 10, whole no. 384 (Oct. 1926): 512.

78. "Naturopathic Law of Florida," *NHH* 42, no. 6, current no. 511 (June 1937): 186–87, 189; "By-Laws of the South Carolina Naturopathic Association and the South Carolina Board of Naturopathic Examiners," ibid. 42, no. 9, current no. 514 (Sept. 1937): 264–65; "Oregon Naturopathic Law: Law, Rules and Regulations Governing the Naturopathic Board of Examiners of the State of Oregon, 1927," ibid., 286–88; Kipnis, State News: Mississippi, *NHH* 44, no. 1, current no. 530 (Jan. 1939): 21; Kipnis, State News: New Mexico, ibid. 44, no. 3, current no. 532 (Mar. 1939): 79; Kipnis, Maryland, ibid.

79. Naturopathic News, "Texas," *NP* 36, no. 7, current no. 441 (July 1931): 207; Benedict Lust, Dr. Lust Speaking, *NHH* 40, no. 3 (Mar. 1935): 67, 90; Walter Seth Kipnis, State News: Connecticut, ibid. 40, no. 6 (June 1935): 180–81; Kipnis, State News: Kansas, ibid. 40, no. 7 (July 1935): 214; Kipnis, State News: Idaho, ibid. 32, no. 3, current no. 508 (Mar. 1937): 80; Kipnis, State News: Pennsylvania, ibid. 44, no. 6, current no. 535 (June 1939): 178; Kipnis, State News: Michigan, ibid. 44, no. 10, current no. 539 (Oct. 1939): 306.

80. Walter Seth Kipnis, State News, "Free Choice of Healing Method Fundamental," *NHH* 45, no. 5, current no. 546 (May 1940): 158.

81. Walter Seth Kipnis, State News: Texas, ibid. 45, no. 7, current no. 548 (July 1940): 223; "Proposed Naturopathic Law for Nevada," ibid. 45, no. 12, current no. 553 (Dec. 1940): 378–80, 382, 384; "Tennessee Naturopathic Act," ibid. 48, no. 2, current no. 579 (Apr. 1943): 53–54; Kipnis, State News: Connecticut, ibid. 48, no. 7, current no. 580 (July 1943): 92; Kipnis, State News: Alaska, ibid. 48, no. 4, current no. 589 (Apr. 1943): 59. If passed, the Alaska law would also be extended to members of the Alaska Drugless Physicians Association. Ongoing efforts in the mid-1940s in Indiana and Maryland, among many others states, were also successful. "Bits of News, Gathered from Here and There, in the Naturopathic Field: Indiana," ibid. 49, no. 5, current

no. 586 (May 1944): 154; "Bits of News, . . . Maryland," ibid. 49, no. 4, current no. 585 (Apr. 1944): 121.

82. "Transcript of Naturopathic Act of the District of Columbia: Public No. 831–70th Congress S. 3936," ibid. 40, no. 6 (June 1935): 178–79, 187; "A Feasible Plan," ibid. 44, no. 10, current no. 539 (Oct. 1939): 307; Walter Seth Kipnis, State News: South Carolina, ibid. 49, no. 5, current no. 586 (May 1944): 156.

83. [Benedict Lust?], "Why Federal and State Naturopathic Laws Should Be Enacted," ibid. 45, no. 10, current no. 551 (Oct. 1940): 312. The basic science laws had also become a problem for allopaths' "licensing reciprocity between states in which the exams tested on different subjects." Whorton, *Nature Cures*, 232.

CHAPTER 9: **Professionalizing and Defining the Nature Cure**

1. Friedhelm Kirchfeld and Wade Boyle, *Nature Doctors: Pioneers in Naturopathic Medicine* (East Palestine, OH: Buckeye Naturopathic, 1993), 193; James C. Whorton saw the term *therapeutic universalism* as emblematic of Lust's philosophy and quoted his use of this term in *Nature Cures: The History of Alternative Medicine in America* (Oxford: Oxford Univ. Press, 2002), 195.

2. NHH 19, no. 4, whole no. 210 (Apr. 1911): 1; Immanuel Pfeiffer, "Our Home Rights," ibid. 5, no. 8, whole no. 56 (Aug. 1904): 192–93. The *Liberator* joined the *NHH* in 1908.

3. [Benedict Lust?], "Naturopathy's Progress," ibid. 5, no. 5, whole no. 54 (May 1904): 110.

4. [Benedict Lust?], "What New Thought Is," ibid. 5, no. 7, whole no. 55 (July 1904): 151; cover, ibid. 8, no. 5, whole no. 89 (Sept. 1907).

5. Edward Earle Purinton, "Who Is Who in Naturopathy," ibid. 9, no. 4, whole no. 99 (Apr. 1908): 101–12, at 101–3; "Information and News Department: To Encourage Honest Advertisers and Eliminate Frauds," ibid. 19, no. 5, whole no. 247 (May 1914), n.p.

6. Margaret Goettler, "The 'Nature Cure,'" ibid. 16, no. 4, whole no. 210 (Apr. 1911): 1; Whorton, *Nature Cures*, 97, 201; Benedict Lust, Correspondence, *NHH* 19, no. 2, whole no. 244 (Feb. 1914): 133.

7. Naturopathic News, *NHH* 19, no. 10, whole no. 252 (Oct. 1914): 682; George Cody, "History of Naturopathic Medicine," in *Textbook of Natural Medicine*, ed. Joseph Pizzorno and Michael Murray (New York: Churchill Livingstone, 1985), 17–40, at 22.

8. "Naturopathic Legislation Series, Part One—Brief I, II and III, Brief II: Practical Ideals Embodied in Naturopathy," *NHH* 19, no. 3, whole no. 245 (Mar. 1914): 145–46; Benedict Lust, ND, DC, "The Chiropractic Fiasco in New York," ibid. 49, no. 5, current no. 586 (May 1944): 150; Naturopathic News, "Announcement of the American Naturopathic Association," ibid. 19, no. 10, whole no. 252 (Oct. 1914): 682.

9. Benedict Lust, ed., *Universal Naturopathic Encyclopedia, Directory and Buyers' Guide: Year Book of Drugless Therapy*, vol. 1, for 1918–19 (New York: privately printed, 1918).

10. Leon Bourgonjon, MD, Correspondence, *NHH* 19, no. 11, whole no. 253 (Nov. 1914): 765; front page, ibid. 20, no. 1, whole no. 255 (Jan. 1915).

11. Edward Earle Purinton, Efficiency in Drugless Healing, "Standardizing the Nature Cure," ibid. 20, no. 3, whole no. 257 (Mar. 1915): 142.

12. Front page, ibid. 20, no. 12, whole no. 266 (Dec. 1915).

13. Benedict Lust, Naturopathic News, ibid., 736; front page, *HHN* 22, no. 6 (Nov.–Dec. 1917). The name change had been adopted earlier, but this issue details the journal's mission.

14. Rene V. Thirion, ND, MD, "Naturopathy, or Nature Cure," *HHN* 23, no. 3, whole no. 283 (Mar. 1918): 275; Benedict Lust, "A Little of the Truth," ibid. 23, no. 10, whole no. 290 (Oct. 1918): 812. The Progressive Naturopathic Physicians were a distinct group from the Association of Naturopathic Physicians of California. See Association of Progressive Naturopathic Physicians of California, "What Is Naturopathy? And What Is the Status of the Naturopathic Physician?," ibid. 25, no. 5, whole no. 309 (June 1920): 239–40.

15. Kirchfeld and Boyle, *Nature Doctors*, 200–201; front page, *Naturopath: formerly Herald of Health* 30, no. 1, whole no. 363 (Jan. 1925); H. M. Shelton and B. Stanford Claunch, "Declaration of Medical Liberty Rights," *Naturopath* 30, no. 3, whole no. 365 (Mar. 1925): 246.

16. Benedict Lust, "Specific Home Study Course of Naturopathy," *Naturopath* 31, no. 4, whole no. 378 (Apr. 1926): 166; Herbert M. Shelton, DP, ND, "The Evils of Condiment Using," ibid. 31, no. 5, whole no. 379 (1926): 227–28.

17. Benedict Lust, "The Natural Life and Healing Methods vs. Medical Piracy," *NP* 36, no. 8, current no. 442 (Aug. 1931): 227; Karl Eric Toepfer, *Empire of Ecstasy: Nudity and Movement in German Body Culture, 1910–1935* (Berkeley: Univ. of California Press, 1997).

18. Benedict Lust, Dr. Lust Speaking, *NHH* 40, no. 1 (Jan. 1935): 3.

19. Front page, ibid. 40, no. 10 (Oct. 1935); "The Science Of Naturopathy," ibid., 307; "Naturopathy—Most Complete Science of Healing," ibid. 40, no. 1 (Dec. 1935): 380; Benedict Lust, "New Prospectus of the American Naturopathic Association," ibid. 42, no. 4, current no. 509 (Apr. 1937): 104; J. Barwick, OD, DSc, PhD, "Cell Science," ibid. 42, no. 6, current no. 511 (June 1937): 167, 179.

20. Benedict Lust, *Collected Works of Dr. Benedict Lust, Founder of Naturopathic Medicine*, ed. Anita Lust Boyd and Eric Yarnell, ND, RH (AHG) (Seattle: Healing Mountain, 2006), 43, 65, 77–78. This is the third published version of Lust's biographical memoir. Whorton, *Nature Cures*, 194; photograph, *NHH* 8, no. 8, whole no. 92 (Sept. 1907): 275; photographs, Lust, *Collected Works*, 98–99; Naturopathic News, *NHH* 19, no. 8, whole no. 250 (Aug. 1914): 537; Naturopathic News, ibid. 19, no. 2, whole no. 244 (Feb. 1914): 119.

21. R. E. Brandman, "A Message to the People: Practical Ideals Embodied in Naturopathy," *NHH* 19, no. 3, whole no. 245 (Mar. 1914): 149; Naturopathic News, ibid. 19, no. 9, whole no. 251 (Sept. 1914): 607.

22. Benedict Lust, Dr. Lust Speaking, ibid. 42, no. 6, current no. 511 (June 1937): 162. The incorporation was different from that under the 1929 and 1931 laws in Washington, DC, which had defined and legitimized naturopathy. Photograph and caption, *Naturopath* 28, no. 9, whole no. 347 (Sept. 1923): 443; advertisement, "American School of Naturopathy and American School of Chiropractic," ibid. 28, no. 2, whole no. 340 (Feb. 1923): 89; "Naturopathic Business Directory and Buyers' Guide: Drugless Schools," ibid., 102.

23. Alzamon Ira Lucas, "The American Drugless University," *HHN* 25, no. 8, whole no. 312 (Sept. 1920): 389–92, at 389; Naturopathic News, ibid. 25, no. 7, whole

no. 311 (Aug. 1920): 356; William F. Havard, "An Open Letter from the Secretary of the ANA," *NNH* 25, no. 9, whole no. 313 (Oct. 1920): 424.

24. Photographs, "Certificate of Attendance and Qualification," *Naturopath* 28, no. 2, whole no. 340 (Feb. 1923): 61; "The American School of Naturopathy Diploma: Doctor of Naturopathy," ibid., 62; "Universal Naturopathic College and Hospital," *HHN* 23, no. 1, whole no. 281 (Jan. 1918): 85.

25. Benedict Lust, "Our Schools and Colleges," *HHN* 23, no. 8, whole no. 288 (Aug. 1918): 709; "Graduates of the Classes of 1928–1929," *NP* 34, no. 5, current no. 415 (May 1929): 219. On a list of a faculty of nine, one woman is named and there is one gender-neutral name; and women made up only a quarter of the alumni association.

26. "Nature School Sued by Senator Straus," *New York Times*, 30 Jan. 1924, 8; Benedict Lust, "New Address of Our School," *Naturopath* 30, no. 9 (Sept. 1925): 847; Lust, *Collected Works*, 82; Albert F. Garcia, "The Senior Class," *Naturopath* 29, no. 7 (July 1924): 615. The sum of $35,000 in 1924 was the equivalent of $464,510 in 2013. Samuel H. Williamson, "Seven Ways to Compute the Relative Value of a U.S. Dollar Amount, 1774 to present," MeasuringWorth, 2015, accessed 26 June 2015, www.measuring worth.com/uscompare/.

27. Benedict Lust, "My Message," *Naturopath* 28, no. 6, whole no. 344 (June 1923): 1; Garcia, "Senior Class," 615.

28. Advertisement, "American School of Naturopathy," *Naturopath* 28, no. 11, whole no. 349 (Nov. 1923): 684; advertisement, "American School of Naturopathy and American School of Chiropractic," ibid. 28, no. 12, whole no. 350 (Dec. 1923): 720; Lust, *Collected Works*, 83.

29. Kirchfeld and Boyle, *Nature Doctors*, 212; "Naturopathic Business Directory and Buyers' Guide," 102; State News: Pennsylvania, *Naturopath* 31, no. 12, whole no. 386 (Dec. 1926): 618–19; photograph, "Graduating Class, 1926, Portland, Maine Branch of American School of Naturopathy," ibid. 31, no. 7, whole no. 381 (July 1926): 357; State News: Connecticut, ibid. 31, no. 8, whole no. 382 (Aug. 1926): 407; "Bits of News, Gathered from Here and There, in the Naturopathic Field: California," *NHH* 49, no. 4, current no. 585 (Apr. 1944): 120, re: Cale College. By the mid-1930s plans were afoot to begin a Natureopathic College in Connecticut, ultimately called the University of the Healing Arts. The Columbia College of Naturopathy in Kansas City, Missouri, was also planned. More were added or planned later in the decade, including the Southern University of Naturopathy and Physiomedicine in Miami (1937); a proposed college of naturopathy in Centerville, Michigan (1940); and the California College of Natural Healing Arts (1944), originally the Cale College of Naturopathy (1927). State News, "Emerson University," ibid. 48, no. 4, current no. 581 (Oct. 1943): 123; State News: Connecticut, ibid. 40, no. 6 (June 1935): 181; State News: Missouri, ibid. 40, no. 9 (Sept. 1935): 281; I. S. Weinger, "Naturopathic Commemoration Day," ibid. 41, no. 1, current no. 506 (Jan. 1937): 33, about the Southern University; "Proposed College of Naturopathy in Centerville," ibid. 45, no. 6, current no. 547 (June 1940): 180; State News: Pennsylvania, ibid. 45, no. 9, current no. 550 (Sept. 1940): 287.

30. Benedict Lust, "Proclamation to the Board of Directors, Trustees, Officers, State Representatives, Delegates and Members Assembled for the 31st Annual Convention of the American Naturopathic Association . . . September 14–18, 1927 Minneapolis, MN," *NP* 32, whole no. 396 (Oct. 1927): 473; Louis S. Reed, *The Healing*

Cults: A Study of Sectarian Medical Practice; Its Extent, Cause, and Control (Chicago: Univ. of Chicago Press, 1932), quoted in Hans A. Baer, "The Sociopolitical Status of US Naturopathy at the Dawn of the 21st Century," *MAQ* 15, no. 3 (2001): 329–46, at 331; "Schools of Chiropractic and of Naturopathy in the United States: Report of Inspections," *JAMA* 90 (May 1928): 1738; Kirchfeld and Boyle, *Nature Doctors*, 201.

31. Lust, *Collected Works*, 66. Lust incorrectly referred to the year Webb-Loomis passed as 1928, however.

32. Ibid., 84–85.

33. An advertisement announcing the school's opening called it the American School of Naturopathy. Yet a photograph of the building refers to it as the American School of Naturopathy and American School of Chiropractic. See Benedict Lust, pamphlet, *Naturopathic Treatment of Disease*, vol. 1, *Lessons 1–9* (New York: Benedict Lust, 1930); "Graduation 1931," *NP* 36, no. 5, current no. 439 (May 1931): 151.

34. Benedict Lust, Dr. Lust Speaking, "Help Finance Our Naturopathic College," *NHH* 40, no. 11 (Nov. 1935): 322.

35. "American College of Naturopathy," *HHN* 25, no. 9, whole no. 313 (Oct. 1920): 422.

36. Advertisement, "Lindlahr College of Natural Therapeutics," ibid. 23, no. 11, whole no. 291 (Nov. 1918): 900; William F. Havard, ND, "A Course in Basic Diagnosis," ibid. 25, no. 1, whole no. 304 (Jan. 1920): 28–33.

37. Helen Wilmans, "Home Course in Mental Science: Practical Healing, Lesson XIX," ibid. 25, no. 1, whole no. 305 (Jan. 1920): 39–46.

38. See Benedict Lust, "Specific Home Study Course of Naturopathy: Lesson I," *Naturopath* 31, no. 2, whole no. 376 (Feb. 1926): 58–61; and monthly lessons after that.

39. "Study Naturopathy at Home!," *NP* 34, no. 3, current no. 413 (Mar. 1929): 146.

40. "Post-Graduate Study of Naturopathy for Students and Practitioners," *NHH* 40, no. 2 (Feb. 1935): 44–48. The first installment's questions focused on vitamins. The examination spanned several issues. See "Post-Graduate Study of Naturopathy for Students and Practitioners," ibid. 41, no. 1, current no. 506 (Jan. 1937): 7–10; and Benedict Lust, Dr. Lust Speaking, ibid. 40, no. 11 (Nov. 1935): 355.

41. Advertisement, "THE MODERN HOME STUDY COURSE: NATUROPATHY," ibid. 49, no. 5, current no. 587 (May 1944): 155; "Complete Index for 1944," ibid. 49, no. 12, current no. 593 (Dec. 1944): 382.

42. Benedict Lust, Dr. Lust Speaking, ibid. 49, no. 6, current no. 535 (June 1939): 162.

43. "Schools of Chiropractic and of Naturopathy in the United States," 1738; Kirchfeld and Boyle, *Nature Doctors*, 214; *Yearbook of the International Society of Naturopathic Physicians and of Emerson University Research Council* (Los Angeles: Schramm, 1945); Kirchfeld and Boyle, *Nature Doctors*, 213. The AMA report claimed that Collins had "graduated" more than four thousand practitioners, many of them World War I veterans whose studies were funded by the government. This contributed to later government refusals to provide funding for soldiers to study at naturopathic schools.

44. F. W. Collins, "Dr. Benedict Lust Lectures at Mecca College of Chiropractic," *HHN* 31 (Feb. 1916): 21; Kirchfeld and Boyle, *Nature Doctors*, 213.

45. Lust, *Collected Works*, 65–66; Kirchfeld and Boyle, *Nature Doctors*, 200.

46. Associated Press, "Connecticut Moves for Wide Exposure of Diploma Frauds," *New York Times*, 19 Nov. 1923, 1.

47. Quinnie H. Cheatam, ND, ed., *The Lighthouse* (Nashville School of Drugless Therapy), reprinted in *NHH* 41, no. 1, current no. 506 (Jan. 1937): 29.

48. Walter Seth Kipnis, State News: Hawaii, *NHH* 45, no. 3, current no. 544 (Mar. 1940): 94.

49. The chart's AMA data was taken from "Laws and Rules of Higher Education in Medicine," *University of the State of New York's Handbook*, no. 9 (June 1926): 59. These were the most recent standards in place in 1940. *NHH* 45, no. 9, current no. 550 (Sept. 1940): 266.

50. Lust claimed that the name change occurred in 1904, yet the masthead of his journal carried the earlier name until 1919. Lust, *Collected Works*, 70. See also Kirchfeld and Boyle, *Nature Doctors*, 202; Naturopathic News, *Naturopath* 31, no. 3, whole no. 377 (Mar. 1926): 143; Whorton, *Nature Cures*, 215; and Lust, *Universal Naturopathic Encyclopedia*.

51. Lust, *Collected Works*, 70; "Complete Route of Dr. Lust's Trip to California," *Naturopath* 31, no. 7, whole no. 381 (July 1926): 349.

52. Edward Earle Purinton, Efficiency in Drugless Healing, "Making It Pay," *NHH* 20, no. 2, whole no. 256 (Feb. 1915): 71–78; insert, *HHN* 23, no. 3, whole no. 283 (Mar. 1918): 286; Benedict Lust, "What Does the A.N.A. Mean to You?," ibid. 23, no. 4, whole no. 284 (Apr. 1918): 368.

53. Chester A. Young, ND, SCT, LSD, "The Color Button of the A.N.A.," *Naturopath* 28, no. 1, whole no. 339 (Jan. 1923): 15; "Drugless Directory of the American Naturopathic Association," ibid. 28, no. 2, whole no. 340 (Feb. 1923): 99–101; "Emblems," ibid. 28, no. 3, whole no. 341 (Mar. 1923): 141; "Directory of the American Naturopathic Association," ibid. 29, no. 7 (July 1924): 91–96; Naturopathic News, ibid. 31, no. 6, whole no. 380 (June 1926): 296.

54. Front page, ibid. 28, no. 10, whole no. 348 (Oct. 1923); Naturopathic News, ibid. 28, no. 11, whole no. 349 (Nov. 1923): 689.

55. "Constitution of the American Naturopathic Association," *NP* 28, no. 11, current no. 421 (Nov. 1929): 520–21; Robert V. Carroll, "An Appeal to All Naturopaths," *NHH* 42, no. 12, current no. 517 (Dec. 1937): 356.

56. Kirchfeld and Boyle, *Nature Doctors*, 204, 226. The ANA underwent a period of uncertain leadership after Lust's death in 1945; Gehman (sometimes spelled Gehmann) was first referred to as president after the Golden Jubilee Congress in 1947. Dr. Jesse Mercer Gehman, "Outstanding Program for Golden Jubilee Congress Practically Completed," *NP* 51, nos. 7–8 (July–Aug. 1947): 410.

57. Harry Finkel, ND, DC, "Status of the Naturopaths," *NHH* 42, no. 5, current no. 510 (May 1937): 133; Walter Seth Kipnis, State News: Connecticut, ibid. 45, no. 2, current no. 543 (Feb. 1940): 62; Kipnis, State News: Hawaii, ibid. 45, no. 3, current no. 544 (Mar. 1940): 92; L. E. Polhemus, "Bits of News, Gathered from Here and There, in the Naturopathic Field," *NHH* 49, no. 8, current no. 589 (Aug. 1944): 249; "Proposed College of Naturopathy in Centerville," 180; Walter Seth Kipnis, State News: Washington, *NHH* 45, no. 6, current no. 547 (June 1940): 192; L. A. Kasparie, "Bits of News, Gathered from Here and There, in the Naturopathic Field: Colorado," ibid. 49, no. 12, current no. 593 (Dec. 1944): 376; Don F. Davis, "Bits of News, Gathered from Here and There, in the Naturopathic Field," ibid. 49, no. 9, current no. 590 (Sept. 1944): 250.

58. Walter Seth Kipnis, State News: California, *NHH* 45, no. 3, current no. 544 (Mar. 1940): 92, on the National Association of Naturopathic Herbalists; William R. Lucas, DN, PhD, Mail Bag, ibid., 79, on joint educational efforts; L. A. Yurman, "Bits of News: New Jersey," ibid. 49, no. 12, current no. 593 (Dec. 1944): 377.

59. State News, "Cuba a Spanish Naturopathic School," *Naturopath* 31, no. 8, whole no. 382 (Aug. 1926): 407; Harry Raitano and Walter Seth Kipnis, State News: Florida, *NHH* 48, no. 4, current no. 581 (Oct. 1943): 124.

60. J. Henry Hallberg, "Ultrashort Wave Selective Therapy," *NHH* 44, no. 1, current no. 530 (Jan. 1939): 9; Cody, "History of Naturopathic Medicine," 22; Harry Riley Spitler, MD, ND, PhD, ed., *Basic Naturopathy* (New York: American Naturopathic Association, 1948); A. W. Kuts-Cheraux, BS, MD, ND, ed., *Naturae Medicina and Naturopathic Dispensatory* (Des Moines: American Naturopathic Physicians and Surgeons Association, 1953).

61. "Program: The American Naturopathic Association and the Pennsylvania State Naturopathic Association in Combined Convention," *Naturopath* 31, no. 7, whole no. 381 (July 1926): 348–49; F. W. Collins, "New Jersey Programme," *NHH* 40, no. 7 (July 1935): 215; M. S. Dantzler, "Address before the First Convention of South Carolina Naturopathic Association," ibid. 42, no. 12, current no. 517 (Dec. 1937): 375–76, on Lust's address. In the photograph of delegates and officers at the thirtieth annual convention, held in Pennsylvania in 1925, 53 people appear; 15 are women, 1 is African American, and 3 to 5 are Latinos. Photograph, "30th Annual Convention of the A.N.A.," *Naturopath* 31, no. 8, whole no. 382 (Aug. 1926): 390; Benedict Lust, "The Twenty-First Annual Convention of the American Naturopathic Association," *HHN* 22, no. 3 (May–June 1917): 1.

62. Announcement, *HHN* 24, no. 10 (Oct. 1919): 1. Beginning in 1914, the annual conventions were held in Butler and Atlantic City, New Jersey (1915), Chicago (1916, 1917), Cleveland (1918), Philadelphia (1919), New York (1920, 1925), Indianapolis (1926), Philadelphia (1927), Minneapolis (1928), Portland, Oregon (1929), New York (1930), Milwaukee (1931), Washington, DC (1932), Chicago (1933), Denver (1934), San Diego (1935), Omaha (1936), Chicago (1937), and so on. Examples of convention reports appeared in *NHH* 42, no. 6, current no. 511 (June 1937): 163; Cody, "History of Naturopathic Medicine," 19; Kirchfeld and Boyle, *Nature Doctors*, 202; and Benedict Lust, "41st Annual Convention of the American Naturopathic Ass'n," *NHH* 42, no. 6, current no. 511 (June 1937): 163.

63. Lust, "41st Annual Convention," 163; Lust, "Gone! Never to Be Forgotten: A Report of the Great Convention," *Naturopath* 28, no. 9, whole no. 347 (Sept. 1923): 485–86; Lust, "41st Annual Convention," 163; Lust, "Report about the Forty-Fourth Annual Congress of the American Naturopathic Convention," *NHH* 45, no. 8, current no. 549 (Aug. 1940): 233; "Officers of the American Naturopathic Association," ibid. 40, no. 6 (June 1935): 168.

64. Naturopathic News, "1918 American Naturopathic Convention," *HHN* 23, no. 5, whole no. 285 (May 1918): 480; Naturopathic News, "The 24th Annual Convention," ibid. 25, no. 7, whole no. 311 (Aug. 1920): 356; "Announcement," *Naturopath* 31, no. 6, whole no. 380 (June 1926): 291, on the Philadelphia convention of 1926; "The 34th National Convention of the American Naturopathic Association (NYC 1932)," *NP* 34, no. 1, current no. 435 (Jan. 1931): 16. The December 1930 program was indeed advertised

in the January 1931 issue. See also "Naturopathic News: The 27th Annual Convention of the ANA, Chicago," *Naturopath* 28, no. 5, whole no. 343 (May 1923): 246.

65. Benedict Lust, "Convention Report," *HHN* 23, no. 8, whole no. 288 (Aug. 1918): 710; "Stenographic Report of the Twenty-second Annual Convention of the American Naturopathic Association, held on June 6th, 7th and 8th, 1918, at Hotel Winton, Cleveland, Ohio," ibid., 716 (the report continued in the September, October, and December issues); "The A.N.A. Convention," *HHN* 32, no. 7, whole no. 287 (July 1918): 681; "Synopsis of Report of the Twenty-fourth Annual Convention of the American Naturopathic Association (New York City, Hotel Commodore, September 23–25, 1920)," ibid. 25, no. 9, whole no. 313 (Oct. 1920): 453–59.

66. "Synopsis of Report of the Twenty-fourth Annual Convention"; E. C. Burton, "Notes on the Proceedings of the Twenty-Eighth Annual Convention of the American Naturopathic Association, Los Angeles, California, November 20–22, 1924," *Naturopath* 30, no. 3, whole no. 365 (Mar. 1925): 205–9; Benedict Lust, "Address of Welcome," *NHH* 40, no. 8 (Aug. 1935) 229, 248–53, at 249; T. M. Schippell (ANA national corresponding secretary), "Proceedings of the 39th Annual Congress of the American Naturopathic Ass'n," ibid., 228, 239–41, 244, 246.

67. Benedict Lust, Dr. Lust Speaking, *NHH* 42, no. 11, current no. 516 (Nov. 1937): 322, 332; T. Louise Nedvidek (convention manager), "The Forty-First Annual Congress of the American Naturopathic Association," ibid., 323, 330; E. C. Burton, "Notes on the Proceedings of the Twenty-Eighth Annual Convention," 207, on John Harvey Kellogg.

68. Harry F. Findeison, "American Naturopathic Association Annual Convention, December 14, 15, 16, 17, 1925: Minutes of the Executive Session, December 17, 1925," *Naturopath* 31, no. 1, whole no. 375 (Jan. 1926): 43–44; Benedict Lust, "Convention Greetings," *NP* 32, whole no. 395 (Sept. 1927): 421.

69. Benedict Lust, "With the Editor in Naturopathy," *NP* 34, no. 11, current no. 421 (Nov. 1929): 492; Byrd Mock, "Echoes of the Naturopathic Convention in the Nation's Capital," *Naturopath* 28, no. 9, whole no. 347 (Sept. 1923): 442; Lust, "41st Annual Convention of the American Naturopathic Ass'n," 163; "Asserts Shower Bath Will Stop Old Age," *New York Times*, 31 Aug. 1923.

70. "Synopsis of Report of the Twenty-fourth Annual Convention," 453; Lust, "41st Annual Convention of the American Naturopathic Ass'n," 163, 175, recalling the difficulties in the 1920s.

71. E. K. Stretch, DO, ND, Osteopathy Department, *NHH* 20, no. 12, whole no. 266 (Dec. 1915): 761; Naturopathic News, *HHN* 23, no. 6, whole no. 286 (June 1918): 582; Naturopathic News, ibid. 23, no. 11, whole no. 291 (Nov. 1918): 901; Naturopathic News, ibid. 25, no. 9, whole no. 313 (Oct. 1920): 460.

72. Edward Earle Purinton, Correspondence, *NHH* 22, no. 12, whole no. 266 (Dec. 1915): 739; Benedict Lust, "Another Case of Chiropractitis," *HHN* 23, no. 7, whole no. 287 (July 1918): 620; Alzamon Ira Lucas to B. J. Palmer, 6 Nov. 1920, in "Chronology of Chiropractic in New Jersey," 7, National Institute of Chiropractic Research, Phoenix, www.chiro.org/Plus/History/Colleges/MeccaNewJersey/new%20jersey_chronology.pdf.

73. Henry Lindlahr, "Welcome to Chicago," *Naturopath* 28, no. 9, whole no. 347 (Sept. 1923): 426.

74. "Drugless Cults Organize to Fight for Recognition," *New York Herald Tribune*, 2 May 1926; Kirchfeld and Boyle, *Nature Doctors*, 257; Walter Seth Kipnis, State News, *Naturopath* 31, no. 8, whole no. 382 (Aug. 1926): 408.

75. Harry F. Findeison, "An Open Letter to the American Naturopathic Association," *NP* 32, whole no. 397 (Nov. 1927): 536–37 (the letter was read at the ANA convention in September 1927); "The 34th National Convention of the A.N.A.: A Glorious Success," ibid. 36, no. 2, current no. 436 (Feb. 1931): 48–49, 55.

76. Walter Seth Kipnis, State News: California and Connecticut, *NHH* 40, no. 5 (May 1935): 154; "Committees Appointed by the Executive Committee of the A.N.A. for the Ensuing Year," ibid. 40, no. 9 (Sept. 1935): 286; F. R. Good, "Organization," ibid. 30, no. 10 (Oct. 1935): 307.

77. Walter Seth Kipnis, State News: Arizona, ibid. 40, no. 11 (Nov. 1935): 338; Benedict Lust, Dr. Lust Speaking, ibid. 40, no. 2 (Feb. 1935): 34; Lust, Dr. Lust Speaking, ibid. 40, no. 8 (Aug. 1935): 227.

78. Kirchfeld and Boyle, *Nature Doctors*, 258.

79. Dr. I. S. Weinger, quoting Benedict Lust, in "Naturopathic Commemoration Day," 25; Louis D. McKenney, PhC, ND, "Naturopaths and Naturopathy," *NHH* 41, no. 2, current no. 507 (Feb. 1937): 35; Cody, "History of Naturopathic Medicine," 20.

80. Israel Shmid, ND, "State Board Examinations," *NHH* 44, no. 1, current no. 530 (Jan. 1939): 4.

81. Front page, ibid. 45, no. 12, current no. 553 (Dec. 1940); "Bits of News, Gathered from Here and There, in the Naturopathic Field: Maryland," ibid. 49, no. 12, current no. 593 (Dec. 1944): 377, scolding those who had not paid their ten-dollar dues and warning them that they would not benefit from a recently submitted licensing bill without a certificate of qualification and a membership card; J. C. Thie, ND, DC, DP, "Why the A.N.A.," ibid. 45, no. 10, current no. 551 (Oct. 1940): 296.

82. Benedict Lust, Dr. Lust Speaking, ibid. 42, no. 10, current no. 515 (Oct. 1937): 290–91.

83. Ibid.

84. Benedict Lust, Dr. Lust Speaking, *NHH* 44, no. 7, current no. 536 (July 1939): 216.

85. Benedict Lust, Dr. Lust Speaking, ibid. 44, no. 12, current no. 541 (Dec. 1939): 354.

86. Walter Seth Kipnis, State News: Connecticut, ibid. 42, no. 7, current no. 512 (July 1937): 208.

87. Walter Seth Kipnis, State News: Arizona, "Action Filed by Naturopath," ibid. 44, no. 2, current no. 531 (Feb. 1939): 47, reprinted from *Arizona Republic* (Phoenix), 15 Dec. 1938. The petitioner was Edward W. Lydic; the board member, W. S. Swank. Kipnis, State News: Pennsylvania, *NHH* 44, no. 8, current no. 537 (Aug. 1939): 246.

88. Mail Bag, *NHH* 44, no. 9, current no. 538 (Sept. 1939): 285; Walter Seth Kipnis, State News: National, ibid. 45, no. 5, current no. 546 (Mar. 1940): 125; Kipnis, State News: Florida, ibid. 45, no. 6, current no. 547 (June 1940): 188; Kipnis, State News: New York, ibid., 190; Hans Zimmerman, Mail Bag, *NHH* 45, no. 8, current no. 549 (Aug. 1940): 234; Benedict Lust, Dr. Lust Speaking, ibid., 226, on osteopathy and the AMA.

89. Kirchfeld and Boyle, *Nature Doctors*, 203–4; "Bits of News, Gathered from Here and There, in the Naturopathic Field: National," *NHH* 45, no. 6, current no. 547 (June 1940): 188; Carl E. Hotchkiss, "Bits of News, . . . California," ibid. 49, no. 3, current

no. 584 (Feb. 1944): 59; R. C. Allred, "Bits of News, . . . Utah," ibid. 49, no. 4, current no. 585 (Apr. 1944): 125.

90. Walter Seth Kipnis, State News: Illinois, *NHH* 45, no. 7, current no. 548 (July 1940): 221, on the United Front for Freedom in Healing and the New Jersey Anti-Medical Trust Federation; Benedict Lust, "Resolution Presented at the National Democratic Convention [in Chicago]," ibid. 45, no. 9, current no. 550 (Sept. 1940): 275, on the Progressive Alliance for Medical Art; Harry Riley Spitler (secretary, International Association of Liberal Physicians), "The International Association of Liberal Physicians, Inc.," ibid. 49, no. 6, current no. 587 (June 1944): 182; Harry Francis (national president, Loyal Liberty Legion), "The Loyal Liberty Legion: Its Relation to the Non-Medical Profession," ibid. 49, no. 9, current no. 590 (Sept. 1944): 267; "Bits of News Gathered from Here and There, in the Naturopathic Field: National," ibid. 49, no. 11, current no. 592 (Nov. 1944): 344, on the National Association of Naturopathic Herbalists; S. J. Singh, MA, BSc, ND (principal, English Naturopathic College, Lucknow, India), "Yoga Therapy," ibid. 42, no. 6, current no. 511 (June 1937): 171; S. S. Goswami, "Naturopathy and Yoga" ibid. 44, no. 11, current no. 540 (Nov. 1939): 327, 338, on a demonstration held at the forty-third annual convention in Pittsburgh.

91. Lucas to Palmer, 6 Nov. 1920; R. B. Jackson, "Willard Carver, LL.B., 1866–1943: Doctor, Lawyer, Indian Chief, Prisoner and More," *Chiropractic History* 14, no. 2 (Dec. 1994): 12–20, describing the ANA takeover attempt by one Dr. Compere and Carver; correspondence between W. W. Shenk, of Minneapolis, J. B. Branyon, of Spartanburg, SC, and Benedict Lust, and correspondence between E. W. Cordingley and Lust, in "There Is Only One A.N.A.," *NHH* 48, no. 3, current no. 580 (July 1943): 84–85.

92. Kirchfeld and Boyle, *Nature Doctors*, 216; Benedict Lust, Dr. Lust Speaking, *NHH* 48, no. 4, current no. 581 (Oct. 1943): 98–99.

93. After Father Divine's death Mrs. S. A. Divine became the spiritual leader. She fought a takeover attempt in 1972 by Jim Jones, leader of the infamous People's Temple, who claimed to be Father Divine reincarnated. The Peace Mission Movement still exists in several American cities and two foreign countries. Benedict Lust, "Father Divine," *NHH* 49, no. 6, current no. 587 (June 1944): 182; Jill Watts, *God, Harlem U.S.A.: The Father Divine Story* (Berkeley: Univ. of California Press, 1992).

94. T. M. Schippell, "Official Bulletin of the American Naturopathic Association," *NHH* 49, no. 11, current no. 592 (Nov. 1944).

95. "Dr. Lust, Advocate of Water Healing," *New York Times*, 6 Sept. 1945.

CHAPTER 10: Deepening Divides, 1945–1969

1. Paul Starr, *The Social Transformation of American Medicine: The Rise of a Sovereign Profession and the Making of a Vast Industry* (New York: Basic Books, 1982), 260, 273, 276.

2. I am indebted to George Cody for his insights on these matters. Personal correspondence, Cody to author, 10 Oct. 2012. See also Cody, "History of Naturopathic Medicine," in *Textbook of Natural Medicine*, ed. Joseph Pizzorno and Michael Murray (New York: Churchill Livingstone), 17–40, at 21–22; Hans A. Baer, "The Sociopolitical Status of US Naturopathy at the Dawn of the 21st Century," *MAQ* 15, no. 3 (2001): 329–46, at 332; and Dr. W. H. Pyott, "Do Naturopathic Physicians Want Hospitals?," *JANA* 3 (Jan. 1950): 8.

3. Starr, *Social Transformation*, 269.

4. Cody, "History of Naturopathic Medicine," 26–28.

5. John Mack Faragher, Mary Jo Buhle, Daniel Czitrom, and Susan Armitrage, *Out of Many: A History of the American People* (Upper Saddle River, NJ: Pearson Prentice Hall, 2006), 678; Starr, *Social Transformation*, 336; Cody, "History of Naturopathic Medicine," 21.

6. Martin J. Blaser, "Who Are We? Indigenous Microbes and the Ecology of Human Diseases," *EMBO Reports* 7, no. 10 (Oct., 2006): 956–60, www.ncbi.nlm.nih .gov/pmc/articles/PMC1618379/.

7. Ibid.; Starr, *Social Transformation*, 336; E. Richard Brown, *Rockefeller Medicine Men: Medicine and Capitalism in America* (Berkeley: Univ. of California Press, 1979), 219–20; Stephen H. Paschen and Leonard Schlup, eds., *Documents Depicting the 1950s: A Decade of Conformity and Dissent*, foreword by William C. Binning (Lewiston, NY: Mellon, 2008), reviewing policies and court opinions on a number of issues and noting the authority of scientific medicine on medical research, smoking, and alcoholism. Another source promoting medical conformity in the 1950s was the television show *Confidential File*, hosted by Paul Coates, which focused on medical quacks. The film *Quacks and Nostrums* (1959) reviewed the dangers of quacks prescribing non-AMA approved remedies. Prelinger Archives, accessed 4 Dec. 2013, https://archive.org/details/medical_quacks_1 and https://archive.org/details/Quacksan1959; Elaine Tyler May, *Homeward Bound: American Families in the Cold War Era* (New York: Basic Books: 2008), 19–20.

8. Cody, "History of Naturopathic Medicine," 21–22; F. Campion, *AMA and U.S. Health Policy since 1940* (Chicago: AMA, 1984).

9. Friedhelm Kirchfeld and Wade Boyle, *Nature Doctors: Pioneers in Naturopathic Medicine* (East Palestine, OH: Buckeye Naturopathic, 1993), 310, as cited in P. Joseph Lisa, *The Assault on Medical Freedom* (Norfolk, VA: Hampton Roads, 1994), 272; Walter I. Wardwell, *Chiropractic: History and Evolution of a New Profession* (Saint Louis: Mosley, 1992), 37–38, 130; Morris Fishbein, "Naturopathy and Its Professors" (1932), Naturowatch, accessed 22 Jan. 2014, www.naturowatch.org/hx/fishbein.html; Hans Baer, *Biomedicine and Alternative Healing Systems in America: Issues of Class, Race, Ethnicity, and Gender* (Madison: Univ. of Wisconsin Press, 2001), 89.

10. "Physicians Opposed Chiropractor Bill: Five County Units Here Fight State Licensing—Chairman for Ban on the 'Cultists,'" *New York Times*, 6 Mar. 1953, 19.

11. Personal correspondence, Cody to author, 10 Oct. 2012; Cody, "History of Naturopathic Medicine," 22; Wardwell, *Chiropractic: History and Evolution*, 39, 105–30, 161–75, 243–47; Walter I. Wardwell, "Chiropractors: Evolution to Acceptance," in *Other Healers: Unorthodox Medicine in America*, ed. Norman Gevitz (Baltimore: Johns Hopkins Univ. Press, 1988), 172.

12. Baer, "Sociopolitical Status of US Naturopathy," 332. According to Walter Wardwell, the trend in chiropractors distancing themselves from naturopaths began in the 1920s, when many chiropractors preferred separate examining boards. Wardwell, *Chiropractic: History and Evolution*, 11; Kirchfeld and Boyle, *Nature Doctors*, 310. The NCNM opened in Portland, Oregon, and moved to Seattle in 1959.

13. Wardwell, "Why Did Chiropractic Survive?," in *Chiropractic: History and Evolution*, 254, 256.

14. Martin Kaufman, "Homeopathy in America: The Rise and Fall and Persistence of a Medical Heresy," in Gevitz, *Other Healers*, 116–17.

15. Cody, "History of Naturopathic Medicine," 28; Wardwell, *Chiropractic: History and Evolution*, 256; Kirchfeld and Boyle, *Nature Doctors*, 258, 203.

16. Kirchfeld and Boyle, *Nature Doctors*, 203, citing George Floden, "The Registry and Its Purpose," *Naturopathic Magazette* 16, no. 2 (1964).

17. "The Golden Jubilee Congress of the American Naturopathic Association, Inc.," *NP* 51, no. 5, current no. 628 (May 1947): 292; George A. Freibott, *The History of Naturopathy or "Pseudomedicalism": Naturopathy's Demise?*, report submitted to the US Department of Education (Priest River, ID: privately printed, 1990), as cited in Kirchfeld and Boyle, *Nature Doctors*, 259.

18. Kirchfeld and Boyle, *Nature Doctors*, 204; Dr. Jesse Mercer Gehman, "Outstanding Program for Golden Jubilee Congress Practically Completed," *NP* 51, nos. 7–8 (July–Aug. 1947), 410; Cody, "History of Naturopathic Medicine," 27.

19. Kirchfeld and Boyle, *Nature Doctors*, 207. At the Golden Jubilee Congress a Woman's Auxiliary of the ANA was formed. There was a presumption in the journals that most women in attendance were accompanying their husbands. "American Naturopathic Association News from the President's Office: 'Women's Auxiliary of the American Naturopathic Association, Inc.,'" *NP* 51, no. 3, current no. 626 (Mar. 1947): 143; Frederick W. Collins, MD, AM, "To All Naturopaths throughout the World," ibid. 51, no. 5, current no. 628 (May 1947): 291; advertisement, "FREDERICK W. COLLINS," ibid., 305.

20. Robert V. Carroll, "Unity and Progress," *JANA* 1 (Apr. 1948): 3; Paul Wendel, *Standardized Naturopathy: The Science and Art of Natural Healing* (Brooklyn: privately printed, 1951); Harry Riley Spitler, MD, ND, PhD, ed., *Basic Naturopathy* (Des Moines: American Naturopathic Physicians and Surgeons Association, 1948); Kirchfeld and Boyle, *Nature Doctors*, 206, citing William A. Turska, "Drugless Physician—A Misnomer?," *Naturopathic Student News* 2 (1976): 5; H. Riley Spitler, "Report of the Council on Schools and Colleges," *JANA* 2 (Sept. 1949): 13; Turska, "Drugless Physician—A Misnomer?," 5.

21. Index-page masthead, *NP* 52, no. 6, current no. 641 (July 1948).

22. Advertisement, ibid. 52, no. 8, current no. 643 (Sept. 1948): 478. *Nature's Path* cost $3.50 annually ($33.98 in 2014 dollars). Samuel H. Williamson, "Seven Ways to Compute the Relative Value of a U.S. Dollar Amount, 1774 to present," MeasuringWorth, 2015, accessed 26 June 2015, www.measuringworth.com/uscompare/; index–cover page, *NP* 51, no. 5, current no. 628 (May 1947); Benedict Lust, Dr. Lust Speaking, ibid. 51, no. 3, current no. 626 (Mar. 1947): 132; Lust, Dr. Lust Speaking, "Three Sources of Life Power," ibid. 51, no. 5, current no. 628 (May 1947): 260; Fulton Ousler, "Personalities: Bernarr Macfadden's True Story," ibid. 51, nos. 7–8, current no. 630 (July–Aug. 1947): 403, 427.

23. Samuel Morrison, ND, "How to Treat Anemia," *NP* 52, no. 4, current no. 639 (May 1948): 262, 286; Dr. Jesse Mercer Gehman (president), "51st Annual Convention of American Naturopathic Association, Inc.," ibid. 52, no. 8, current no. 643 (Sept. 1948): 473; Eugene A. Wimmershoff, PhT, "Colonic Irrigation and Good Health," ibid. 52, no. 6, current no. 641 (July 1948): 361, 378; Nathaniel Williams, "Exercise a Vital Adjunct to Every-Day Living," ibid., 360, 386; "Quick Reference

Vitamin Table," *NP* 52, no. 4, current no. 639 (May 1948): 271; advertisement, "Men, Women over 40 Rundown—Listless," ibid. 51, no. 5, current no. 628 (May 1940): 294; Paul C. Bragg, "The New Science of Healthier, Longer Life for Those Past 40," ibid. 51, nos. 7–8, current no. 630 (July–Aug. 1947): 390, 391, 409; Neil Nelson, "Sugar Drunkards," ibid. 51, no. 5, current no. 628 (May 1947): 320.

24. E. W. Cordingley, AM, ND, "Fibroid Tumors of the Uterus," ibid. 52, no. 4, current no. 639 (May 1948): 298; Dr. Max Rosenfeld, "How to Fight Cancer," ibid. 52, no. 8, current no. 643 (Sept. 1948): 468, 490; Irma Goodrich Mazza, "How to Grow Herbs," ibid. 52, no. 6, current no. 641 (July 1948): 370. The December 1928 issue of *Nature's Path* referred to the work of Herman J. DeWolff, a Dutch epidemiologist; ibid. 33, no. 12 (Dec. 1928). Cody, "History of Naturopathic Medicine," 21.

25. John B. Lust, "Sickness Is Nature's Warning," *NP* 58, no. 8 (Aug. 1953): 4; M. A. Brandon, DO, "The Chemical Poisoning of Our Drinking Water," ibid. 59, no. 2 (Feb. 1954): 10–13; advertisement, "Filter-All," ibid. 67, no. 3 (Fall 1963): 19; Barbara Lust, "Poisons Formed by Aluminum Cook Ware," ibid., 18; Jeannette Druce, "A New Look At . . . Doctors, Drugs and Patients," *NP* 61, no. 3 (Mar. 1956): 12, 18–19; Evelyn Carnell, "Getting Rid of Varicose Veins," ibid. 58, no. 8 (Aug. 1953): 8; Eugene A. Bergholz, MD, "One Cure for Stomach Ulcers," ibid., 11–12, 28–29; Ruth Laverty, "The 'Coughs' Mean Bronchitis," *NP* 59, no. 2 (Feb. 1954): 5, 16–18, 22; John B. Lust, ed., "Fifty, the Age of Opportunity," ibid., 4; O. M. Richardson, "How to Relieve Bursitis," ibid., 12, 35–36.

26. John B. Lust, *Lust for Living: How to Grow Younger, Healthier, Happier* (New York: Benedict Lust, 1953); and Lust, *Drink Your Troubles Away* . . . (New York: Benedict Lust, 1967), on raw-juice therapy; "Directory of Health Aids," *NP* 58, no. 8 (Aug. 1953): 31; Dr. Jules Lemoine, "How to Use Yoga for Restful Sleep," ibid. 59, no. 2 (Feb. 1954): 9, 29–31; Edgar J. Saxon, ND, "Foot Health . . . Pays Dividends," ibid. 61, no. 3 (Mar. 1956): 10, 14–15; "El Rancho Adolphus Home of Scientific Living, Inc. of Scranton, PA," ibid. 52, no. 6, current no. 641 (July 1948): 353; "Directory of Health Aids," 31–32; Philip S. Cascio, "Nature's Miracle Healers—Citrus Bioflavonoids," *NP* 61, no. 3 (Mar. 1956): 5, 16; Barbara Lust, "Vegetable Juices and Their Uses," ibid. 67, no. 3 (Fall 1963): 15, 54–55; advertisement, "THE CHAMPION JUICER," ibid., 13.

27. Virginia S. Lust, "Salads for Health," *NP* 58, no. 8 (Aug. 1953): 7, 25; Alice Chase, "Your Health Problems—A Case of Gynecological Problems," ibid. 59, no. 2 (Feb. 1954): 6, 15, 19.

28. Kirchfeld and Boyle, *Nature Doctors*, 218, 289–90; Mario T. Campanella, editorial, *JNM* 3 (Sept. 1952): 3; advertisement, "International Society of Naturopathic Physicians," *JN* 61 (Nov. 1956): 32; Campanella, ND, "Dear Doctor," undated letter to members of the ISNP on ISNP letterhead.

29. Iva Lloyd, *History of Naturopathic Medicine* (Toronto: McArthur, 2009); Arno Koegler, "Comments on the Merits of Homeopathic Therapeutics," *JNM* 3 (Feb. 1952): 6–7. On membership, see ibid. 3 (Aug. 1952): 3; and Mario T. Campanella, ND, "We are 15 years . . . young!," ibid. 4 (Jan. 1953): 3–4. Arthur Schramm Speaking, "Fifteen years ago . . . ," ibid., 5–6. Schramm specialized in botanicals and had been dean of the Naturopathic Department of the California University of Liberal Physicians in Los Angeles.

30. Mario T. Campanella, editorial, *JNM* 3 (Mar. 1952): 3–4; Cody, "History of Naturopathic Medicine," 28; frontispiece, ibid. 2 (May 1951).

31. Kirchfeld and Boyle, *Nature Doctors*, 207; Wendel, *Standardized Naturopathy*; "Naturopathic Activities," *JNM* 3 (Jan. 1952): 22.

32. Arthur Schramm (president Emeritus, ISNP), "Message from Dr. Schramm," *JNM* 1 (July–Aug. 1950): 13.

33. "Diagnosis of Paralysis of the Median Nerve," ibid. 1 (Oct. 1950): 11; W. Kelly, "Technic [*sic*] for the Expression of Pus from the Ducts of Skene's Glands," ibid. 1 (Nov. 1950): 10; A. L. Ranney, "Diagnosis of Paralysis of the Median Nerve," *JN*, June 1957, 22; "Technique for the Expression of Pus from the Ducts of Skene's Glands," ibid., July 1957, 8; William A. Turska, "Botanics in Diseases of the Heart," *JNM* 5 (Apr. 1954): 16–17.

34. Kirchfeld and Boyle, *Nature Doctors*, 289–91; A. Christensen, "The Allopathic Adaptation Trend," *JNM* 2 (Apr. 1951): 8.

35. Turska articulated five natural methods of living ad treatment: elementary remedies, chemical remedies, mechanical remedies, mental remedies, and constructive thought. William A. Turska, "Catechism of Naturopathy," *JNM* 4 (Oct. 1953): 9–11; "Naturopathic Activities," ibid. 4 (Nov. 1953): 18–19.

36. William A. Turska, ND, "Naturae Medicine Therapeusis and Rationale," *Journal of the American Association of Naturopathic Physicians* 8 (Jan. 1956): 5.

37. William A. Turska, "Some Observations of Naturopathic Philosophy," *JN*, Oct. 1956, 10–13; Turska, "Some Observations of Naturopathic Philosophy," ibid., Nov. 1956, 10–13.

38. Henry Bodewein, "Treating Arthritis and Rheumatism by Old Forgotten Naturopathic Methods," *JNM* 3 (Mar. 1952): 7; Ellen Schramm, "Obstetrics and Pre-Natal Care," ibid. 18.

39. "Naturopathic Activities," ibid. 4 (Nov. 1953): 18.

40. Kirchfeld and Boyle, *Nature Doctors*, 291–92.

41. JANA's board of trustees hailed from Washington State, Kansas, Colorado, Pennsylvania, Texas, Utah, Michigan, and Missouri. "1948 National Convention and Symposium of the American Naturopathic Association," *JANA* 1 (Apr. 1948): 6; "Prominent Naturopath Passes," *JNM* 2 (May 1951): 6. Henry Schlichting Jr., The President's Message, *JANA* 3 (Jan. 1950): 5. Robert Carroll died in 1951.

42. Arthur Schramm, ND, ed., Department of International Relations, *JANA* 3 (July 1950): 7.

43. "Nature's Way Public Health Association," ibid. 2 (July 1949): 17. Harry F. Bonnelle (chairman), "Public Relations and Publicity Department," ibid. 2 (Dec. 1949): 10. The size of the new association's membership was not reported.

44. Odessa Whiting (secretary), "The Ladies' Auxiliary of the A.N.A.," ibid. 2 (July 1949): 14.

45. R. C. Allred, "Childbirth without Fear," ibid. 2 (May 1949): 8, 21; "Sanitarium—Hospital Department," ibid., 10; Membership Committee, Sanitarium Foundation, "Sanitarium—Hospital Department," *JANA* 1 (Apr. 1948): 5; Pyott, "Do Naturopathic Physicians Want Hospitals?," 8.

46. Advertisement, "Send Students," *JANA* 2 (May 1949): 22; "Resolutions Passed by the A.N.A. Convention," ibid. 2 (July 1949): 18. By 1955 the American Association of

Naturopathic Physicians recognized only two schools of naturopathic medicine: the Central State College of Physiatrics in Eaton, Ohio, under H. Riley Spitler's leadership, and the Western States College of Chiropractic and Naturopathy, located near Portland, Oregon, led by W. A. Budden. Both of these schools were affiliated with the values of the western ANA. Cody, "History of Naturopathic Medicine," 22.

47. Henry Schlichting Jr., ND, The President's Message, *JANA* 3 (May 1950): 5; "New Naturopathic Definition," ibid., 17; "Naturopathic Definition Changed," *JANA* 4 (Mar. 1951): 12; Harold Wellington Jones, Norman L. Hoerr, and Arthur Osol, eds., *Blakiston's New Gould Medical Dictionary* (New York: Blakiston, 1949).

48. Phil North, "Scene of the A.N.A. Convention: Houston's Shamrock Hotel Opens 'Mid Scenes of Lucullean Grandeur," as reprinted in *JANA* 2 (May 1949): 9.

49. H. Riley Spitler, "*Naturae Medicina and Naturopathic Dispensatory*," *Journal of the American Naturopathic Physicians and Surgeons Association* 6, no. 9 (Dec. 1953): 5–11; A. W. Kuts-Cheraux, BS, MD, ND, ed., *Naturae Medicina and Naturopathic Dispensatory* (Des Moines: American Naturopathic Physicians and Surgeons Association, 1953).

50. Henry Schlichting Jr., ND, editorial, *Journal of the American Association of Naturopathic Physicians* 7 (Aug. 1954): 4; "A. R. Hedges Re-Elected President," in Harry Riley Spitler, ND, "Annual Report of the Legal and Legislative Committee," ibid. 4, 6–7.

51. Henry Schlichting Jr., The President's Message, *JANA* 2 (July 1949): 5; W. H. Pyott, "Does the Naturopathic Profession Need Organization?," ibid. 2 (May 1949): 11, 14; Schlichting, The President's Message, ibid. 1 (Dec. 1948–Jan. 1949): 5; Schlichting, The President's Message, ibid. 2 (Dec. 1949): 5, on Gehman's reason for cancellation; Carroll, "Unity and Progress," 1.

52. Carroll, "Unity and Progress," 1; Henry Schlichting Jr., "Convention Highlights," *JANA* 2 (May 1949): 5; "Dr. Henry Schlichting, Jr., Re-Elected President of the A.N.A." ibid. 2 (June 1949): 6; "Program Recommended by the Planning Committee," ibid. 2 (Apr. 1949): 7, 16. On insignia, see A. R. Hedges (member, of Signs and Insignia Committee), "A.N.A. Sign," ibid. 2 (Apr. 1949): 21.

53. "A.N.A. Unification Convention," ibid. 3 (July 1950): 8; Kirchfeld and Boyle, *Nature Doctors*, 203.

54. Waldo H. Jones, PhD, "What Do You Know about Cancer? . . . Very Little," *JANA* 1 (Apr. 1948): 20; advertisement, ibid., 19. Authors tended to admit that little was known about a disease, or they held true to traditional naturopathic beliefs.

55. Cody, "History of Naturopathic Medicine," 28, 22; Kirchfeld and Boyle, *Nature Doctors*, 205; Kuts-Cheraux, *Naturae Medicina and Naturopathic Dispensatory*; Spitler, *Basic Naturopathy*; Wendel, *Standardized Naturopathy*; Leon Chaitow, *Naturopathic Physical Medicine: Theory and Practice for Manual Therapists and Naturopaths* (Philadelphia: Elsevier Health Sciences, 2008), 13.

56. *Joint Naturopathic Convention Report*, pamphlet (Aug. 1957), 1; W. A. McMillan, ND, "Jurisdiction and Training," *JN*, Nov. 1957, 14–15.

57. Harry F. Bonelle, ND (president, NANP), "Highlights of Denver Convention," *JNM*, Sept.–Oct. 1959, 3–5; Campanella, "Dear Doctor."

58. Baer, "Sociopolitical Status of US Naturopathy," 332, citing Lisa, *Assault on Medical Freedom*; *AANPQN* 1, no. 3 (1986): 4; ibid. 2, no. 5 (1987): 3.

59. Bureau of Economic and Business Research, "Survey of Naturopathic Schools Prepared for the Utah State Medical Association," Dec. 1958, 5, NCNM, 3. The phrase *desolate state* was a handwritten notation on a copy of the survey from Friedhelm Kirchfeld to this author, 1992.

60. Kirchfeld and Boyle, *Nature Doctors*, 201. A letter from Lust to R. C. Allred saying the DC incorporation was his "charter" was not included in the survey but was said to be "on file in the department of business registration." Bureau of Economic and Business Research, "Survey of Naturopathic Schools," 58–59, on the American School of Naturopathy. The 1958 survey also detailed ten naturopaths trained at the American School of Naturopathy then practicing in Utah. They had graduated in the years 1919 through 1942. None were interviewed for the report. US Congress, House, Committee on the District of Columbia, *An Act to Amend Meaning And Intent of Healing Art Practice Act of February 27, 1929: Hearing Before the Committee On the District of Columbia, House of Representatives, Seventy-First Congress, Third Session, On H.R. 12169, a Bill to Amend the Meaning And Intention of an Act of Congress Entitled "An Act to Regulate the Practice of the Healing Art to Protect the Public Health of the District of Columbia," Approved February 27, 1929. January 28, 1931* (Washington, DC: GPO, 1931).

61. Mario T. Campanella, "Some Observations and Comments," *JN* 1 (Jan. 1957): 7. The nine affiliated states were Connecticut, Missouri, Nevada, Wyoming, Florida, Georgia, Maryland, Texas, and Washington.

62. Harry F. Bonnelle, "National and International Naturopathic Activities: NANP Summer Report—1960," *JNM*, Aug. 1960, 28–32.

63. Bill Becker, "Explosives Found," *New York Times*, 22 Jan. 1960, 13; Associated Press, "Naturopath Defended: Wife Quotes Spears as Denying He Placed a Bomb on Airliner," reprinted in ibid., 23 Jan. 1960, 1, 10; Associated Press, "F.B.I. Remains Silent," 24 Jan. 1960, reprinted in ibid., 25 Jan. 1960, 24; Associated Press, "Naturopath Defended," 1; Associated Press, "F.B.I. Questions Turska," 23 Jan. 1960, reprinted in *New York Times*, 24 Jan. 1960, 43; "Plane Search Slated," ibid., 16 Feb. 1960, 17.

64. "Book Guide," *JN*, Feb. 1957, 29; William A. Turska, "Wet Blood Drop Test (Original Research)," ibid., 16–21; J. H. Tilden, MD, *Toxemia Explained: The True Interpretation of the Cause of Disease*, 5th ed. (Denver: privately printed, 1935), first published in 1926.

65. Louis A. LaVine, "What We Know through Research," *JN*, Aug. 1957, 20; V. L. Fernandez (vice president, ISNP), "Poliomyelitis," ibid., 9; Thomas T. Lake, "Naturopathic Treatment of Tonsillitis," *JN*, Nov. 1957, 9–12; I. John Williams, "Psychotherapy as a Specialty in Naturopathic Medicine," ibid., May 1957, 9–13; J.A.S., "Allergies and Food Adulteration," ibid., 19; William Dunkler, "Physical Therapy and Progressive Naturopathy," ibid., Aug. 1957, 10–13; ANA of Pennsylvania, "Classification, Etiology, Endogenous Causes and Diagnosis of Obesity," ibid., Sept. 1957, 16–17; study panel, Washington State Naturopathic Association, "Calcium," ibid., Oct. 1957, 17–22. "Products and Their Merits" names specific companies that sold valuable and reputable products. Among these were Dartell Laboratories, the Physical Therapy Company, and Vitaminerals Laboratories (all of Los Angeles); Ethical Specialties (Kalamazoo, Michigan); the Natural Sales Company and the Homeopathic Pharmacy (both in Pittsburgh); Lawrence Laboratories (Philadelphia); and Metabolic Products

(Boston). Each company's products are detailed, their prices given, and at times their efficacy is praised. "Products and Their Merits," ibid., June 1957, 29–30.

66. "National Naturopathic Activities: Dr. Lorna M. Murray Elected President of the N.A.N.P.," *JN*, Sept. 1957, 29–30, at 30; Lorna Mae Murray, ND, "Unity and Cooperation," ibid., May 1957, 7; Mario T. Campanella, ND, editorial, *JNM*, Feb. 1961, 2–4; "Report of the 26th Annual Meetings of the International Society of Naturopathic Physicians," ibid., Aug.–Sept. 1964, 2–10.

67. Lisa Bolton Stevenson, "Does Your Wife Know?," *JN* 1 (Jan. 1957): 19.

68. Kenneth C. Hitchcock, "Another Milestone of Progress," ibid. 1 (Sept. 1957): 18–22.

69. Arno Koegler, "A Message from Dr. Koegler," *JN*, Oct. 1956, 3; Kirchfeld and Boyle, *Nature Doctors*, 258, 287–301, 310; "A Brief History of NCNM," accessed 12 Jan. 2014, www.ncnm.edu/about-ncnm/getting-to-know-ncnm/history.php; Friedhelm Kirchfeld and Wade Boyle, "Father of Modern Naturopathic Medicine: John Bastyr (1912–)," in *Nature Doctors*, 303; "About Dr. John Bastyr," accessed 12 Jan. 2012, www .bastyr.edu/about/about-our-university/history-heritage#Namesake-Dr-Bastyr.

70. "National Naturopathic Activities," *JN* 1 (Jan. 1957): 25; "Treasurer's Report: The Naturopathic Foundation Report," ibid. 1 (July 1957): 31; "Good News on the Educational Front: National Naturopathic Activities," ibid. 1 (May 1957): 27; Kuts-Cheraux, *Naturae Medicina and Naturopathic Dispensatory*.

71. "College of Naturopathic Missionary Medicine Established," *JNM*, June 1959, 28–29; D. E. McArthur, "Greetings Fellow Naturopathic Physicians," ibid., Aug. 1959, B, C, reprinted from NANP-ISNP Convention program; Bonnelle, "Highlights of Denver Convention," 3–4. In photographs of basic and advanced nutrition classes, nearly all students are female, reflecting again the gender norms of the period. Sierra was in compliance with the California State Department of Education, approved and accredited by the National Association of Naturopathic Physicians and approved by the US Department of Justice for students without visas. It also had a graduate school. George Floden, ed., *Naturopathic Magazette* 13 (Spring 1960).

72. *The Naturopath* 1 (Nov. 1963): 2–3.

73. See, e.g., the 8-page November 1963 issue of *The Naturopath*, including R. L. Sanders, "Did Grandma Practice Medicine Without a License?," 2; Old Doc Cherokee, "How to Treat a Cow," 3; and Dewey Pleak, "Diagram of a Compost Heap," 5. For the currency information, see "CPI Inflation Calculator."

74. Lisa Bente, MS, RD, and Shirley A. Gerrior, PhD, RD, "Selected Food and Nutrient Highlights of the 20th Century: U.S. Food Supply Series," *U.S. Department of Agriculture Center for Nutrition Policy and Promotion* 14, no. 1 (2002): 43, 44–51, 97, 106–7.

75. *An American Feast: Food, Dining, and Entertainment in the United States from Simmons to Rombauer*, exhibition, Hugh M. Morris Library, University of Delaware, 21 June–30 Sept. 1994, section titled "Dining, Etiquette, and Social Meaning," p. 14 of 22, accessed 3 May 2005, www.lib.udel.edu/ud/spec/exhibits/american.htm; William H. Young with Nancy K. Young, "Food: The American Diet," in *The 1930s* (Westport, CT: Greenwood, 2002), 19–22, 97–117; Sandra Opdyke, *The Routledge Historical Atlas of Women in America* (New York: Routledge, 2000), 132.

76. Lowell K. Dyson, "American Cuisine in the 20th Century," *Food Review* 23, no. 1 (2000): 5–6; Rachel Carson, *Silent Spring* (New York: Houghton, 1962).

77. Dewey Pleak, "Organic Gardening," *The Naturopath* 1 (Nov. 1963): 5; Pleak, "What to Grow and Why," ibid. 2 (May 1964): 10; Pleak, Organic Gardening the Natural Way, ibid. 2 (Jan. 1964): 9.

78. "Pesticide Kills Rodents, Insects—*and Children*; Deadly Poison Arouses Health Authorities, Parents," ibid. 1 (May–June 1963): 1–2.

79. Old Doc Cherokee, "The Story of Herbs and the Lore of the Blue Ridge Mountains," ibid. 1 (Nov. 1963): 3, 6; "Conservation Story: Forest Facts and Fancies," ibid. 2 (Dec. 1963): 9; "Iroquois 'Folk Remedy' Saves Explorers' Lives," ibid. 2 (May 1964): 4; "Healthful Delicacies from Your Herb Garden," ibid. 2 (June 1964): 11; Clarence Bly, Your Foods and Your Health, appearing monthly in *The Naturopath*; Linda Clark, "(author of 'Stay Young Forever')," The Question Box, *The Naturopath* 3 (Jan. 1965): 4.

80. J. A. Boucher, "Naturopaths Must Meet Rigid School Requirements," *The Naturopath* 1 (Feb. 1963): 1–2; "Second National College Symposium Slated Jan. 18–19," ibid. 2 (Jan. 1964): 5; "Naturopaths to Convene at Portland, Ore., in May; Challenge of the Future Faced," ibid. 2 (Mar. 1964): 7; "Look to Nature for Arthritis Answer Says Noted Doctor," ibid. 1 (Feb. 1963): 1–2; John W. Noble, "Doctor's Notebook: Pulmonary Emphysema," ibid. 2 (Jan. 1964): 2, 5; M. O. Garten, "Quick Fact about Your Organs of Elimination," ibid. 2 (Apr. 1964): 3, 7; W. J. McCormick, MD, "Cancer: A Preventable Disease, Secondary to Nutritional Deficiency," ibid. 2 (May 1964): 4, 13; "The Rise and Decline of the Age of Miracles," ibid. 2 (Mar. 1964): 3; "Drugs Now in Use May Never Be Tested," ibid.; "Dr. John W. Noble to Speak on Medical Monopoly," *The Naturopath* 3 (Mar. 1965): 8.

81. "Chemical Poisons Pose Potent Threat to Man," *The Naturopath* 1 (May–June 1963): 2; J. A. Boucher, "The Nature Doctor: On the Subject of Quackery," ibid. 2 (Aug. 1964): 8.

CHAPTER 11: The 1970s and Beyond

1. J. A. Roth, *Health Purifiers and Their Enemies* (New York: Prodist, 1976), as cited in Hans Baer, "Partially Professionalized and Lay Heterodox Medical Systems within the Context of the Holistic Health Movement," in *Biomedicine and Alternative Healing Systems in America: Issues of Class, Race, Ethnicity, and Gender* (Madison: Univ. of Wisconsin Press, 2001), 104–5; Friedhelm Kirchfeld, personal correspondence with author, July 1993.

2. Cash Asher, *Bacteria, Inc.* (Milwaukee: Lee Foundation for Nutritional Research, 1955); Jean A. Oswald, *Yours for Health: The Life and Times of Herbert M. Shelton* (Franklin, WI: Franklin Books, 1989); *Dr. Shelton's Hygienic Review* (1939–80).

3. Michael Murray, ND, and Joseph Pizzorno, ND, "The Naturopathic Revolution," *Venture Inward*, July–Aug. 1992, 13–18, at 17; Hans A. Baer, "The Sociopolitical Status of US Naturopathy at the Dawn of the 21st Century," *MAQ* 15, no. 3 (2001): 329–46, at 333.

4. Victoria Moran, "The Natural Doc," *Vegetarian Times*, Aug. 1990, 40, 42–47.

5. Arnold DeVries, *Therapeutic Fasting* (Los Angeles: Chandler, 1963); Henry G. Bieler, MD, *Food Is Your Best Medicine* (New York: Random House, 1965); Harry

Benjamin, ND, *Everybody's Guide to Nature Cure* (1936; repr., Surrey: Health For All, 1967); Boston Women's Health Book Collective, *Our Bodies, Ourselves* (Boston: New England Free Press, 1971); *Books That Shaped America*, exhibition, Library of Congress, 2012; Mark Bricklin (executive editor, *Prevention* magazine), *The Practical Encyclopedia of Natural Healing* (Emmaus, PA: Rodale Press, 1976): xix, as cited in Baer, "Partially Professionalized," 107, where Baer identifies and discusses the following seminal texts and their impact: Norman Cousins, *Anatomy of an Illness* (New York: Norton, 1979); Ken Pelletier, *Mind as Healer, Mind as Slayer* (New York: Delta, 1977); Pelletier, *Towards a Science of Consciousness* (New York: Delacorte, 1978); Pelletier, *Holistic Medicine: From Stress to Optimum Health* (New York: Delacorte & Delta, 1981); Paavo O. Airola, ND, *Health Secrets from Europe* (West Nyack, NY: Parker, 1970); Airola, *How to Get Well: Therapeutic Uses of Foods, Vitamins, Food Supplements, Juices, Herbs, Fasting, Baths and Other Ancient and Modern Nutritional and Biological Modalities in Treatment of Common Ailments* (Sherman, OR: Health Plus, 1974), in its twenty-third printing by 1989; Andrew Weil, MD, *Health and Healing: Understanding Conventional and Alternative Medicine* (Boston: Houghton Mifflin, 1983); Weil, *The Natural Mind: A Revolutionary Approach to the Drug Problem* (New York: Houghton Mifflin, 1972); Bernie Siegal, *Love, Medicine and Miracles* (New York: HarperCollins, 1986); Dolores Krieger, *The Therapeutic Touch: How to Use Your Hands to Help or Heal* (New York: Prentice Hall, 1979).

6. Murray and Pizzorno, "Naturopathic Revolution," 14.

7. In 2013 Cayce's centers had regional representatives throughout the United States and in 37 other countries and individual members in more than 70 countries. "Who Was Edgar Cayce? Twentieth Century Psychic and Medical Clairvoyant," Edgar Cayce's Association for Research and Enlightenment, accessed 24 Nov. 2013, www.edgarcayce.org/are/edgarcayce.aspx; Murray and Pizzorno, "Naturopathic Revolution," 17.

8. Baer, "Partially Professionalized," 107. For an excellent overview of New Age practices and ideas, see ibid., 110–13.

9. Keki Sidhwa, "Don't Hand Out Remedies: You Must Teach Your Patients How to Live," *International Journal of Alternative and Complementary Medicine*, June 1993, 19–21, at 19, reprinted from ibid., July 1983; Kirchfeld, personal correspondence with author, July 1993.

10. Mary Walker, "Choosing a Naturopath: How to Rate Qualifications of N.D.'s," *EastWest: The Journal of Natural Health & Living*, Nov. 1990, 38–39; Nancy A. Ruhling, "The Spa Experience . . . Or, How to Take the Watercure without Getting in over Your Head: Here's a Guide to Naturopathy, Rolfing Seaweed Ways, and the Ins and Outs of Spa-Going," *USAIR Magazine* 14 (Dec. 1992): 50–56, 88; "Naturopathy," *Healing Currents: A Journal for the Healing Arts*, Dec. 1993; Lauri Aesoph, "Sweet Buzz or Lullaby: How Does Sugar Really Affect Behavior?," *Health Store News*, Apr.–May 1991, 17–18; Skye Weintraub, ND, "Naturopathic Medicine and Naturopathic Physicians . . . Does Health Insurance Cover Naturopathic Care?," *Healing Currents*, n.d., 24–25, "Naturopathy, History" file, NCNM; Dana Ullman, "Getting beyond Wellness Macho: The Promise and Pitfalls of Holistic Health," *Utne Reader*, Jan.–Feb. 1988, 68–72; Sharon Cohen, "Doctor, Doctor, Give Me the News," *SHAPE*, Aug. 1996, 98–103, 114, 116; Nancy Serrell, "Naturopathic Medicine: Nontraditional

Health Care Moves toward the Mainstream," *Connecticut Valley News*, 10 Feb. 1992, 17, 24, at 17; Rep. William Kidder and Rep. Kathleen Ward to Paul Bergner, editor, *Naturopathic Physician*, 1 Sept. 1994, American Naturopathic Medical Association, www.anma.org/hampsh.html.

11. Joshua A. Perper and Stephen J. Cina, "It's All Natural! Alternative and Complementary Medicine," in *When Doctors Kill: Who, Why, and How* (New York: Springer Science & Business Media, 2010), 179–86, as cited in Baer, "Partially Professionalized," 119.

12. Murray and Pizzorno, "Naturopathic Revolution," 14; "Bastyr University History and Heritage: Founding of Bastyr," accessed 27 Oct. 2012, www.bastyr.edu /about/about-our-university/history-heritage.

13. Murray and Pizzorno, "Naturopathic Revolution," 16. Physical medicine included "physiotherapy equipment such as ultrasound, diathermy and other electro-magnetic techniques; therapeutic exercise, massage, joint mobilization (manipulation) and immobilization techniques, and hydrotherapy."

14. Friedhelm Kirchfeld to author, 3 Aug. 1993, including the letter sent to the old-timers, "National Treasure: Homecoming '93 Celebration," n.d., in author's possession.

15. L. S. McGinnis, "Alternative Therapies: An Overview," *Cancer* 67 (Mar. 15 1991): 1788–92; G. D. Appelt, "Toward an Appreciation of Cross-Cultural Perspectives in Modern Health Care," *Journal of Clinical Pharmacy and Therapeutics* 13 (June 1988): 175–78, at 175, 177. Bernie S. Siegel, a pediatric and general surgeon, described cancer patients who employed meditation and spirituality to generate greater healing powers. Siegel, *Love, Medicine and Miracles: Lessons Learned about Self-Healing from a Surgeon's Experience with Exceptional Patients* (New York: HarperCollins, 1990). Another author concluded that integrating a patient's beliefs and practices, including prayer, with his or her physical healing produced valuable results. Chandy C. John, MD, "Faith, Hope and Love in Medicine," *Pharos* 51 (Fall 1988): 12–17.

16. Alan C. Mermann, MD, "The Doctor's Critic: The Unorthodox Practitioner," *Pharos* 53 (Winter 1990): 9–13; Miriam K. Feldman, "Patients Who Seek Unorthodox Medical Treatment," *Minnesota Medicine* 73 (June 1990): 19–25, at 19; Michael H. Kottow, "Classical Medicine v Alternative Medical Practices," *Journal of Medical Ethics* 18 (1992): 18–22, at 18; Sid Kemp, "The Case for Reporting on Medical Alternatives," *Nieman Reports* 47 (Winter 1993): 40–43, at 41.

17. Cathy Rogers, ND, "President's Letter: Rising Demand for Naturopathic Doctors across U.S.," *AANPQN* 5, no. 1 (1990): 1–2. Rogers also cautioned practitioners about the problems they would face in states that refused them a license.

18. "Referral Calls Skyrocket after Good Housekeeping Article," *NPhys* 9 (Spring 1994): 17.

19. Kemp, "Case for Reporting on Medical Alternatives"; Perper and Cina, "It's All Natural!"; Baer, "Partially Professionalized," 119; P. M. Barnes, B. Bloom, and R. Nahin, "Complementary and Alternative Medicine Use among Adults and Children: United States, 2007," CDC National Health Statistics Report 12, 10 Dec. 2008, http://nccam.nih.gov/news/camstats/2007; W. Weber, J. A. Taylor, R. L. McCarty, et al., "Frequency and Characteristics of Pediatric and Adolescent Visits in Naturo-pathic Medical Practice," *Pediatrics* 120 (July 2007): 142–46; D. P. Albert and

D. Martinez, "The Supply of Naturopathic Physicians in the United States and Canada Continues to Increase," Complementary Health Practice Review 11 (2006): 120–22; Hans A. Baer, "The Potential Rejuvenation of American Naturopathy as a Consequence of the Holistic Health Movement," MAQ 13, no. 4 (1992): 337.

20. "Considering Alternative Medicine? Arm Yourself with Information," Taking Care (Travelers Insurance Company), no. 11 (Nov. 1993): 4–5.

21. Weintraub, "Naturopathic Medicine and Naturopathic Physicians," 25; Steven Bratman, The Alternative Medicine Sourcebook: A Realistic Evaluation of Alternative Healing Systems (Chicago: Contemporary Books, 1997), 214; Rena J. Gordon and Gail Silverstein, "Marketing Channels for Alternative Health Care," in Alternative Thera-pies: Expanding Options in Health Care, ed. Rena J. Gordon, Barbara Cable Nien-stedt, and Wilbert M. Gesler (New York: Springer, 1998), 87–103, at 94, as cited in Baer, "Partially Professionalized," 119–20; P. M. Herman, O. Szczurko, K. Cooley, et al., "Cost-Effectiveness of Naturopathic Care for Chronic Low Back Pain," Alternative Therapies in Health and Medicine 14, no. 2 (2008): 32–39.

22. Baer, "Partially Professionalized," 106–7. Publications in which mindbodyspirit ideas were exchanged included the Journal of Holistic Health, Alternative Medicine, the Holistic Health Review (published by the American Holistic Medical Associa-tion), and the Journal of Alternative and Complementary Medicine: Research on Paradigm, Practice, and Policy. See also Jeanne Achterberg, PhD, "What Is Medicine?," Alternative Therapies 2, no. 3 (May 1996): 58–61, at 58; and Roger Newman Turner, ND, DO, BAC, "Naturopathy: Baseline for Alternative Medicine," International Journal of Alternative and Complementary Medicine, July 1993, 15–19.

23. Baer, "Potential Rejuvenation," 369; Susan M. Fitzgerald, "Naturopathy: An Age-Old Medicine for the 'New Age,'" My Health: American Association of Alternative Medicine Magazine, July 1997, www.ncnm.edu/article.htm (accessed 19 Aug. 2000; URL no longer valid); Joseph E. Pizzorno Jr. and Pamela Snider, "Naturopathic Medicine," in Fundamentals of Complementary and Alternative Medicine, ed. Marc S. Micozzi, 2nd ed. (New York: Churchill Livingston, 2001), 189.

24. Mary Bisceglia, "Naturopathy—Putting Nature Back into Medicine," Earth-tone, no. 16 (July–Aug. 1981): 12.

25. George Cody, "History of Naturopathic Medicine," in Textbook of Natural Medicine, ed. Joseph Pizzorno and Michael Murray (New York: Churchill Living-stone), 17–40, at 29; Baer, "Sociopolitical Status of US Naturopathy," 333; "Founding of Bastyr," Bastyr University, accessed 10 July 2015, www.bastyr.edu/about/about-our -university/history-heritage; "Academics," Bastyr University, accessed 10 July 2015, www .bastyr.edu/academics. Another school, the Pacific College of Naturopathic Medicine in Monte Rio, California, was created in 1979, but it was short-lived. It closed quickly because there was no licensure for naturopathic schools in California and because of poor management. Baer, "Sociopolitical Status of US Naturopathy," 333.

26. "I see no change in the foreseeable future," wrote Robert H. Hayes, MD (chief, Office of Medical Services, South Dakota Department of Health) to Davies on 21 Oct. 1980. Copy sent to author by Friedhelm Kirchfeld, founding NCNM library director (1978–2006). Other responses to Davies came from the following (in chronological order): Lillian S. Golovin (administrative assistant, Department of Health, Vermont), 23 Oct. 1980; Curtis G. Power III (assistant attorney general, West Virginia), 24 Oct.

1980; Eugenia K. Dorson (executive secretary, Virginia Department of Health Regulatory Boards), 27 Oct. 1980; Hal L. Nelson (chief, Legal and Claims Services, Texas Department of Health), 29 Oct. 1980; Stanley A. Miller (commissioner, Pennsylvania Bureau of Professional and Occupational Affairs), 5 Nov. 1980; and Robert W. McClanaghan (administrator, Professional Regulation, Rhode Island and Providence Plantations Department of Health), 18 Nov. 1980.

27. Southwest College opened in Scottsdale, Arizona, and then relocated to Tempe, maintaining a clinic in Scottsdale. At Bridgeport, the donations from the Professors World Peace Group totaled by some reports $98 million or $110 million. Baer, "Sociopolitical Status of US Naturopathy," 333, citing Frederick Clarkson, *Eternal Hostility: The Struggle between Theocracy and Democracy* (Monroe, ME: Common Courage, 1997), 56–59; Joseph Berger, "U. of Bridgeport Honors Rev. Moon, Fiscal Savior," *New York Times*, 8 Sept. 1995; Associated Press, "University of Bridgeport Memorializes Rev. Sun Myung Moon," CBS New York, 2 Sept. 2012, http://newyork.cbslocal.com/2012/09/02/university-of-bridgeport-memorializes-rev-sung-myung-moon/.

28. Friedhelm Kirchfeld, letter to author, 1998; *JNM* 6 (Aug. 1996): 1–76 (entire issue); Leanna J. Standish, ND, PhD (director of Research, Bastyr College), "A Simple Method for Doing Research in Your Medical Practice," n.d., NCNM; "What Is Naturopathic Medicine? Philosophy," Naturopathic Medicine Network, accessed 1998, www.pandamedicine.com:80/natmed.html. The current webpage is "Naturopathic Medicine," Naturopathic Medicine Network, accessed 10 July 2015, www.pandamedicine.com/naturopathic_medicine.html.

29. Baer, *Biomedicine and Alternative Healing Systems in America*, 93; "Naturopathic Medicine FAQ," Association of Accredited Naturopathic Medical Colleges, accessed 10 July 2015, http://aanmc.org/naturopathic/faq/.

30. Association of Accredited Naturopathic Medical Colleges webpage, accessed 12 Apr. 2013, www.aanmc.org; "Accreditation," Bastyr College, San Diego Campus, accessed 6 Dec. 2013, www.bastyr.edu/about/accreditation. The National University of Health Sciences was founded in 1906 by John Fitz Alan Howard as the National School of Chiropractic in Davenport, Iowa. In 1963 it moved to Lombard, Illinois, and in 2009 it began offering the ND. "Our History: National University of Health Sciences," National University of Health Science, accessed 27 Oct. 2012, www.nuhs.edu/extras/historytimeline/index.html. The University of Bridgeport opened in 1927 as the Junior College of Connecticut, the first junior college in the Northeast. See "University of Bridgeport History and Mission," accessed 27 Oct. 2012, www.bridgeport.edu/welcome/history-and-mission; and D. M. Eisenberg, M.H. Cohen, A. Hrbek, et al., "Credentialing Complementary and Alternative Medical Providers," Annals of Internal Medicine 137, no. 12 (2002): 965–73. On the NCNM's name change, see the NCNM website, accessed 12 Jan. 2014, www.ncnm.edu/about-ncnm/getting-to-know-ncnm/history.php.

31. "Canadian College of Naturopathic Medicine, Accreditation," accessed 27 Oct. 2012, www.ccnm.edu; Boucher Institute of Naturopathic Medicine website, accessed 24 Jan. 2014, www.binm.org/; Canadian College of Naturopathic Medicine website, accessed 24 Jan. 2014, www.ccnm.edu/; Council on Naturopathic Medical Education website, accessed 27 Oct. 2012, www.cnme.org.

32. Baer, "Potential Rejuvenation," 379; Friedhelm Kirchfeld and Wade Boyle, *Nature Doctors: Pioneers in Naturopathic Medicine* (East Palestine, OH: Buckeye Naturopathic, 1993), 226. SisterSong is a vital part of the women's health networks. It brings together local, regional, and national grass-roots organizations in the United States that represent the reproductive needs and health social justice issues and concerns of women of color. See SisterSong website, accessed 6 Dec. 2013, www .SisterSong.net.

33. See "Table 200. Total fall enrollment in degree-granting institutions, by attendance status, sex, and age: Selected years, 1970 through 2020," *Digest of Educational Statistics*, National Center for Education Statistics, accessed 2012, http://nces.ed .gov/programs/digest/d11/tables/dt11_200.asp. For histories showing health care as a culturally acceptable role for women as nurturers and caretakers, see Susan E. Cayleff, *Wash and Be Healed: The Water-Cure Movement and Women's Health* (1987; repr., Philadelphia: Temple Univ. Press, 1992); and Regina Morantz Sanchez, *Sympathy and Science: Women Physicians in American Medicine* (Chapel Hill: Univ. of North Carolina Press, 2000). On women's issues since the second wave of feminism that resonate with naturopathy, see Nancy C. Unger, *Beyond Nature's Housekeepers: American Women in Environmental History* (Oxford: Oxford Univ. Press, 2012); Robert K. Musil, *Rachel Carson and Her Sisters: Extraordinary Women Who Have Shaped America's Environment* (New Brunswick, NJ: Rutgers Univ. Press, 2014); Barbara Seaman, ed., *Voices of the Women's Health Movement*, with Laura Eldridge (New York: Seven Stories, 2012); Francine Rota, "50 Ways to Better Women's Health," *EastWest: The Journal of Natural Health & Living*, Nov. 1990, 46, 48–54, 74–78; Cathy N. Rogers, ND, *Women and Work Healing Seminars Renewal Retreats*, pamphlet (privately printed, ca. Spring 1990); Rogers, "President's Letter," *AANPQN* 4 (Early Winter 1989): 2–3, discussing applicable strategies for naturopaths given in Cayleff, *Wash and Be Healed*; Jamison Starbuck, JD, ND, "From the President," *NPhys* 7 (Winter 1992): 2; Jim Sensenig, "From the Founding President," ibid. 8 (Fall 1993): 2. Tom Kruzel succeeded Starbuck as AANP president, and Starbuck became the director of national affairs. In 1993 the manager for the Portland convention was Dr. Teri Davis. See Teri Davis, "From the Convention Manager," ibid. 8 (Spring 1993): 29. The AANP manager was Sheila Quinn, credited on the homepage of AANP for her instrumental role in managing the new Bastyr College in Washington State. See Sheila Quinn, "Inside-out: The View from the Office," ibid. 8 (Fall 1993): 5, 9–10. Of the nine student volunteers in the AANP office, seven were female students from Bastyr and the NCNM. Robin Spiegel (AANP office manager), "Student Volunteers," ibid. 8 (Spring 1993): 12. In 1993 Vida Fassler, secretary of the board of the Institute of Naturopathic Medicine, an independent nonprofit organization that helps maximize the benefits of naturopathic medicine for the general public, had the honor of symbolically launching the institute in Portland. Paul Bergner, "Naturopathic Fund-raising Institute Launched," ibid. 8 (Fall 1993): 1. Additionally, the president of the Washington (State) Association of Naturopathic Physicians was Dr. Patricia Hastings. Ibid., 35. One year later, in 1994, Vida Fassler and Dr. Harry Swope, INM board members, met privately with Surgeon General Dr. Jocelyn Elders to discuss naturopathic medicine and provided her with written materials about it. Harry Swope, MBA, ND, "INM Board Members Meet with

Surgeon General," *NPhys* 9 (Spring 1994): 1; Hans A. Baer, "The Status of Women in Naturopathy: A View from the United States and Australia," *Health Sociology Review* 22 (June 2013): 194; "About," California Naturopathic Doctors Association, accessed 12 Dec. 2014, www.calnd.org/about.

34. Lauri Aesoph, "When Doctors Are Patients: Naturopathic Physicians and Cancer," *NPhys* 7 (Winter 1992): 23–27; Teri Davis, "Naturopathic Medicine: Healing the Person, Healing the Planet," ibid. 9 (Spring 1994): 15; Lauri Aesoph, "Dr. Jeanne Albin: First Native American Naturopathic Physician," ibid. 6 (Fall 1991): 23; "Women's Health Care," *JNM* 7 (Winter 1997): 1–129; Eileen Stretch, ND, "Clinical Manifestations of HIV Infection in Women," ibid. 3, no. 1 (1992): 12–19; Jill E. Stansbury, ND, *Botanical Medicines Acting on the Female Reproductive System* (n.p., n.d.), pamphlet acquired by author from Bastyr, late 1990s; Rota, "50 Ways to Better Women's Health," 46, 48–54, 74–78, containing an interview with Cathy Rogers, ND, and 5 other holistic female practitioners on their treatments for women's ailments and well-being, which are an unselfconscious melding of naturopathic and natural medical healing methods.

35. Baer, *Biomedicine and Alternative Healing Systems*, 185.

36. "AANP Invited to Testify at Congressional Hearing," *NPhys* 8 (Fall 1993): 13; Cody, *Sociopolitical Status*, 334; National Center for Complementary and Alternative Medicine, National Institutes of Health, "What Is Complementary and Alternative Medicine?," accessed 27 Oct. 2012, www.nccam.nih.gov/health/naturopathy; James C. Whorton, *Nature Cures: The History of Alternative Medicine in America* (Oxford: Oxford Univ. Press, 2002), 294–95. The 2013 budget was down from its peak of $128 million in 2012. National Center for Complementary and Alternative Medicine, "NCCAM Funding: Appropriations History." The requested 2014 NIH budget was $31.3 billion. National Institutes of Health, "Fact Sheet: Impact of Sequestration on the National Institutes of Health," accessed 6 Dec. 2102, www.nih.gov/news/health/jun2013/nih-03.htm; National Institutes of Health Executive Summary FY 2014 Budget Request, accessed 6 Dec. 2013, http://officeofbudget.od.nih.gov/pdfs/FY14/Tab%201%20-%20Executive%20Summary_final.pdf; "National Institutes of Health: Actual Total Obligations by Institute and Center, FY 2000–2014," accessed 12 July 2015, http://officeofbudget.od.nih.gov/pdfs/FY16/Actual%20Obligations%20By%20IC%20FY%202000%20-%20FY%202014%20%28V%29%20%282%29.pdf, for a breakdown of NIH budgets; Baer, *Biomedicine and Alternative Healing Systems*, 185.

37. Irvine Loudon, "A Brief History of Homeopathy," *Journal of the Royal Society of Medicine* 99 (Dec. 2006): 607–10, at 607, on CAM as a game changer; Whorton, *Nature Cures*, 299; Elizabeth Austin, "Looking for Cures," *Self* 18 (1 Sept. 1996): 192–93, at 192. On medical sciences' ridicule of NCCAM's regulatory effectiveness, see Tom Delbanco, "Leeches, Spiders and Astrology: Predilections and Predictions," *JAMA* 280 (Nov. 1998): 1560–62, as quoted in Whorton, *Nature Cures*, 297; and Stephen Barrett, MD's Quackwatch website, accessed 16 Dec. 2014, www.quackwatch.com/, containing skewed and biased—and debatably documented—diatribes about a variety of topics in alternative medicine.

38. Brian Inglis, "Osteopathy and Chiropractic," in *The Case for Unorthodox Medicine* (New York: Putman, 1965), 128–29; Norman Gevitz, *The D.O.'s: Osteopathic Medicine in America* (Baltimore: Johns Hopkins Univ. Press, 1982).

39. Francis Helminski, MA, JD, "That Peculiar Science: Osteopathic Medicine and the Law," *Law, Medicine and Health Care* 12 (Feb. 1984): 32–36; Hans A. Baer, "The Organizational Rejuvenation of Osteopathy: A Reflection of the Decline of Professional Dominance in Medicine," *Social Science and Medicine* 15A (1981): 701–11.

40. C. J. Denbow and R. Feeck, "The Many Meanings of 'Holistic Medicine,'" *Journal of the American Osteopathic Association* 88 (Jan. 1988): 15–18.

41. Irvin M. Korr, PhD, "Osteopathic Research: The Needed Paradigm Shift," ibid. 91, no. 2 (Feb. 1991): 156–71; George W. Northup, "Stop, Look, and Listen When Your Patients 'Complain,'" ibid. 91 (Sept. 1991): 857–58; Thomas Wesley Allen, DO, "Osteopathic Research: Where Have We Been and Where Are We Going?," ibid. 91, no. 2 (1991): 122.

42. Inglis, "Osteopathy and Chiropractic," 127–28, 131–37; "Massachusetts Board of Registration of Chiropractors," Federation of Chiropractic Licensing Boards, accessed 15 Jan. 2014, http://directory.fclb.org/US/Massachusetts.aspx.

43. Ronald L. Caplan, "Health Care Reform and Chiropractic in the 1990s," *Journal of Manipulative and Physiological Therapeutics* 14 (July–Aug. 1991): 341–53, at 344–46; David Coburn, PhD, "Legitimacy at the Expense of Narrowing of Scope of Practice: Chiropractic in Canada," ibid. 14 (Jan. 1991): 14–21.

44. American Holistic Medical Association, "Training," accessed 15 Jan. 2013, www .holisticmed.org/. In 2014 the AHMA became the Academy of Integrative Health & Medicine, and members are developing "a new paradigm in integrative medicine. AIHM, accessed 12 July 2015, http://aihm.org/fellowship/.

45. The four schools offering holistic degrees are Tennessee State, New York College of Health Professions, Canyon College (California), and New York University. www.ahna.org, accessed 10 Jan. 2013. L. W. Brallier, "Biofeedback and Holism in Clinical Practice," *Holistic Nurse Practitioner*, May 1988, 26–33; Dolores Gorton, "Holistic Health Techniques to Increase Individual Coping and Wellness," *Journal of Holistic Nursing* 6, no. 1 (1988): 25–30; Joyce E. White, "Is Holistic Care Possible in the 'Real Life' Clinical Setting?," ibid., 1–5.

46. *JAMA* 280 (11 Nov. 1998); Baer, *Biomedicine and Alternative Healing Systems*, 180.

47. Terri A. Winnick, "From Quackery to 'Complementary' Medicine: The American Medical Profession Confronts Alternative Therapies," *Social Problems* 52 (Feb. 2005): 38–61, at 38. The journals Winnick examined in her study were *JAMA*, the *New England Journal of Medicine*, the *American Journal of Medicine*, the *Annals of Internal Medicine*, and the *Archives of Internal Medicine*.

48. H. Bostrom and S. Rossner, "Quality of Alternative Medicine—Complications and Avoidable Deaths," *Quality Assurance Health Care* 2, no. 2 (1990): 111–17; Jennifer R. Jamison, "Counteracting Nutritional Misinformation: A Curricular Approach," *Journal of Manipulative and Physiological Therapeutics* 13 (Oct. 1990): 454–62, at 456.

49. Hans A. Baer, "Naturopathy, Acupuncture, and the Holistic Health Movement," in *Biomedicine and Alternative Healing Systems*, 97–102, at 101.

50. Associated Press, "Americans Give Acupuncture a Shot," *Pawtucket (RI) Evening Times*, 10 Nov. 1988, 23; Betsy Johnson, "Easing the Pain," *Attleboro (MA) Sun Chronicle*, 3 Oct. 1991, 33–34; Jason C. Shu, MD, FICAE, Diplomate in Acupuncture, Pennsylvania State Board of Medicine, "Overview of Acupuncture in the United

States," *Acupuncture & Electro-Therapeutics Research: The International Journal* 13 (1988): 59–62, at 60.

51. Program, "Fifth Annual International Symposium on Acupuncture and Electro-Therapeutics: October 19–22, 1989"; T. W. Wong, "Acupuncture: From Needle to Laser," *Family Practitioner* 8 (June 1991): 168–70; Dahong Zhuo, "Traditional Chinese Medicine: Rehabilitative Therapy in the Process of Modernization," *International Disability Studies* 10, no. 3 (1988): 140–42; David R. Aylin (general practitioner), "Using Acupuncture in General Practice," *Practitioner* 22 (Apr. 1988): 431–32.

52. National Council Against Health Fraud, "Acupuncture: The Position of the National Council Against Health Fraud," *Clinical Journal of Pain* 7, no. 2 (1991): 162–66; James A. Merolla, "Acupuncture, the Ancient Eastern Healing Art," *Bay Windows*, Sept. 1990, 2–3; David H. Freedman, "The Triumph of New-Age Medicine," *Atlantic*, July–Aug. 2011, 90, 92, 94, 97–100; Baer, "Naturopathy, Acupuncture," 101.

53. Brian Inglis, "Naturopathy and Herbalism," in *Case for Unorthodox Medicine*, 67–78; Jane Jones, "Bringing Plants Closer to Humans: A Philosophy," *Midwife Health Visitor and Community Nurse* 24 (July 1988): 257; Health Center for Better Living in Florida, *A Useful Guide to Herbal Health Care* (Naples, FL, n.d.), cover. The center's credibility cannot be determined. The *Herb Quarterly*, published by Long Mountain Press in San Anselmo, California, was one example of missed opportunities for collaboration in the 1990s. It covered such topics as the value of individual plants, medicinal herb gardens, fragrancy, herbal cuisine and teas, and recipes. See, e.g., *Herb Quarterly*, no. 64 (Winter 1994).

54. Leigh Fenly, "Green Medicine: Scientists Are Racing against Time to Uncover Nature's Healing Secrets," Currents, *San Diego Union*, 7 Nov. 1990, C-1, C-5; Steven Hollifield, AP, *A Review of Traditional Rainforest Botanicals: According to Traditional Chinese Medicine*, pamphlet (Jupiter, FL: privately printed, 1991).

55. *Medical Herbalism: A Clinical Newsletter for the Herbal Practitioner* (Portland, OR) 3 (May–June 1991); pamphlet, *GAIA Herb Symposium, 1994: Naturopathic Herbal Wisdom Expanding Our Knowledge of Nature's Curative Plants* (Welches, OR, 1994).

56. The basic sciences are biochemistry, human physiology, histology, anatomy, macro- and microbiology, immunology, human pathology, neuroscience, and pharmacology. "Naturopathic Medicine," Association of Accredited Naturopathic Medical Colleges, accessed 17 Dec. 2013, http://aanmc.org/naturopathic/rigorously _trained/.

57. Moran, "Natural Doc," 44.

58. Joseph Pizzorno, "State of the College," *JNM* 3 (Winter 1985): 111, as quoted in Whorton, *Nature Cures*, 290. An example of a review article in this period is Tori Hudson, ND, "Some Thoughts on Research," *AANPQN* 6 (Spring 1991): 1, 31, at 1. See also Leanna J. Standish, PhD, ND, "A Crisis in Research," ibid., 1, 33–35, at 35. On monetary prizes, see "In-Office Research Awards," *NPhys* 6 (Fall 1991): 41; and Paul Berger and Emily Kane, "Whistler Convention Emphasizes Naturopathic Research," ibid., 1, 18. On presenting findings at the next AANP conference, see Carlo Calabrese, ND, MPH, chair of Bastyr College Research Department, "Outcomes Evaluation Form for Naturopathic Therapeutic Interventions," handout, Feb. 1991; Standish, "Simple

Method for Doing Research." On research scholarships, see Carlo Calabrese, "Research News from Bastyr College," *NPhys* 8 (Winter 1993): 28.

59. "Non-Standard HIV/ARC/AIDS Management," *JNM* 3 (1992): 12–29; Francis Brinker, "Inhibition of Endocrine Function by Botanical Agents I. Boraginaceae and Labiatae," ibid. 1 (1990): 10; Leanna J. Standish, Carlo Calabrese, and Pamela Snider, "The Naturopathic Medical Research Agenda: The Future and Foundation of Naturopathic Medical Science," *Journal of Alternative and Complementary Medicine* 12, no. 3 (2006): 341–45; Heather S. Boon, Daniel C. Cherkin, Janet Erro, et al., "Practice Patterns of Naturopathic Physicians: Results from a Random Survey of Licensed Practitioners in Two U.S. States," *BMC Complementary and Alternative Medicine* 4 (2004): 14; Holly J. Hough, Catherine Dower, and Edward H. O'Neil, *Profile of a Profession: Naturopathic Practice* (San Francisco: Center for the Health Professions, University of California, San Francisco, 2001), www.futurehealth.ucsf .edu/Content/29/2001-09_Profile_of_a_Profession_Naturopathic_Practice.pdf; "Vitamin D Fact Sheet for Health Professionals," National Institutes of Health Office of Dietary Supplements, accessed 3 Dec. 2014, www.ods.od.nih.gov/factsheets /vitamind-healthprofessional/; Mark P. Mattson, David B. Allison, et al., "Meal Frequency and Timing in Health and Disease," *Proceedings of the National Academy of Sciences* 111 (17 Nov. 2014): 16647–53; Dianna Douglas, "Should More Women Give Birth Outside the Hospital?," NPR, accessed 13 July 2015, www.npr.org/sections/health -shots/2015/07/13/419254906/should-more-women-give-birth-outside-the-hospital?sc=17 &f=1001&utm_source=iosnewsapp&utm_medium=Email&utm_campaign=app.

60. Goldie Blumenstyk, "Obama Singles Out For-Profit Colleges and Law Schools for Criticism," *Chronicle of Higher Education*, 23 Aug. 2013, www.chronicle.com/article /article-content/141253/; "Naturopathic Medicine FAQ, Accreditation & Licensure: Q: Why doesn't the AANMC represent any online degree programs?," Association of Accredited Naturopathic Medical Colleges, accessed 10 Jan. 2014, http://aanmc.org /naturopathic/faq/.

61. "About," Trinity School of Natural Health, accessed 2 Dec. 2013, www .trinityschool.org/page.php?id=a. The site changed its language slightly in 2015; see www.trinityschool.org/page.php?id=1. The American Naturopathic Medical Accreditation Board also accredits the University of Natural Medicine in San Dimas, California, where one can earn an ND degree; the Naturopathic Institute of Therapies & Education in Mt. Pleasant, Michigan; the International Institute of Original Medicine in Smithfield, Virginia, where one can earn a Doctor of Naturopathy in Original Medicine degree; and the International College of Healing Arts in Conroe, Texas, where one can earn a diploma in natural medicine or an ND degree. Other programs similarly accredited are also listed on the board's website. "Accredited Member Schools," American Naturopathic Medical Accreditation Board, accessed 7 Dec. 2013, www.anmab.org/ANMAB_MEMBER_SCHOOLS.html.

62. "Southern College of Naturopathic Medicine Sued by Arkansas Attorney General" thread, DegreeInfo Forum, accessed 7 Dec. 2013, www.degreeinfo.com /general-distance-learning-discussions/5860-southern-college-naturopathic -medicine-sued-arkansas-attorney-general.html; the consent decree in the State of Arkansas case against the Herbal Healer Academy is online at www.quackwatch.org /02ConsumerProtection/AG/AR/hha2.html.

63. Baer, *Biomedicine and Alternative Healing Systems*, 189–90.

64. "Licensed States & Licensing Authorities," accessed 7 Dec. 2013, www
.naturopathic.org/content.asp?contentid=57. There are also five Canadian licensing
associations for naturopaths. "State and Federal Advocacy," accessed 7 Dec. 2013, www
.naturopathic.org/article_section.asp?edition=101&section=156; "Advocacy," American
Association of Naturopathic Physicians website, accessed 12 Dec. 2013, www.naturopathic
.org/content.asp?contentid=12; Adam Silberman, "Students Lobby Congress to Support
Naturopathic Medicine," accessed 10 Jan. 2014, www.bastyr.edu/blogs/california
-campus/students-lobby-congress-support-naturopathic-medicine; American Association of Naturopathic Physicians, "Advocacy Priorities in 2015," accessed 18 Dec. 2014,
www.naturopathic.org/article_content.asp?edition=101&section=156&article=965.

65. "Health Care Reform Extends Reach of Naturopathic Medicine," accessed 8
Dec. 2013, www.bastyr.edu/news/general-news-home-page/2013/01/health-care-reform
-extends-reach-naturopathic-medicine.

66. Kaiser Health News and Ankita Rao, "Alternative Treatments Could See Wide
Acceptance Thanks to Obamacare," *The Rundown* (blog), *PBS Newshour*, 29 July 2013,
www.pbs.org/newshour/rundown/2013/07/how-the-health-reform-law-will-impact-alt
ernative-medicine-access.html.

67. Ibid.; Mark Jawer, "Including NDs in Health Insurance—the Affordable Care
Act," 16 Aug. 2013, http://naturopathic.org/article_content.asp?edition=101&section=156
&article=827.

68. Ibid.

69. Kaiser Health News and Rao, "Alternative Treatments"; Krishnadev Calamur,
"Supreme Court Rules Obamacare Subsidies Are Legal," NPR, accessed 11 July 2015,
www.npr.org/sections/thetwo-way/2015/06/25/417425091/supreme-court-rules-obama
care-subsidies-are-legal.

70. "Visions and Goals," accessed 12 Dec. 2014, www.naturopathic.org/content.asp
?contentid=19.

71. "Demographics and Enrollment," accessed 8 Dec. 2013, www.bastyr.edu/about
/about-our-university/demographics-enrollment. Several leaders at the NCNM are
female. All the members of the board of directors and all the committee chairs of the
California Naturopathic Doctors Association are women. The first naturopath to be
elected to a public (allopathic) hospital board was Jane Guiltinan, at Seattle's
Harborview.

The student population of the NCNM is not made public, but information on
faculty numbers is available. Both the dean and associate dean of the program in
naturopathic medicine are women, and 4 of the 16 full-time faculty are women.
Among adjunct faculty, women predominate: 38 are women, 19 and men, and the
gender of 7 is unclear. "School of Naturopathic Medicine Faculty," accessed 8 Dec.
2013, www.ncnm.edu/academic-programs/school-of-naturopathic-medicine/nd-faculty
.php; "About," California Naturopathic Doctors Association, accessed 15 Jan. 2014, www
.calnd.org/about; "Jane Guiltinan, ND," accessed 15 Jan. 2014, www.bastyr.edu/people
/faculty-researcher/jane-guiltinan-nd. Further or more extensive alliances could
include reproductive choice groups, women's studies departments and programs, local
women's museums, LGBTQ centers, gender-conscious politicians and businesses,
sustainability initiatives, the Green Movement, foods co-ops, and animal rights groups.

72. Paul Sisson, "Natural Medical School Latest Development in Growing Trend: Bastyr University, First Accredited Postgraduate Institution in State, Opened in S.D. in September," *U-T San Diego*, 9 Dec. 2012. An online search of the *Los Angeles Times* reveals no coverage. "Branding Push Makes Big Splash in Key Media," 3 Oct. 2013, http://ucsdnews.ucsd.edu/feature/branding_push_makes_big_splash_in_key_media. The University of California, San Diego, dominates media coverage through its well-organized "branding splash."

73. *Natural Medicine Journal* website, accessed 10 Dec. 2013, www.natural medicinejournal.com; "Corporate Partners: Program Participants," accessed 12 Dec. 2013, www.naturopathic.org/content.asp?pl=9&sl=656&contentid=656; Sussanna Czeranko, "Homeopathy Meets Naturopathy," *Naturopathic Doctor News & Review* 9 (Feb. 2013): 26–28; Czeranko, "Yungborn," ibid. 10 (May 2014): 27–29. Czeranko is also preserving naturopathic history by publishing the foundational writings of Benedict Lust and others. Czeranko, ed., *Philosophy of Naturopathic Medicine: In Their Own Words* (Portland, OR: NCNM Press, 2013); Czeranko, ed., *Origins of Naturopathic Medicine: In Their Own Words* (Portland, OR: NCNM Press, 2013).

74. "Mission and Vision," Patient-Centered Outcomes Research Institute, accessed 12 Dec. 2013, www.pcori.org/about-us/mission-and-vision; "Research and Support," Patient-Centered Outcomes Research Institute, accessed 12 Dec. 2013, www.pcori.org /research-we-support/landing/; Kaiser Health News and Rao, "Alternative Treatments"; Whorton, *Nature Cures*, 295; John Weeks, *The Integrator Blog*, accessed 8 Dec. 2013, http://theintegratorblog.com/. Weeks has honorary degrees from Bastyr University (1992), the National University of Health Sciences (2011). and the Canadian College of Naturopathic Medicine (2012).

75. Janet McKee, "Holistic Health and the Critique of Western Medicine," *Social Science and Medicine* 26 (1988): 775–84, at 778, as cited in Baer, "Naturopathy, Acupuncture," 102.

76. Other social justice groups focus on climate change and green living; the environment; gender and racial identity and sexual preference equality; animal rights; and a livable wage. They face dilemmas similar to those of naturopaths and often share common goals, making them potential powerful allies. These goals are clearly articulated by the naturopathic profession. Bastyr was chosen as a green campus in 2013. Goal 2(D) of the American Association of Naturopathic Physicians is that "people from diverse socioeconomic groups and cultures will be served by NDs in a manner that is culturally competent and respectful." Goal 5(C) is that "NDs will advocate for global health, sustainable human communities and a healthy planetary ecosystem." See www.naturopathic.org/content.asp?pl=9&sl=19&contentid=19, accessed 9 Feb. 2014.

Index